S0-DUE-199

DIGITAL PRINCIPLES AND APPLICATIONS

DIGITAL PRINCIPLES AND APPLICATIONS

ALBERT PAUL MALVINO

Vice-president
Director of Research
Time Systems Corporation
Mountain View, California

DONALD P. LEACH

Professor, Electronics Department
Foothill College
Los Altos Hills, California

McGRAW-HILL BOOK COMPANY

New York · St. Louis · San Francisco
Toronto · London · Sydney
Mexico · Panama

DIGITAL PRINCIPLES AND APPLICATIONS

Copyright © 1969 by McGraw-Hill, Inc. All Rights Reserved. Printed in the United States of America. No part of this publication may be reproduced, stored in a retrieval system, or transmitted, in any form or by any means, electronic, mechanical, photocopying, recording, or otherwise, without the prior written permission of the publisher.
Library of Congress Catalog Card Number: 68-31663

39849

1 2 3 4 5 6 7 8 9 10 −MAMM− 7 6 5 4 3 2 1 0 6 9

PREFACE

Computers exert such a great influence that digital circuits and concepts permeate all areas of electronics. Because of this, every engineer and technician should know at least the rudiments of digital electronics.

This book introduces the exciting field of digital electronics. The earlier chapters impart principles by discussing binary numbers, codes, Boolean algebra, logic circuits, and other important concepts. The later chapters describe applications, as in counters, digital clocks, computing circuits, and storage devices. After completing the book, the reader will have a basic, up-to-date knowledge of digital electronics. Furthermore, he will have a solid base upon which to do advanced study in this important branch of electronics.

The length and level of the book make it suitable for a beginning course in digital electronics for engineers and technicians. Glossaries, review questions, and problems appear at the end of each chapter as study aids.

A. P. MALVINO
D. P. LEACH

CONTENTS

1. DIODES AND TRANSISTORS   1

    1-1. Diode Characteristics   1-2. The Ideal Diode   1-3. The Second Approximation of a Diode   1-4. The Reverse Resistance of a Diode   1-5. Zener Diodes   1-6. Transistor Characteristics   1-7. The Ideal Transistor   1-8. The Second Approximation of a Transistor   1-9. The Transistor Used as a Switch   1-10. Switching Time   1-11. The Emitter Follower

2. BINARY AND OCTAL NUMBERS   30

    2-1. Binary Numbers   2-2. Binary Addition   2-3. Binary-to-Decimal Conversion   2-4. Decimal-to-Binary Conversion   2-5. Binary Subtraction   2-6. The 9's and 10's Complements   2-7. Binary Multiplication and Division   2-8. Octal Numbers   2-9. Octal-Binary Conversion

3. BINARY CODES   57

    3-1. The 8421 Code   3-2. The Excess-3 Code   3-3. Other 4-bit BCD Codes   3-4. The Parity Bit   3-5. 5-bit Codes   3-6. Codes with More than 5 Bits   3-7. The Gray Code

4. BOOLEAN ALGEBRA 81

4-1. The OR Gate 4-2. The AND Gate 4-3. Positive and Negative Logic Systems 4-4. The NOT Circuit 4-5. OR Addition 4-6. AND Multiplication 4-7. The NOT Operation 4-8. De Morgan's Theorems 4-9. The Universal Building Block 4-10. Laws and Theorems of Boolean Algebra

5. ARITHMETIC CIRCUITS 116

5-1 The Exclusive-OR Gate 5-2. The Half-adder 5-3. The Full-adder 5-4. A Parallel Binary Adder 5-5. An 8421 Adder 5-6. An Excess-3 Adder 5-7. Half- and Full-subtractors 5-8. Classifying Logic Systems

6. MULTIVIBRATORS 142

6-1. The T Flip-flop 6-2. The RS and RST Flip-flops 6-3. The JK Flip-flop 6-4. The Schmitt Trigger 6-5. The Astable Multivibrator 6-6. The Monostable Multivibrator

7. BASIC ELECTRONIC COUNTERS 167

7-1. Using Flip-flops to Count 7-2. A Decade Counter 7-3. Decoding a Counter 7-4. Cascading Decade-counter Units 7-5. Gating a Counter 7-6. A Frequency Counter 7-7. A Digital Voltmeter

8. COUNTER TECHNIQUES 183

8-1. Binary Ripple Counter 8-2. Modified Counter Using Feedback 8-3. Parallel Counter 8-4. The Race Problem 8-5. Series-Parallel Combination Counters 8-6. Binary Decade Counter with Decoding Gates 8-7. Higher-modulus Counters 8-8. A BCD Counter

9. SPECIAL COUNTERS AND REGISTERS 213

9-1. A Serial Shift Register 9-2. A Ring Counter 9-3. A Shift Counter 9-4. A Moduilo-10 Shift Counter with Decoding 9-5. A Digital Clock 9-6. Up-Down Counter 9-7. Shift-register Operations

10. INPUT-OUTPUT DEVICES 248

10-1. Punched Cards 10-2. Paper Tape 10-3. Magnetic Tape 10-4. Digital Recording Methods 10-5. Other Peripheral Equipment 10-6. Encoding and Decoding Matrices

11. D/A AND A/D CONVERSION 275

11-1. Variable-resister Network 11-2. Binary Ladder 11-3. D/A Converter 11-4. D/A Accuracy and Resolution 11-5. A/D Converter—

Contents ix

Simultaneous Conversion 11-6. A/D Converter—Counter Method 11-7. Advanced A/D Techniques 11-8. A/D Accuracy and Resolution 11-9. Electromechanical A/D Conversion

12. MAGNETIC DEVICES AND SYSTEMS 316

12-1. Magnetic Cores 12-2. Magnetic-core Logic 12-3. Magnetic-core Shift Register 12-4. Coincident-current Memory 12-5. Memory Addressing 12-6. Apertured Plate 12-7. Multiaperture Devices 12-8. Magnetic-drum Storage

13. DIGITAL ARITHMETIC 358

13-1. Number Representations 13-2. Fundamental Processes 13-3. Serial Binary Adder 13-4. Parallel Binary Adder 13-5. BCD Addition 13-6. A Parallel BCD Adder 13-7. Binary Multiplication 13-8. Binary Division

14. CLOCK AND CONTROL 398

14-1. Basic Clocks 14-2. Clock Systems 14-3. D/A Converter Control 14-4. Multiple D/A Conversion 14-5. A/D Converter Control 14-6. Ring-counter Control 14-7. MPG Computer

BIBLIOGRAPHY 417
APPENDIX A: STATES AND RESOLUTION FOR BINARY NUMBERS 418
APPENDIX B: DECIMAL-OCTAL-BINARY NUMBER CONVERSION TABLE 419
ANSWERS TO SELECTED ODD-NUMBERED PROBLEMS 420
INDEX 426

DIGITAL PRINCIPLES AND APPLICATIONS

CHAPTER 1

DIODES AND TRANSISTORS

Diodes and transistors are normally covered in a basic electronics course; nevertheless, in this chapter we make a brief review of these devices. We will discuss ideas that are most useful in our later work, and will stress approximations because these are adequate for most digital analysis.

1-1 Diode Characteristics

A semiconductor diode lets current pass easily in one direction, but not in the other. Because of this, we can think of a diode as an automatic switch — it is closed when current tries to go one way, but open when it tries to go the other way.

We have shown the schematic symbol for a semiconductor diode in Fig. 1-1a. Note that *conventional* current flows easily from the anode to the cathode.

To get a concrete notion of how a diode works, let us engage in a hypothetical experiment. Imagine that we increase the d-c supply of Fig. 1-1a

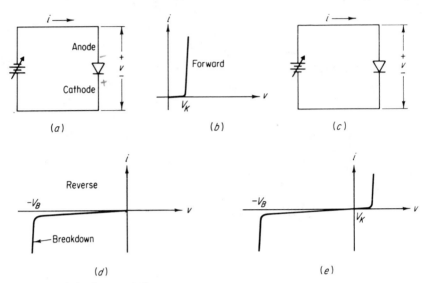

Fig. 1-1 Diode characteristics.

from 0 volts to higher values. How much current flows? At first there would be very little current. After a few tenths of a volt, however, the current would increase sharply as shown in Fig. 1-1b. The approximate voltage V_K where the current increases rapidly is called the "knee voltage." For germanium diodes the knee voltage is around 0.3 volt; for silicon diodes it is around 0.7 volt.

The graph of Fig. 1-1b is called the forward characteristic of the diode. This graph shows the relation between current and voltage for the circuit of Fig. 1-1a. With the plus terminal of the battery connected to the anode, the diode is forward-biased.

What happens if we reverse the battery connection as in Fig. 1-1c? As we start to increase the battery voltage, there is very little current as shown in Fig.1-1d. But if we increase the voltage to a large enough value, the current suddenly increases. We call this point where the current suddenly increases the "breakdown point," and we designate V_B as the breakdown voltage. Normally, a rectifier diode operates at voltages that are much less than the breakdown voltage.

By combining the forward and reverse characteristics, we get the graph of Fig. 1-1e. This graph is typical of semiconductor diodes. The most important points to remember about this graph are:

• In the forward direction, current flows easily after reaching the knee voltage.

- In the reverse direction very little current flows below the breakdown point.

1-2 The Ideal Diode

To analyze diode circuits quickly and easily, we will often use the ideal-diode approximation. This approximation views the diode as a perfect switch — there is no voltage drop across it when closed, and no current through it when open.

Figure 1-2a displays and summarizes the properties of the ideal diode. When the diode is forward-biased, there is zero voltage drop. When the diode is back-biased, there is no current through it.

It is always helpful to have an equivalent circuit for an ideal device. In Fig. 1-2b we have shown our circuit viewpoint of an ideal diode. As indicated, when the conventional current tries to flow from anode to cathode (in the direction of the triangle), the diode is like a closed switch. But if the conventional current tries to flow the other way — against the diode triangle — the diode is like an open switch. This *switch* viewpoint of a diode is a very simple and crude way to look upon diode action. Yet for preliminary analysis of diode circuits it is an excellent and practical tool.

EXAMPLE 1-1

Sketch the output voltage for the circuit of Fig. 1-3a. Use the ideal-diode approximation.

SOLUTION

During each positive half-cycle the input voltage is +20 volts. Conventional current flows in the direction of the diode triangle; therefore, the diode is shorted. As a result, +20 volts appears across the output during each positive half-cycle.

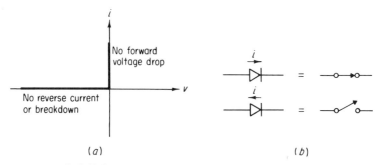

Fig. 1-2 Ideal diode

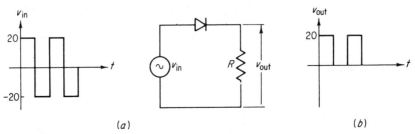

Fig. 1-3 Example 1-1.

During each negative half-cycle the diode is reverse-biased. The diode therefore looks like an open switch. With no current flow, no voltage can be developed across the output.

Figure 1-3b shows the output voltage. Note that the negative parts of the source voltage have been clipped off. From basic electronics we recognize this circuit as a simple half-wave rectifier.

EXAMPLE 1-2

What is the output waveform of the circuit in Fig. 1-4a?

SOLUTION

During each positive half-cycle conventional current flows in the direction of the diode triangle; therefore, the diode is like a closed switch. Since no voltage can appear across a short, the output voltage must be zero throughout each positive half-cycle.

During each negative half-cycle, however, conventional current tries to flow opposite the diode triangle, so that the diode is open. With no current in the circuit there can be no voltage drop across the resistor, and all the input voltage must appear across the output terminals.

Figure 1-4b shows the output waveform.

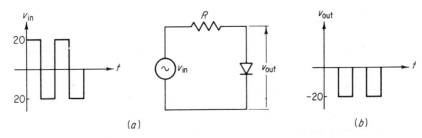

Fig. 1-4 Example 1-2.

Diodes and Transistors

The circuit of Fig. 1-4a is often called a "positive clipper" because it clips off all positive parts of the input signal.

1-3 The Second Approximation of a Diode

The ideal diode is the simplest and most useful approximation of a rectifier diode. There are times, however, when it leads to considerable error. One way to improve our approximation of a diode is to allow for the knee voltage. In other words, what we propose to do is to approximate the i-v characteristic of a real diode by the simple i-v characteristic of Fig. 1-5a. This graph tells us that the diode has a voltage drop of V_K when it is conducting. (Remember: 0.3 volt for germanium and 0.7 volt for silicon.)

As far as an equivalent circuit is concerned, in our second approximation we think of an ideal diode in series with a battery of V_K volts as shown in Fig. 1-5b. Thus, in our second approximation of a real diode we are merely saying that a few tenths of a volt drop occurs across the real diode when it is conducting.

When should we use the second approximation? As a simple guide, we use it whenever an error of a few tenths of a volt is objectionable. For instance, in a circuit where the source voltage is only 1 or 2 volts, a few tenths of a volt is important, and we should use the second approximation.

To summarize diode-circuit analysis:

- Use the ideal-diode approach to get a basic idea of how the circuit operates. This is often adequate.
- If an error of a few tenths of a volt is too much, use the second approximation.

EXAMPLE 1-3

Analyze the circuit of Fig. 1-6a by using the second approximation.

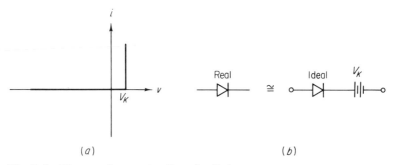

Fig. 1-5 The second approximation of a diode.

6   **Digital Principles and Applications**

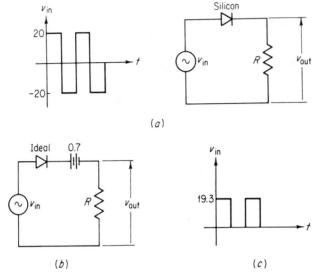

Fig. 1-6 Example 1-3

SOLUTION

We recognize this circuit as a half-wave rectifier. If the diode were ideal, the output would simply have positive half-cycles with 20-volt peaks.

To use the second approximation, think of the circuit as shown in Fig. 1-6b. During each positive half-cycle the output voltage is 19.3 volts (0.7 volt less than 20 volts). In other words, we are saying that about 0.7 volt is dropped across the silicon diode.

During each negative half-cycle the diode is off; therefore, the output voltage is zero.

Figure 1-6c displays the output waveform. Note that this waveform is only slightly different from what we would get by using the ideal-diode approach.

EXAMPLE 1-4

In the positive clipper of Fig. 1-7a, use the second approximation to find the output waveform and the current during the positive half-cycle.

SOLUTION

During the positive half-cycle the diode is on; therefore, the voltage across it is about 0.7 volt. During the negative half-cycle the diode is off, so that all the input voltage appears across the output. Thus, we can sketch the output waveform as in Fig. 1-7b.

Diodes and Transistors

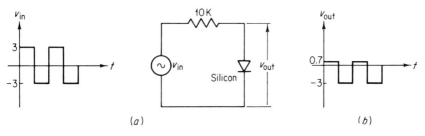

Fig. 1-7 Example 1-4.

To find the current during the positive half-cycle, note that the voltage across the 10-kilohm resistor is the difference between the source voltage and the diode voltage. That is,

$$V_R = 3 - 0.7 = 2.3 \text{ volts}$$

The current through the resistor is simply

$$I = \frac{V_R}{R} = \frac{2.3}{10(10^3)} = 0.23 \text{ ma}$$

1-4 The Reverse Resistance of a Diode

The simplest way to account for the reverse current below breakdown is to think of the diode as a large resistance. To find the value of this reverse resistance we need only take the ratio of any reverse voltage to the corresponding current. That is,

$$R_R = \frac{V_R}{I_R} \tag{1-1}$$

where R_R = reverse resistance
V_R = any reverse voltage below the breakdown voltage
I_R = corresponding reverse current

As an example, suppose a diode has a current of 100 na for a reverse voltage of 50 volts. The reverse resistance would be

$$R_R = \frac{V_R}{I_R} = \frac{50}{100(10^{-9})} = 500 \text{ megohms}$$

The use of reverse resistance is a crude approximation. The behavior of a diode in the reverse direction is not really so simple as that of an ordinary resistor. Yet, as a first approximation we can use the concept of reverse resistance to simplify circuit analysis. When using the reverse-resistance concept, remember that the diode must *not* be in the breakdown region.

Example 1-5

The diode of Fig. 1-8a has a current of 1 μa for a reverse voltage of 100 volts. If we allow 0.3 forward-voltage drop, what is the output waveform?

Solution

First, we get the reverse resistance.

$$R_R = \frac{V_R}{I_R} = \frac{100}{10^{-6}} = 100 \text{ megohms}$$

Now we can draw the equivalent circuit for the negative half-cycle (Fig. 1-8b). The 100-megohm resistance represents the diode, so that the circuit behaves like a voltage divider during the negative half-cycle. With a 100:1 ratio, about 1 percent of the source voltage appears across the output terminals.

During the positive half-cycle there is about 0.3 volt across the diode; therefore, the output is around 19.7 volts.

Figure 1-8c shows the output waveform.

The reverse resistance becomes important only when the circuit resistance is comparable in size to the reverse resistance of the diode. For instance, in Fig. 1-8b, only about 1 percent of the source voltage appears across the output during the negative half-cycle. However, if we were to raise the 1-megohm resistor to a 100-megohm resistor, half the source

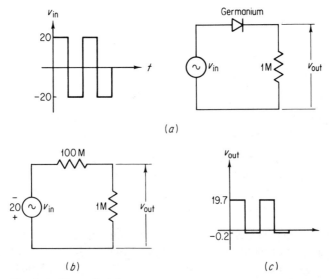

Fig. 1-8 Example 1-5.

Diodes and Transistors

voltage would then appear across the output terminals when the diode was reverse-biased. Naturally, this is undesirable, and as a rule, we try to make the reverse resistance of the diode much larger than the resistance in series with the diode.

1-5 Zener Diodes

Zener diodes (sometimes called avalanche diodes) are silicon diodes that have been specially processed to have a sharp breakdown characteristic. For instance, in Fig. 1-9a, note that at the breakdown point, the diode current increases very sharply and the breakdown region is almost vertical.

When a zener diode enters the breakdown region, it will not burn out unless there is enough current to cause excessive power dissipation. For instance, suppose a zener diode can dissipate 1 watt before burning out, and that it has a breakdown voltage of 50 volts. If this diode operates in the breakdown region, it will not burn out unless the current exceeds

$$I = \frac{P}{V} = \frac{1 \text{ watt}}{50 \text{ volts}} = 20 \text{ ma}$$

Zener diodes are constant-voltage devices; that is, in the breakdown region the voltage remains constant even though the current can change. To bring this out clearly, examine Fig. 1-9a. At breakdown the voltage across the diode is V_B. When the diode current increases, the voltage increases only slightly. In other words, *in the breakdown region the voltage across the zener diode is almost constant.*

To simplify analysis of zener-diode circuits, we can use the ideal-zener-diode approximation. The i-v graph of such a diode is shown in Fig. 1-9b. In the forward direction we neglect the small voltage drop; in the reverse direction there is no current until we reach the breakdown point. We will denote the breakdown voltage by V_Z, which simply stands for the zener voltage.

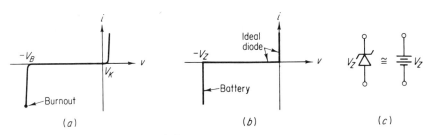

Fig. 1-9 Zener-diode characteristics.

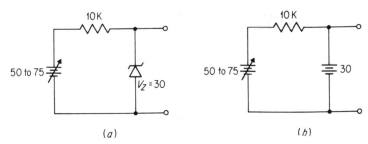

Fig. 1-10 Example 1-6.

Our circuit model of an ideal zener diode is quite simple. When the diode is *not* in the breakdown region, it acts like an ideal diode. But when it is in the breakdown region, it is like a *battery* of V_Z volts. We have shown the latter equivalent circuit in Fig. 1-9c. Note the schematic symbol used to indicate a zener diode.

EXAMPLE 1-6

What is the voltage across the zener diode of Fig. 1-10a?

SOLUTION

For any source voltage between 50 and 75 volts, there is enough voltage to break down the diode. If we like, we can visualize the diode as shown in Fig. 1-10b. The output voltage is constant, being equal to the zener voltage.

This is an example of simple voltage regulation. Even when the source voltage changes, the output remains at 30 volts.

EXAMPLE 1-7

What is the output waveform in Fig. 1-11a? Use the ideal-zener-diode approximation.

SOLUTION

During each positive half-cycle the upper diode is like a short and the lower diode is like a battery of 10 volts. During each negative half-cycle the situation is reversed; that is, the upper diode is in the breakdown region and the lower diode is forward-biased.

Figure 1-11b shows the output waveform. The circuit has merely clipped the waveform at $+10$ and -10 volts.

EXAMPLE 1-8

Suppose we use a sine wave instead of a square wave to drive the circuit of Fig. 1-11a. What will the output waveform be?

Diodes and Transistors

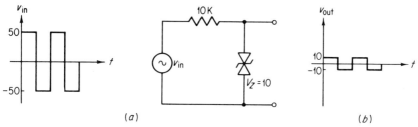

Fig. 1-11 Examples 1-7 and 1-8.

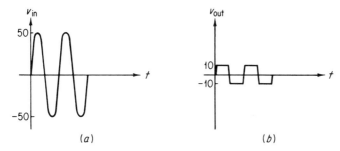

Fig. 1-12 Example 1-8.

SOLUTION

The circuit simply clips the sine wave at $+10$ and -10 volts. The output waveform is shown in Fig. 1-12.

1-6 Transistor Characteristics

The common types of transistors are n-p-n and p-n-p. We have shown the schematic symbol for an n-p-n transistor in Fig. 1-13a. The terminal with the arrow is the emitter, the upper terminal is the collector, and the other

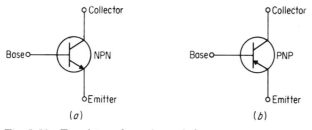

Fig. 1-13 Transistor schematic symbols.

terminal is the base. For an *n-p-n* transistor the arrow points from the base to the emitter. The schematic symbol for a *p-n-p* transistor is different, as shown in Fig. 1-13b. Note that the arrow points from the emitter to the base. (Incidentally, the arrow points in the easy direction of conventional current.)

The transistor behaves like a controlled-current source. To bring this concept out clearly, consider the common-emitter circuit of Fig. 1-14a, where we have shown voltage sources driving the base and collector. We denote the base current by I_B and the collector current by I_C. Suppose we adjust the base current to exactly 0.01 ma. In the collector circuit this is what we find: when we increase the collector voltage V_{CE}, the collector current increases very sharply at first, but after a few tenths of a volt, the collector current reaches an *almost fixed* value (see Fig. 1-14b). For a typical transistor the collector current is much larger than the base current. Arbitrarily, we have shown a collector current of about 1 ma above the knee of the curve.

Suppose we change the base current from 0.01 to 0.02 ma. In the collector circuit we find an almost constant value of collector current, once the collector voltage is more than a few tenths of a volt (see Fig. 1-14c). In this case, the collector current is around 2 ma, about 100 times larger than the base current.

If we repeat this procedure for other values of base current and draw all graphs on the same set of axes, we get the typical *i-v* characteristic of Fig.

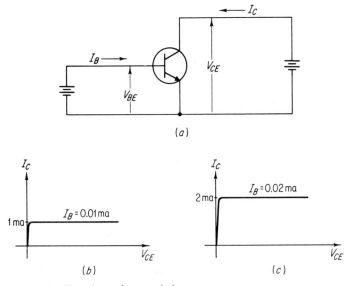

Fig. 1-14 Transistor characteristics.

Diodes and Transistors

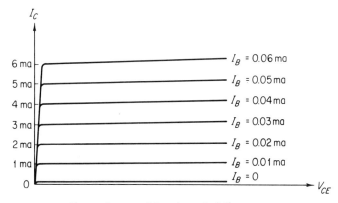

Fig. 1-15 Composite transistor characteristics.

1-15. Note how the collector current reaches an almost fixed value above the knee of each curve. Also, the ratio of collector current to base current is around 100 for each curve.

The beta of a transistor is simply the ratio of the collector current to the base current. We are most concerned with the d-c beta; it is defined as

$$\beta = \frac{I_C}{I_B} \tag{1-2}$$

where I_C = total collector current
I_B = total base current

In our previous discussion we assumed a transistor with a beta of 100. When we examine a large number of transistors, however, we typically find that beta lies in the range of about 20 to 200. Betas between 50 and 100 are very common.

The really important ideas to get from Fig. 1-15 are:

- Above the knee, the collector current is almost constant.
- The base current controls the collector current.

Another important point about transistor action is that the base-emitter part of a transistor acts almost like a single diode. Typically, the graph of base current versus base voltage looks like Fig. 1-16a. This shape is like an ordinary diode graph, and it should be, because the base-emitter part of a transistor is almost the same as a *p-n* diode. If we were to reverse the base supply in Fig. 1-14a, the base diode would be reverse-biased and very little current would flow until we reached the breakdown voltage of the diode (see Fig. 1-16b). Thus, to a first approximation we can think of the base-emitter part of a transistor as being an ordinary diode.

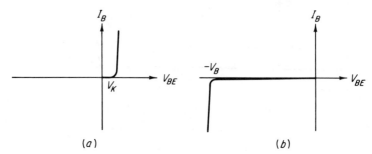

Fig. 1-16 Base characteristics.

The p-n-p transistors operate the same as n-p-n transistors except that all currents and voltages are reversed.

Let us summarize the key ideas of this section.

- The base-emitter part of a transistor behaves almost like an ordinary diode.
- If the collector voltage is more than a few tenths of a volt, the collector current remains almost constant even though the collector voltage changes.
- The base current controls the collector current.
- The d-c beta is the ratio of the total collector current to the total base current.

1-7 The Ideal Transistor

To simplify the analysis of transistor circuits, we will often use the ideal-transistor approximation. We know that the base-emitter part of a transistor acts almost like an ordinary diode; therefore, as an ideal approximation we will visualize the i-v characteristic as shown in Fig. 1-17a.

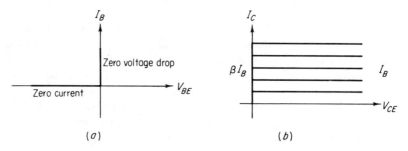

Fig. 1-17 Ideal-transistor characteristics.

Diodes and Transistors

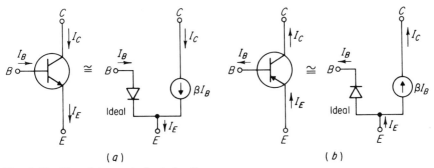

Fig. 1-18 Transistor equivalent circuits.

As far as the collector is concerned, we will use the ideal i-v graph of Fig. 1-17b. This graph tells us that the collector current is constant for any value of collector voltage greater than zero; also, the collector current equals beta times the base current.

Figure 1-18a shows the model we will use for the ideal n-p-n transistor. There is an ideal diode in the base circuit and an ideal current source in the collector. The value of the current source is beta times the base current.

Figure 1-18b shows the equivalent circuit of a p-n-p transistor. As indicated earlier, the action in a p-n-p transistor is basically the same as in an n-p-n except that all currents and voltages are in opposite directions.

Realize that we are approximating. We are not saying the equivalent circuits of Fig. 1-18 exactly describe transistor behavior; we are saying only that these circuits are a first approximation for transistor behavior. These circuits are quite useful. In spite of their simplicity, we can adequately analyze many digital circuits with these ideal approximations.

EXAMPLE 1-9

Find the collector-emitter voltage in the circuit of Fig. 1-19a. Assume the transistor is ideal and use a beta of 100.

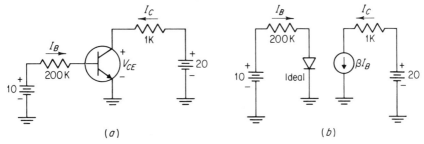

Fig. 1-19 Example 1-9.

Solution

We can replace the transistor by its ideal equivalent circuit (Fig. 1-19b). Clearly, the ideal diode in the base is *on* because conventional current tries to flow in the direction of the diode triangle. Therefore, the base current is

$$I_B = \frac{10}{200(10^3)} = 0.05 \text{ ma}$$

The collector current is controlled by the base current, and equals

$$I_C = \beta I_B = 100(0.05 \text{ ma}) = 5 \text{ ma}$$

In Fig. 1-19a, the collector-emitter voltage V_{CE} must equal the supply voltage less the drop across the 1-kilohm resistor. That is,

$$V_{CE} = 20 - 5(10^{-3})(10^3) = 15 \text{ volts}$$

Example 1-10

Find the approximate value of beta in Fig. 1-20.

Solution

The voltage from collector to ground is given as 10 volts. Since the supply is 30 volts, there must be a 20-volt drop across the 20-kilohm load resistor. But if there is 20 volts across the 20-kilohm resistor, the current must be

$$I_C = \frac{20}{20(10^3)} = 1 \text{ ma}$$

In the base circuit it is clear that the base diode is forward-biased and that the base current is

$$I_B = \frac{10}{2(10^6)} = 5 \text{ μa}$$

We can find beta very easily. It is the ratio of collector current to base current; therefore,

$$\beta = \frac{I_C}{I_B} = \frac{1 \text{ ma}}{5 \text{ μa}} = 200$$

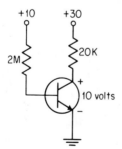

Fig. 1-20 Example 1-10.

1-8 The Second Approximation of a Transistor

The results we get with the ideal-transistor approximation are adequate in some circuits, but somewhat inaccurate for others. Therefore, in addition to the ideal-transistor approximation, we will sometimes use a second approximation of the transistor.

Figure 1-21 shows the equivalent circuit for the second approximation of a transistor. The only change is in the base circuit. Instead of treating the base diode as ideal, we are now using the second approximation for the base diode. The collector circuit remains the same as before. In essence, in the second approximation of a transistor, we simply allow 0.3 or 0.7 volt across the base diode, depending on the transistor type.

EXAMPLE 1-11

Find the collector current in Fig. 1-22a. Use the second approximation.

SOLUTION

The base diode is on. Because the transistor is made of silicon, we allow about a 0.7-volt drop across the base diode. So the base current is

$$I_B = \frac{5 - 0.7}{10^6} = 4.3 \; \mu a$$

With a beta of 75, the collector current is

$$I_C = \beta I_B = 75(4.3 \; \mu a) = 0.322 \; ma$$

EXAMPLE 1-12

Find the collector current in Q2 of Fig. 1-22b. The transistors are made of silicon. Use the second approximation.

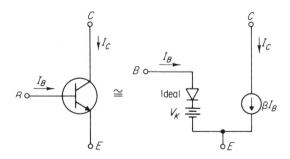

Fig. 1-21 The second approximation of a transistor.

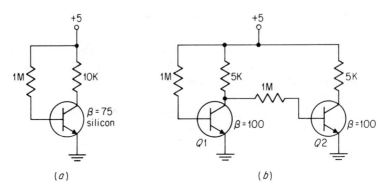

Fig. 1-22 Examples 1-11 and 1-12.

SOLUTION

The base current in Q1 is

$$I_B = \frac{5 - 0.7}{10^6} = 4.3 \ \mu a$$

and the collector current is

$$I_C = 100(4.3 \ \mu a) = 0.43 \ ma$$

The voltage from the collector of Q1 to ground approximately equals the supply voltage less the drop across the 5-kilohm resistor.

$$V_{C1} = 5 - 0.43(10^{-3})5(10^3) = 2.85 \text{ volts}$$

The collector of Q1 drives the base resistor of Q2; that is, about 2.85 volts is applied to the 1-megohm base resistor of Q2. This means that the base current in Q2 is

$$I_B = \frac{2.85 - 0.7}{10^6} = 2.15 \ \mu a$$

The collector current of Q2 therefore equals

$$I_C = 100(2.15 \ \mu a) = 0.215 \ ma$$

1-9 The Transistor Used as a Switch

In digital circuits we almost always use transistors as switches to change from one voltage level to another. As a simple example of what we mean by switching, look at Fig. 1-23a. We can either short or open terminals A and B. A mechanical switch is too slow in some applications; therefore, we sometimes will use the transistor as an electronic switch between A and B.

Diodes and Transistors

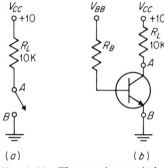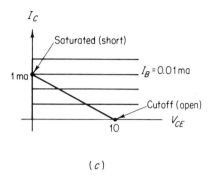

Fig. 1-23 The transistor switch.

Here is how it is done. We connect a transistor as shown in Fig. 1-23b. The base current controls the collector current; if there is no base current, there is no collector current (neglect leakage). In other words, if there is no base current, A and B are open. To prevent base current, we need only back-bias the base diode by making V_{BB} negative.

To short A and B together, we must saturate the transistor. Recall that saturation occurs when there is enough collector current to drop V_{CE} to about 0 volts. Figure 1-23c shows this saturation condition. When the transistor works at the upper end of the load line, the collector-emitter voltage has dropped to zero. Under this condition the collector current is maximum and the transistor is saturated. Since the collector-emitter voltage is around zero at saturation, we can say that terminals A and B are shorted.

Thus, by holding the transistor at either cutoff or saturation, we are either opening or shorting A and B together. In effect, the transistor acts like a switch. Whether the switch is open or closed depends upon the base drive. To open the switch, we merely back-bias the base diode; this ensures no base current, which in turn means no collector current. To close the switch, we forward-bias the base diode enough to cause saturation. This means that the base current must be equal to or greater than

$$I_{B(\text{sat})} = \frac{I_{C(\text{sat})}}{\beta} \tag{1-3}$$

where $I_{B(\text{sat})}$ = minimum value of base current needed to saturate the transistor
$I_{C(\text{sat})}$ = saturation value of collector current

Equation (1-3) gives us a threshold value for base current; at this value of base current, the transistor has just saturated; if the base current is more than this value, the transistor stays saturated; if the base current is less than this value, the transistor comes out of saturation.

When V_{CE} drops to zero in Fig. 1-23b, all the supply voltage appears across the load resistor. Thus, the saturation current is simply

$$I_{C(\text{sat})} = \frac{10}{10(10^3)} = 1 \text{ ma}$$

In general, the saturation current is simply

$$I_{C(\text{sat})} \cong \frac{V_{CC}}{R_L} \qquad (1\text{-}4)$$

Equation (1-4) is approximate because at saturation V_{CE} is almost, but not quite, zero. In other words, at saturation, the voltage across R_L is a few tenths of a volt less than V_{CC}.

EXAMPLE 1-13

Determine if the transistor in Fig. 1-24a is saturated. Use a beta of 50.

SOLUTION

The saturation value of collector current is

$$I_{C(\text{sat})} \cong \frac{V_{CC}}{R_L} = \frac{20}{10(10^3)} = 2 \text{ ma}$$

and the threshold value of base current is

$$I_{B(\text{sat})} = \frac{I_{C(\text{sat})}}{\beta} = \frac{2 \text{ ma}}{50} = 0.04 \text{ ma}$$

We must now find out if the actual value of base current is more or less than the threshold value. It is clear in Fig. 1-24a that the base diode is forward-biased. Treating the base diode as ideal, we get a base current of

$$I_B = \frac{V_{BB}}{R_B} = \frac{10}{100(10^3)} = 0.1 \text{ ma}$$

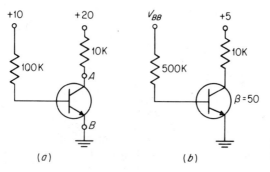

Fig. 1-24 Examples 1-13 and 1-14.

Diodes and Transistors

The actual base current is 0.1 ma; the threshold value needed to just saturate is 0.04 ma. Clearly, there is more than enough base current to saturate the transistor.

EXAMPLE 1-14

In the circuit of Fig. 1-24b what is the minimum value of V_{BB} that saturates the transistor?

SOLUTION

The saturation value of collector current is simply

$$I_{C(\text{sat})} \cong \frac{V_{CC}}{R_L} = \frac{5}{10(10^3)} = 0.5 \text{ ma}$$

and the threshold value of base current is

$$I_{B(\text{sat})} = \frac{I_{C(\text{sat})}}{\beta} = \frac{0.5 \text{ ma}}{50} = 0.01 \text{ ma}$$

If we neglect the base-emitter knee voltage, the minimum base voltage needed to saturate is

$$V_{BB} = I_{B(\text{sat})} R_B = 0.01(10^{-3}) 500(10^3) = 5 \text{ volts}$$

In other words, when V_{BB} is 5 volts or more, the transistor will saturate.

If a silicon transistor were used, a drop of about 0.7 volt would appear across the base diode. Therefore, as a refinement, we would say that about 5.7 volts or more will saturate the transistor.

1-10 Switching Time

By controlling the base current, we can control the collector current and make the transistor act like a switch. Ideally, as soon as the base current exceeds the threshold value, the transistor saturates. For example, in Fig. 1-25a, a pulse drives the base circuit. Before point A in time the base diode is back-biased so that the transistor is cut off; under this condition the collector voltage to ground equals V_{CC}. At point A in time the base drive suddenly changes from $-V$ to $+V$. Let us assume that $+V$ volts is enough to saturate the transistor. Ideally, if there were no capacitive effects or other time delays, the collector voltage would immediately drop to 0 volts because the transistor is saturated (see Fig. 1-25b). Between A and B the base diode remains forward-biased and the collector voltage stays at zero. At point B in time the base drive reverses so that the base diode is reverse-biased. Ideally, the collector current would immediately cut off and the

22 Digital Principles and Applications

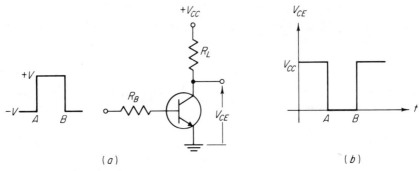

Fig. 1-25 Pulse operation.

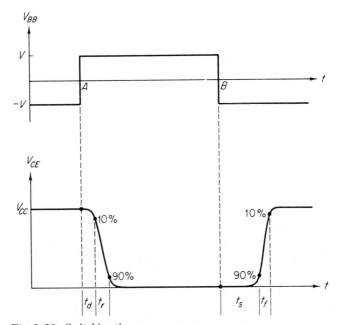

Fig. 1-26 Switching times.

collector voltage would suddenly rise to V_{CC}. To a first approximation we would get the collector waveform of Fig. 1-25b.

Actually, the collector voltage cannot change instantaneously; it always takes time to change from one voltage level to another. Figure 1-26 shows the base drive and the resulting collector waveform. The transitions have been exaggerated to bring out the different times. These are:

Diodes and Transistors

- Delay time t_d is the amount of time that elapses from the instant the base drive is turned on until the collector voltage has changed 10 percent. Two factors contribute to this delay: (1) the time needed for carriers to pass from emitter to collector; (2) the base-emitter capacitance.
- Rise time t_r is the time that elapses between the 10 and 90 percent point in Fig. 1-26. The transistor cutoff frequency and the external capacitance determine the rise time.*
- Storage time t_s (also called saturation delay time) is the time that elapses after the base drive is reversed and the collector voltage returns to the 90 percent point. Storage time arises because a large number of carriers are stored at the collector junction when the transistor is saturated. These excess carriers must first be cleared out of the junction before the transistor can cut off.
- Fall time t_f is the time needed for the transistor to change from the 90 percent to the 10 percent point. Like rise time, fall time is related to the transistor cutoff frequency and the external capacitances.

These various times set an upper limit on how fast we can switch a transistor on and off. For instance, suppose that the transistor circuit of Fig. 1-25a has the following switching data: $t_d = 5$ nsec, $t_r = 10$ nsec, $t_s = 40$ nsec, and $t_f = 10$ nsec. The sum of these times is 65 nsec. Thus, to turn a transistor on and off, we need slightly more than 65 nsec. In this case, the highest frequency of operation is about

$$f_{\max} \cong \frac{1}{65(10^{-9})} \cong 15 \text{ MHz}$$

Turn-on time is the sum of delay time t_d and rise time t_r; *turn-off time* is the sum of storage time t_s and fall time t_f. In high-speed digital circuits we need to make the turn-on and turn-off times as small as possible. One way to reduce the turn-on time is by overdriving the base diode, that is, using enough base drive to make I_B much larger than the threshold value $I_{B(\text{sat})}$. Similarly, we can reduce the turn-off time by using a large reverse voltage, thereby driving the base diode well into the reverse region.

One of the best ways to reduce the switching time is to *prevent the transistor from saturating*. As mentioned earlier, when a transistor saturates a large number of carriers are stored at the collector junction because the collector diode becomes forward-biased. When the base drive suddenly reverses, the stored charges flow for a while, temporarily keeping the collector current from cutting off. By simply avoiding saturation, the collector diode remains back-biased, so that storage time is almost eliminated.

---

*This part of the waveform is called the rise because the current actually rises during this time.

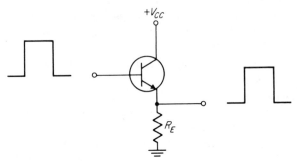

Fig. 1-27 Emitter follower.

For our purposes, the key points to remember are:

• Even though the base drive is suddenly changed, the collector voltage cannot suddenly respond — some time must elapse before the collector voltage changes from one level to another.

• The turn-on and turn-off times place a limit on the maximum switching frequency.

1-11 The Emitter Follower

Recall one more type of transistor circuit — the emitter follower. Figure 1-27 shows a simple emitter follower. The input signal drives the base; the output is taken from the emitter. To a first approximation the output signal is a replica of the input signal. For instance, if we drive the base with a rectangular pulse of amplitude V, we get a rectangular output pulse of the same size. This is an ideal approximation; actually, there is a small amount of signal dropped across the base-emitter diode. However, in a well-designed emitter follower, at least 90 percent of the input signal appears at the output.

Like the common-emitter circuit of previous sections, the emitter follower has turn-on and turn-off times; this results in an upper limit on the maximum switching frequency.

SUMMARY

The ideal-diode approximation is the simplest and most practical viewpoint of a rectifier diode. In this approximation we think of the diode as an automatic switch — closed when the conventional current flows in the direction of the diode triangle, but open when current tries to flow the other way.

In the second approximation of a diode we allow for a forward voltage drop equal to the knee voltage V_K. In germanium diodes V_K is about 0.3 volt; in silicon, about 0.7 volt.

In the reverse region below breakdown we can think of the diode as a reverse resistance R_R that equals the ratio of a reverse voltage to the corresponding current. When the diode is in the breakdown region, we approximate it by a battery.

Two very important transistor properties are:

- Above the knee the collector current is almost constant.
- The base current controls the collector current.

In the ideal-transistor approximation we visualize the base-emitter junction as an ideal diode and the collector as a current source. In the second approximation of a transistor we allow for a voltage drop of V_K across the base-emitter junction.

In digital circuits the transistor is almost always used as a switch to change from one voltage level to another. The transistor cannot immediately respond to a change in the input signal. The turn-on and turn-off times establish an upper limit on the maximum switching frequency.

GLOSSARY

beta (β) The d-c beta is the ratio of the total collector current to the total base current.

conventional current A mathematical concept of current that flows opposite to the electron flow.

current source An ideal device whose current is independent of the voltage across it. It is the dual of voltage source, which is a device whose voltage is independent of the current through it.

ideal diode An approximation of real diode in which the diode is visualized as a perfect switch.

knee voltage The approximate value of forward voltage above which diode current increases sharply.

positive clipper A circuit that removes all positive parts of an input signal.

reverse resistance The ratio of a reverse voltage across a diode to the corresponding current.

turn-off time The sum of storage time and fall time.

turn-on time The sum of delay time and rise time.

REVIEW QUESTIONS

1. What are the knee voltages for germanium and silicon?
2. Describe the ideal-diode approximation.
3. What is a positive clipper?
4. What is the second approximation of a diode?
5. How do we find the reverse resistance of a diode?
6. When a zener diode is operating in the breakdown region, what determines if it burns out?
7. What is the equivalent circuit for a zener diode that is operating in the breakdown region?
8. How do we define the d-c beta of a transistor?
9. In the ideal-transistor approximation, how do we view the base-emitter part of the transistor? The collector-base part?
10. How do we find the base current that just saturates the transistor using $I_{C(sat)}$ and beta?
11. Define the delay time and the rise time.
12. Define the storage time and the fall time.
13. What are the turn-on and turn-off times?
14. In an emitter follower, how is the output voltage related to the input?

PROBLEMS

1-1 In Fig. 1-28a, use the ideal-diode approximation to find the value of direct current I.

1-2 In Fig. 1-28b, use the ideal-diode approximation to find I.

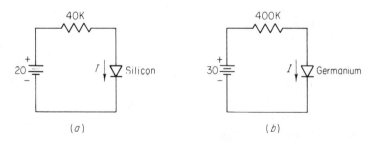

Fig. 1-28

Diodes and Transistors

1-3 Sketch the waveform of v_{out} in Fig. 1-29a.

1-4 In Fig. 1-29b, what is the output waveform?

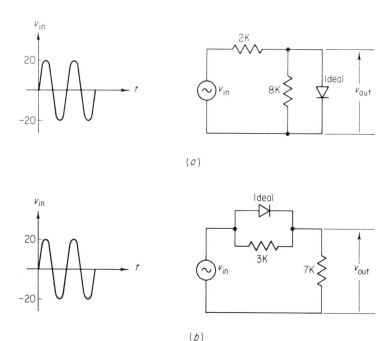

Fig. 1-29

1-5 What is the output waveform in Fig. 1-30a?

1-6 Sketch v_{out} in the circuit of Fig. 1-30b.

1-7 In Fig. 1-29b, replace the ideal diode by the second approximation of a silicon diode, and sketch v_{out}.

1-8 In Fig. 1-30a, use the second approximation of a silicon diode and sketch v_{out}.

1-9 Sketch v_{out} in Fig. 1-30b, using the second approximation of a silicon diode in the place of the ideal diode.

1-10 In Fig. 1-31, sketch the waveform of v_{out} for the following diodes:
  (a) A germanium diode with a reverse current of 0.125 ma for a reverse voltage of 50 volts. Neglect the forward voltage drop.
  (b) A silicon diode with a reverse current of 1.25 μa for a reverse voltage of 50 volts. Neglect the forward voltage drop.

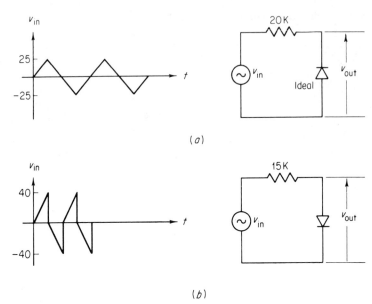

Fig. 1-30

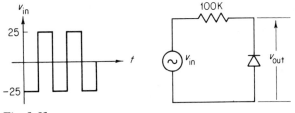

Fig. 1-31

1-11 Find the minimum and maximum values of current through the zener diode of Fig. 1-32a. Use the ideal-zener-diode approximation.

1-12 In Fig. 1-32b, find the minimum and maximum values of zener-diode current for an $R = 10$ kilohms.

1-13 Find the minimum value of beta that just saturates the transistor of Fig. 1-33a.

1-14 In Fig. 1-33a, the d-c beta equals 50. Find the value of I_B. Is the transistor saturated? What is the value of V_{CE} and I_C?

1-15 In Fig. 1-33b, the d-c beta is 100. Find the base current, the collector current, and the voltage from collector to ground. Use both the ideal-transistor approximation and the second approximation.

1-16 What is the minimum value of beta that saturates the transistor of Fig. 1-33b?

Diodes and Transistors

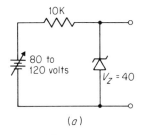
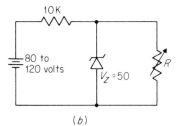

Fig. 1-32

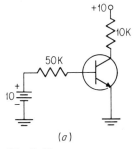
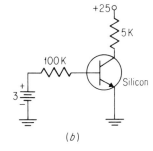

Fig. 1-33

1-17 In Fig. 1-22b, what is the voltage of the second collector if the beta is 50 instead of 100 for each transistor?

1-18 A transistor circuit has the following switching characteristics: $t_d = 10$ nsec, $t_r = 35$ nsec, $t_s = 90$ nsec, and $t_f = 35$ nsec. Find the approximate value of the maximum switching frequency. The transistor is alternately saturated and cut off.

1-19 Suppose that the transistor in the previous problem is *not* saturated. What is the approximate value of the maximum switching frequency?

CHAPTER 2

BINARY AND OCTAL NUMBERS

When we hear the word "number," most of us immediately think of the familiar decimal number system with its 10 digits: 0, 1, 2, 3, 4, 5, 6, 7, 8, and 9. It may seem strange and even useless to talk about number systems with only two digits, or five, or perhaps eight. But using 10 basic symbols as we do today is really a historical accident. A study of anthropology shows that 10 basic symbols arose because we have 10 fingers. No doubt, had we had a different number of digits on our hands, there would now be a different number of digits in our number system.

Besides decimal numbers, two other systems have been widely used in the past. The quinary system with its five basic symbols arose because some people counted with the fingers of only one hand; in the vigesimal system, which has 20 basic symbols, its users counted with the digits of both hands and feet.

In this chapter we will discuss the binary and octal number systems (binary uses two digits, octal uses eight). These two systems are essential to the study of digital circuits; almost all counting and computing circuits use binary numbers (or binary-coded-decimal), whereas devices for getting information into and out of a digital system often use octal numbers.

Binary and Octal Numbers

2-1 Binary Numbers

A number system is nothing more than a code representing quantity. For each distinct quantity, there is an assigned symbol for the quantity. After we memorize the code, we can count, and this leads to arithmetic and higher forms of mathematics.

The most familiar number system is the decimal system, whose digits we have symbolized in Table 2-1. The black circles indicate pebbles which we will use to represent different quantities. In the table are 10 basic symbols or digits: 0 through 9. Each of these symbols stands for a certain number of pebbles. Since we long ago memorized the meaning of each symbol, we immediately know that 4 represents ●●●●, 7 denotes ●●●●●●●, and so on. But these symbols have the meaning they do only because we have thoroughly *memorized them*. Other symbols could just as easily be used. For instance, instead of 0, 1, 2, ... , 9, we can use $A, B, C, ... , J$. In this case, C would mean ●●, D would denote ●●●, and so on.

The use of *10 digits*, 0 through 9, *is really unnecessary*. After all, since a number system is only a code, we can use any number of code symbols we want.

Table 2-1 The Decimal Digits

Pebbles	Symbol
None	0
●	1
●●	2
●●●	3
●●●●	4
●●●●●	5
●●●●●●	6
●●●●●●●	7
●●●●●●●●	8
●●●●●●●●●	9

A binary number system is a code that uses only two basic symbols. These digits can be any two distinct characters like A and B, · and -, or the customary 0 and 1. Table 2-2 shows the basic symbols of the binary number system.

Table 2-2 The Binary Digits

Pebbles	Symbol
None	0
●	1

Probably the first question that comes to mind at this point is how do we represent quantities that are greater than ●? What can we use for ●●, ●●●, and so on?

After we reach 9 in the decimal number system, we form combinations of decimal digits to get 10, 11, 12, etc. In other words, the next decimal number after 9 is obtained by using *the second digit followed by the first* to get 10. The decimal number after 10 is obtained by using the second digit followed by second to get 11, and so forth.

In the binary number system we use the same approach. After we reach 1, we have run out of binary digits (there is no 2, 3, . . . in the binary number system). To represent ●●, we merely use the second binary digit followed by the first to get 10. To represent ●●●, we use 11. Thus we count in binary as follows: 0, 1, 10, 11. To avoid confusion with decimal numbers, it helps to read these binary numbers as zero, one, one-zero, and one-one.*

What is the next binary number after 11? It is not 12 because 2 is not a binary digit.

After we reach 99 in the decimal number system, we have exhausted all the two-digit numbers; we then use three digits at a time to get 100, 101, 102, 103, etc.

Table 2-3 Three Ways of Counting to Seven

Quantity	Binary number	Decimal number
None	0	0
●	1	1
●●	10	2
●●●	11	3
●●●●	100	4
●●●●●	101	5
●●●●●●	110	6
●●●●●●●	111	7

*A tribe of the Torres Straits counts in binary as follows: urapun, okosa, okosa-urapun, okosa-okosa.

Binary and Octal Numbers 33

By following this same pattern with binary numbers, we count as follows: 0, 1, 10, 11, 100, 101, 110, 111. Table 2-3 summarizes these numbers. To remember how to count with binary numbers, the following method is quite useful.

1. Think of the decimal numbers.
2. Eliminate any number that contains a digit greater than 1.
3. The numbers that remain are the binary numbers.

For instance, if we think of the decimal numbers and cross out all numbers with a digit greater than 1, we get

0, 1, ~~2~~, ~~3~~, ..., ~~9~~, 10, 11, ~~12~~, ~~13~~, ..., ~~99~~, 100, 101, 1~~0~~2, 1~~0~~3, ..., 1~~0~~9, 110, 111, ...

Collecting the numbers that remain, we have the binary number system:

0, 1, 10, 11, 100, 101, 110, 111, ...

The binary number system of Table 2-3 is really nothing more than a code. Once we are accustomed to working with it, it becomes almost as familiar as the ordinary decimal number system.

Of course, we can form many other number systems by selecting a different number of basic symbols or digits.

The *base* or *radix* of a number system refers to the number of basic symbols used. For instance, 10 is the base of the decimal number system because this system uses 10 digits, 0 through 9. The binary number system has a base of 2 because the only digits used are 0 and 1. The octal number system, which we will study shortly, has a base of 8.

EXAMPLE 2-1

Show how to count in a number system with a base of 3. Use the digits 0, 1, 2.

SOLUTION

The first three numbers are simply 0, 1, 2. At this point we have run out of basic symbols, and we now form two-digit combinations as we do in the decimal number system. Thus, in a base-3 system the next number after 2 is 10 (second digit followed by first). The number after 10 is 11 (second followed by second), and the number after 11 is 12 (second followed by third). Up to this point, we count as follows: 0, 1, 2, 10, 11, 12. Since 2s are allowed, the next two-digit number after 12 is 20. Then come 21 and 22.

Now we have exhausted the allowable two-digit combinations (the remaining two-digit numbers are from 23 to 99 and contain digits greater

Fig. 2-1 Example 2-3.

than 2). We now can form three-digit combinations, then four, five, and so forth. Therefore, here is how we count in a base-3 system:

0, 1, 2, 10, 11, 12, 20, 21, 22, 100, 101, 102, 110, 111, 112, 120, 121, 122, 200, 201, 202, 210, 211, 212, 220, 221, 222, 1000, ...

EXAMPLE 2-2

Instead of using digits 0 and 1 for our binary numbers, let us use A and B, respectively. Show how to count to seven.

SOLUTION

$A, B, BA, BB, BAA, BAB, BBA, BBB$

EXAMPLE 2-3

Suppose that there are three lamps as shown in Fig. 2-1, and we are told that an on-lamp stands for binary number 1, whereas an off-lamp denotes binary number 0. Reading from left to right, what binary number do the lamps symbolize? What is the corresponding decimal number?

SOLUTION

The lamps are on-off-on, which stands for 101. From Table 2-3 we know that this represents decimal number 5.

EXAMPLE 2-4

In Fig. 2-2a we have a transistor switching circuit. Whenever a voltage is approximately 5 volts, we will use binary number 1 to indicate this condition; whenever a voltage is zero, we will use binary number 0 for this. What binary numbers do the input and output waveforms of Fig. 2-2b and c represent? (Read from left to right during *each* second of time.)

SOLUTION

Since +5 volts denotes binary 1, and 0 volts stands for binary 0, we read the input waveform as 10110.

The output waveform indicates binary number 01001.

Incidentally, note that the transistor is either saturated (shorted) or cut off (open) so that the output voltage is either +5 or 0 volts.

Binary and Octal Numbers

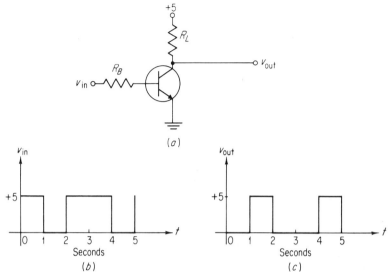

Fig. 2-2 Example 2-4.

2-2 Binary Addition

Addition is a manipulation of numbers that represents the combining of physical quantities. For example, in the decimal number system $2 + 3 = 5$ symbolizes the combining of •• with ••• to get a total of •••••.

To discover the rules for binary addition, we need to discuss four simple cases.

Case 1: When nothing is combined with nothing, we get nothing. The binary representation of this is $0 + 0 = 0$.

Case 2: When nothing is combined with •, we get •. Using binary numbers to denote this, we write $0 + 1 = 1$.

Case 3: Combining • with nothing gives •. The binary equivalent of this is $1 + 0 = 1$.

Case 4: When we combine • with •, the result is ••. Using binary numbers, we symbolize this by $1 + 1 = 10$.

The last result is sometimes very disturbing because of our long-time association with decimal numbers. But it is correct and makes perfect sense *because we are using binary numbers*. Binary number 10 stands for •• and not for ••••••••••(ten). Naturally, it is important to know when we are using binary numbers and when we are using decimal numbers. Ordinarily, it is clear from the context which kind of numbers we are talking about.

But when there is doubt about the kind of numbers involved, we will use a subscript to indicate the base. For instance, we can write 11_2 to denote binary number 11. Or we can write 45_{10} to mean decimal number 45. Thus, both the following equations make good sense:

$$1_{10} + 1_{10} = 2_{10}$$

$$1_2 + 1_2 = 10_2$$

To summarize our results for binary addition, we have

$$0 + 0 = 0$$
$$0 + 1 = 1$$
$$1 + 0 = 1$$
$$1 + 1 = 10 \quad \text{(zero, carry one)}$$

To add larger binary numbers, we carry into higher-order columns as we do with decimal numbers. As an example, suppose we want to add 10 to 10. We do it as follows:

$$\begin{array}{r} 10 \\ +\ 10 \\ \hline 100 \end{array}$$

In the first column, zero plus zero is zero. In the second column, one plus one is zero, carry a one. (Note that in decimal numbers the foregoing addition was $2 + 2 = 4$.)

As another example, suppose we want to add $1 + 1 + 1$. We can add two of the 1s to get $10 + 1$. By adding again, we get 11, that is,

$$1 + 1 + 1 = 10 + 1$$
$$= 11$$

If four 1s are to be added, we can cumulatively add two numbers each time.

$$1 + 1 + 1 + 1 = 10 + 1 + 1$$
$$= 11 + 1$$
$$= 100$$

EXAMPLE 2-5

(a) Add 101 and 110
(b) Add 111 and 110.

SOLUTION

(a) $\quad \begin{array}{r} 101 \\ +110 \\ \hline 1011 \end{array}$ first column: $1 + 0 = 1$
second column: $0 + 1 = 1$
third column: $1 + 1 = 10 \quad$ (zero, carry one)

Binary and Octal Numbers

(b) 111   first column: $1 + 0 = 1$
    $+110$ second column: $1 + 1 = 10$ (zero, carry one)
    ────   third column: $1 + 1 + 1 = 10 + 1 = 11$
    1101

EXAMPLE 2-6

Add the following:
(a) $1010 + 1101$
(b) $1011 + 1010$

SOLUTION

(a)   1010
     $+1101$
     ─────
     10111

(b)   1011
     $+1010$
     ─────
     10101

2-3 Binary-to-Decimal Conversion

How do we convert a binary number into its decimal counterpart? For instance, what decimal number does binary number 1010011 stand for? In this section we learn how to convert a binary number quickly and easily into its decimal equivalent.

We can express any decimal integer (a whole number) in units, tens, hundreds, thousands, and so on. For instance, we can express decimal number 2945 as

$$2945 = 2000 + 900 + 40 + 5$$

In powers of 10, this becomes

$$2945 = 2(10^3) + 9(10^2) + 4(10^1) + 5(10^0)$$

In a similar way, we can partition any binary number into its constituent parts. For instance, binary number 111 becomes

$$111 = 100 + 10 + 1 \tag{2-1}$$

This decomposition is valid because the right side of the equation does indeed equal the left side.

Something interesting occurs when we convert Eq. (2-1) into its decimal counterpart. Using Table 2-3,

$$\begin{array}{ccccccc} 111 & = & 100 & + & 10 & + & 1 \quad \text{binary} \\ \downarrow & & \downarrow & & \downarrow & & \downarrow \quad \downarrow \\ 7 & = & 4 & + & 2 & + & 1 \quad \text{decimal} \end{array}$$

Breaking a binary number into its parts is equivalent to splitting its decimal equivalent into units, twos, fours, eights, and so on. In other words, each digit position in a binary number has a value or weight. The least significant digit (the one on the right) has a value of 1. The second position from the right has a value of 2, the next 4, then 8, 16, 32, etc. Diagrammatically, the digit positions of a binary number have the following decimal weights:

etc. ← | 32 | 16 | 8 | 4 | 2 | 1 |

As a matter of fact, these numbers are in ascending powers of 2.

etc. ← | 2^5 | 2^4 | 2^3 | 2^2 | 2^1 | 2^0 |

The important point about these weights is this: whenever we look at a binary number, we can find its decimal equivalent as follows:

1. When there is a 1 in a particular digit position, add the weight of that position.
2. When there is a 0 in a digit position, disregard the weight of that position.

For example, binary number 101 has a decimal equivalent of

$$101 = 4 + 0 + 1 = 5$$
Binary = decimal

As another example,

$$1011 = 8 + 0 + 2 + 1 = 11$$
Binary = decimal

Still another example is

$$11001 = 16 + 8 + 0 + 0 + 1 = 25$$
Binary = decimal

We can streamline this binary-to-decimal conversion by the following procedure:

1. Write the binary number.
2. Directly under the binary number write 1, 2, 4, 8, 16, ... , working from right to left.
3. If a zero appears in any digit position, cross out the decimal weight for that position.
4. Add the remaining weights to get the decimal equivalent. As an example of this approach, let us convert binary 101 to its decimal equivalent.

Binary and Octal Numbers

Step 1: 1 0 1
Step 2: 4 2 1
Step 3: 4 ~~2~~ 1
Step 4: 4 + 1 = 5

As another example, notice how quickly 10101 is converted to its decimal equivalent.

$$1\ 0\ 1\ 0\ 1$$
$$16\ \not{8}\ 4\ \not{2}\ 1 = 21$$

So far, we have discussed binary integers (whole numbers). How do we convert binary fractions into corresponding decimal equivalents? For instance, what is the decimal equivalent of 0.101? In this case, the weights of each digit position to the right of the binary point are given by

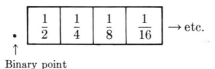

Binary point

or in powers of 2, we have

.| 2^{-1} | 2^{-2} | 2^{-3} | 2^{-4} | → etc.

Binary point

Thus 0.101 would have a decimal equivalent of

0.1 0 1
$\frac{1}{2}\ \not{\frac{1}{4}}\ \frac{1}{8}$ → 0.5 + 0.125 = 0.625

Another example, the decimal equivalent of 0.1101 is

0.1 1 0 1
$\frac{1}{2}\ \frac{1}{4}\ \not{\frac{1}{8}}\ \frac{1}{16}$ → 0.5 + 0.25 + 0.0625 = 0.8125

As far as mixed numbers are concerned (numbers that have both an integer and a fractional part), we handle each part according to the rules just developed. The weights for a mixed number are

etc. ← | 2^3 | 2^2 | 2^1 | 2^0 | 2^{-1} | 2^{-2} | 2^{-3} | → etc.

Binary point

With practice, the conversion of binary numbers into decimal numbers becomes a fast and easy procedure.

EXAMPLE 2-7

Convert binary 110.001 to a decimal number.

SOLUTION

$$1\ 1\ 0\ .\ 0\ 0\ 1$$
$$4\ 2\ \cancel{1}\ \cancel{\tfrac{1}{2}}\ \cancel{\tfrac{1}{4}}\ \tfrac{1}{8} \rightarrow 6.125$$

EXAMPLE 2-8

Solve for x in the following equation:

$$1011.11_2 = x_{10}$$

SOLUTION

This is just another way of asking for the decimal equivalent of a binary number. (Remember that the subscript tells us the base or radix of the number system.)

$$1\ 0\ 1\ 1\ .\ 1\ 1$$
$$8\ \cancel{4}\ 2\ 1\ \tfrac{1}{2}\ \tfrac{1}{4} = 11.75$$

Therefore, $1011.11_2 = 11.75_{10}$.

2-4 Decimal-to-Binary Conversion

One way to convert a decimal number into its binary equivalent is to reverse the process described in the preceding section. For instance, suppose we want to convert decimal number 9 into the corresponding binary number. All we need to do is express 9 as a sum of powers of 2, and then write 1s and 0s in the appropriate digit positions. Thus,

$$\text{Decimal } 9 = 8 + 1 = 8 + 0 + 0 + 1$$
$$= 1001 \text{ binary}$$

As another example,

$$\text{Decimal } 25 = 16 + 8 + 1 = 16 + 8 + 0 + 0 + 1$$
$$= 11001 \text{ binary}$$

A more popular method for converting decimal numbers to binary numbers is the double-dabble method. We will first illustrate how to convert a decimal integer to a binary integer. In the double-dabble method we progressively divide the decimal number by 2, writing down the remainder after each division. The remainders taken in reverse order form the binary number. We can understand this method better after going through some

Binary and Octal Numbers

examples. Suppose we start by converting 9 to a binary number. We progressively divide by 2 as shown in the following:

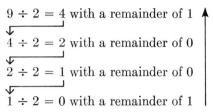

In the first step we divide 9 by 2 to get 4 with a remainder of 1. Next we divide the result of the first division (4) by 2 to get 2 with a remainder of 0. We continue in this fashion, progressively dividing the result of each division by 2, and writing down the remainder after each division. In the last step, $1 \div 2 = 0$ with a remainder of 1. In other words, 2 is not contained in 1; so we write 0 with a remainder of 1. The remainders taken in reverse order (from the bottom, up) give us the answer, which is 1001.

Let us convert decimal 25 to its binary equivalent using the double-dabble method.

$$25 \div 2 = 12 \text{ with a remainder of } 1$$
$$12 \div 2 = 6 \text{ with a remainder of } 0$$
$$6 \div 2 = 3 \text{ with a remainder of } 0$$
$$3 \div 2 = 1 \text{ with a remainder of } 1$$
$$1 \div 2 = 0 \text{ with a remainder of } 1$$

The remainders taken in reverse order (from the bottom, up) give the binary equivalent, which is 11001. The foregoing process can be simplified by writing down the numbers in an orderly fashion like

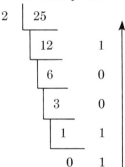

In other words, starting at the top, we divide 25 by 2 and write 12 with a remainder of 1. Then we divide 12 by 2 and write 6 with a remainder of 0, and so forth. With a little practice we can quickly and easily convert decimal integers to binary integers.

As far as fractions are concerned, we do not divide by 2; we *multiply* by 2 and record a carry into the integer position. The carries taken in the *forward* order are the binary fraction. As an example, convert 0.625 to a binary fraction.

$0.625 \times 2 = 1.25 = 0.25$ with a carry of 1
$0.25 \times 2 = 0.5$ with a carry of 0
$0.5 \times 2 = 1.0 = 0$ with a carry of 1

By taking the carries in forward order, we get 0.101, the binary equivalent of 0.625.*

As another example, take 0.85 and convert to binary.

$0.85 \times 2 = 1.7 = 0.7$ with a carry of 1
$0.7 \times 2 = 1.4 = 0.4$ with a carry of 1
$0.4 \times 2 = 0.8 = 0.8$ with a carry of 0
$0.8 \times 2 = 1.6 = 0.6$ with a carry of 1
$0.6 \times 2 = 1.2 = 0.2$ with a carry of 1
$0.2 \times 2 = 0.4 = 0.4$ with a carry of 0

Taking the carries in forward order, we get binary fraction 0.110110. In this case, we stopped the process after getting 6 binary digits; our answer is a good approximation. If more accuracy is needed, we continue multiplying by 2 until we have as many digits as necessary.

Incidentally, the word "bit" is very common in digital work. It is simply a short way of saying "binary digit." In other words, a binary number like 1101 has 4 binary digits, or 4 bits. 111010 has 6 binary digits, or 6 bits.

EXAMPLE 2-9

Convert 21.6_{10} to a binary number.

SOLUTION

Split 21.6 into an integer of 21 and a fraction of 0.6, and apply double-dabble to each part.

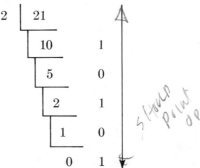

*Note that in this example the last product is 1.0. Whenever 1.0 is a product, we can terminate the conversion because all succeeding carries must be 0s.

Binary and Octal Numbers

and

$$0.6 \times 2 = 1.2 = 0.2 \text{ with a carry of } 1$$
$$0.2 \times 2 = 0.4 \text{ with a carry of } 0$$
$$0.4 \times 2 = 0.8 \text{ with a carry of } 0$$
$$0.8 \times 2 = 1.6 = 0.6 \text{ with a carry of } 1$$
$$0.6 \times 2 = 1.2 = 0.2 \text{ with a carry of } 1$$

The binary number is 10101.10011. This 10-bit number is an approximation of 21.6_{10} because we terminated the conversion of the fractional part after 5 bits.

2-5 Binary Subtraction

To subtract binary numbers, we first need to discuss four simple cases.

Case 1: $0 - 0 = 0$
Case 2: $1 - 0 = 1$
Case 3: $1 - 1 = 0$
Case 4: $10 - 1 = 1$

The last result represents

$$\bullet\bullet - \bullet = \bullet$$

which is obviously correct.

To subtract larger binary numbers, we use a method that is similar to that used with decimal numbers — we subtract column by column, borrowing from the adjacent column when necessary. For instance, in subtracting 101 from 111, we proceed as follows (in all our examples we will also show the equivalent decimal subtraction):

```
    7     111     first column:  1 − 1 = 0
   −5    −101     second column: 1 − 0 = 1
   ──    ────     third column:  1 − 1 = 0
    2     010
```

Here is another example: subtract 1010 from 1101.

```
   13    1̶1̶01     first column:  1 − 0 = 1
  −10   −1010     second column: 10 (after borrow) − 1 = 1
  ───   ─────     third column:  0 (after borrow) − 0 = 0
    3    0011     fourth column: 1 − 1 = 0
```

Binary numbers can also be negative, just like decimal numbers. As an illustration, subtract 111 from 100. We get

```
    4     100     first column:  1 − 0 = 1
   −7    −111     second column: 1 − 0 = 1
   ──    ────     third column:  1 − 1 = 0
   −3    −011
```

Note that we subtracted the *smaller magnitude from the larger*, and prefixed the *sign of the larger magnitude;* this is identical to the procedure used with decimal numbers.

For another example, let us subtract 11011 from 10111.

$$\begin{array}{rr} 23 & 10111 \\ -27 & -11011 \\ \hline -\ 4 & -00100 \end{array}$$

Again note that we subtract the upper number from the lower because we must subtract the smaller magnitude from the larger, just as we do with decimal numbers.

The foregoing examples illustrate one method of doing subtraction. Many digital computers subtract in this way. However, there are other methods for doing subtraction which require less circuitry. Before we describe these other ways to subtract, we need to define the 1's and 2's complements.

The 1's complement of a binary number is the number we get when we change each 0 to a 1, and each 1 to a 0. In other words, the 1's complement of 100 is 011 (each 1 is changed to 0, and each 0 to 1). The 1's complement of 1001 is 0110. Thus we examine each bit in the binary number, and change the 1s to 0s and the 0s to 1s. More examples of 1's complements are

1010 has a 1's complement of 0101
1110 has a 1's complement of 0001
0011 has a 1's complement of 1100

The 2's complement is the binary number we get when we *add* 1 to the 1's complement. That is,

2's complement = 1's complement + 1

For instance, to find the 2's complement of 1011, we first get the 1's complement, which is 0100. Next, we add 1 to 0100 to get 0101 — the 2's complement of 1011. For another example, find the 2's complement of 11001. First, get the 1's complement, which is 00110. After adding 1 to this, we have 00111. Therefore, 00111 is the 2's complement of 11001. Here are some more worked-out 2's complements.

Number	*2's complement*
1110	0001 + 1 = 0010
0001	1110 + 1 = 1111
10110	01001 + 1 = 01010

Binary and Octal Numbers

How do we use complements to do subtraction? Instead of subtracting a number, we can add its 2's complement, and disregard the last carry. For instance, to subtract 101 from 111, we proceed as follows:

Decimal	Conventional	2's complement	
7	111	111	
−5	−101	+011	(2's complement of 101)
2	010	̶1010	

Note that we added the 2's complement, and crossed out the carry. What remains is the same answer that we get by conventional subtraction.

Take another example: subtract 1010 from 1101.

13	1101	1101
−10	−1010	+0110
3	0011	̶10011

Again, by forming the 2's complement and adding, we get the same answer as with conventional subtraction, provided we disregard the carry.*

Some digital computers do subtraction by using the 2's-complement method. The advantage of this approach is a reduction in hardware; instead of having digital circuits that add and subtract, we need only adding-type circuits when we use the 2's complement.

Still another approach is the 1's-complement method of subtraction. Instead of subtracting a number, we add the 1's complement of the number. The last carry is then added to the number to get the final answer. The method is best understood by doing an example. Let us subtract 101 from 111. We proceed as follows:

```
    111
   +010      (the 1's complement of 101)
  ┌─1001
  │  001
  └→ +1
     010
```

Here is what we did: (1) We formed the 1's complement of 101 to get 010. (2) We added 010 to 111 to get 1001. (3) The last carry is 1. Whenever there is a 1 carry in the last position, we *remove* it and *add* it onto the remainder as shown. This is called "end-around carry."

---

*A thorough study of the 2's complement shows that the carry indicates whether the answer is positive or negative. When the carry is 1, the answer is positive. When there is no carry, however, the answer is negative and is the 2's complement of the actual magnitude.

Take another example. Subtract 1010 from 1101.

```
  13      1101        1101
 -10     -1010       +0101     (the 1's complement of 1010)
 ---     -----      ┌─10010
   3      0011      │  0010
                    └→  +1
                       ────
                       0011
```

Whenever there is an end-around carry, the answer is positive and the magnitude is expressed in binary form. But when there is no end-around carry, this indicates that the answer is negative and that it is in the 1's-complement form. For instance, when we subtract 1101 from 1010, we get

```
  10      1010        1010
 -13     -1101       +0010     (the 1's complement of 1101)
 ---     -----        ────
 - 3     -0011        1100 → -0011
                       ↑
```

(There is no carry.)

Whenever there is no end-around carry, as in this case, the answer is negative and in 1's-complement form. Therefore, we take the 1's complement of 1100 to get 0011 and prefix a minus sign for a final answer of −0011.

The general proofs for using the 1's and 2's complements must be left to advanced books in finite mathematics.

The use of the 1's complement to do subtraction is popular in digital computers because only adders are needed; besides, it is quite easy with digital circuits to get the 1's complement of a number (this will be shown in a later chapter).

EXAMPLE 2-10

Subtract 01101 from 11011. Use the 1's complement.

SOLUTION

```
          11011
        +10010      (the 1's complement of 01101)
       ┌─101101
       │  01101
       └→   +1
           ─────
           01110
```

EXAMPLE 2-11

Subtract 11011 from 01101. Use the 1's complement. Also, show the decimal subtraction and the conventional binary subtraction.

Binary and Octal Numbers

SOLUTION

```
   13      01101     01101
  -27     -11011    +00100     (1's complement)
  ---     ------    ------
  -14     -01110    10001  →  -01110
                    ↑
                    └─ (No carry)
```

Since there is no carry, we take the 1's complement and prefix a minus sign.

2-6 The 9's and 10's Complements

We have just seen that the 1's and 2's complements are useful in binary addition. There also are complements in the decimal system. The 9's complement (similar to the 1's complement) is found by subtracting each decimal digit from 9. For instance, the 9's complement of 25 is obtained as follows:

$$\begin{array}{r} 99 \\ -25 \\ \hline 74 \end{array}$$ the 9's complement of 25

We subtracted each decimal digit from 9. As another example, the 9's complement of 6291 is

$$\begin{array}{r} 9999 \\ -6291 \\ \hline 3708 \end{array}$$ the 9's complement of 6291

Again, we subtracted each decimal digit from 9.

The 10's complement (similar to the 2's complement) of a decimal integer is one greater than the 9's complement,* that is,

$$10\text{'s complement} = 9\text{'s complement} + 1$$

As an example, we find the 10's complement of 25 as follows:

$$\begin{array}{r} 99 \\ -25 \\ \hline 74 \end{array} + 1 = 75$$ the 10's complement of 25

*If mixed numbers are used, we add 1 to the least significant digit. For instance, for 4.52, the 9's complement is 5.47, and the 10's complement is 5.48.

As another example, the 10's complement of 6291 is

$$\begin{array}{r} 9999 \\ -6291 \\ \hline 3708 \end{array} + 1 = 3709 \qquad \text{the 10's complement of 6291}$$

We can use 9's and 10's complements to subtract decimal numbers just as we did with binary numbers. For instance, to subtract 25 from 83 using the 9's complement we proceed as follows:

$$\begin{array}{r} 83 \\ -25 \\ \hline 58 \end{array} \qquad \begin{array}{r} 83 \\ +74 \\ \hline \lceil 157 \\ 57 \\ \hookrightarrow 1 \\ \hline 58 \end{array} \qquad \begin{array}{l} \text{the 9's complement of 25} \\ \\ \text{end-around carry} \end{array}$$

We get the same answer with the 10's complement:

$$\begin{array}{r} 83 \\ +75 \\ \hline \cancel{1}58 \end{array} \qquad \text{the 10's complement of 25}$$

We disregard the last carry, the same as we do with the 2's complement.

2-7 Binary Multiplication and Division

Binary multiplication is much easier than decimal because the multiplication table for binary has only four cases.

Case 1: $0 \times 0 = 0$
Case 2: $0 \times 1 = 0$
Case 3: $1 \times 0 = 0$
Case 4: $1 \times 1 = 1$

To handle larger numbers, that is, numbers with several bits, we follow the pattern used with decimal numbers — we form partial products and add these. For instance, in multiplying 111 by 101, we proceed as follows:

$$\begin{array}{r} 7 \\ \times 5 \\ \hline 35 \end{array} \qquad \begin{array}{r} 111 \\ \times 101 \\ \hline 111 \\ 000 \\ 111 \\ \hline 100011 \end{array}$$

Binary and Octal Numbers 49

Consider another example: 10110 times 110.

```
       22            10110
      ×6            ×110
      ---           -----
      132           00000
                   10110
                  10110
                  --------
                  10000100
```

Digital circuits for multiplying are straightforward, and we will discuss them in a later chapter.

Division follows the same pattern in binary as in decimal. To divide 1100 by 10, we proceed as follows:

```
       6            110
    ------       --------
   2 | 12      10 | 1100
                   10
                   --
                    10
                    10
                    --
                    00
```

With practice, binary multiplication and division become easier than decimal multiplication and division.

EXAMPLE 2-12

Multiply 101.1 by 11.01.

SOLUTION

```
       5.5            101.1
     ×3.25          ×11.01
     -----          ------
       275            1011
       110            0000
       165           1011
     ------         1011
      17875         --------
      17.875        10001111
                    10001.111
```

2-8 Octal Numbers

The octal numbers are quite important in digital work. To begin with, the octal number system has a base of eight, meaning that it has eight basic symbols. Although we can use any eight distinct symbols, it is customary to use the first eight decimal digits. In other words, the digits of the octal number system are

$$0, 1, 2, 3, 4, 5, 6, 7$$

(There is no 8 or 9.) These digits, 0 through 7, have exactly the same physical meaning as decimal symbols; that is, 2 stands for ••, 5 symbolizes •••••, and so on.

How do we count beyond 7 with octal numbers? As with binary and decimal numbers, after running out of basic symbols we merely form two-digit combinations, taking the second digit followed by the first digit, then the second followed by the second, and so forth. Therefore, after we reach 7 in octal numbers, the next number is 10 (second digit followed by first), then 11 (second followed by second), and so on. We count as follows with octal numbers:

$$0, 1, 2, 3, 4, 5, 6, 7,$$
$$10, 11, 12, 13, 14, 15, 16, 17,$$
$$20, 21, 22, 23, 24, 25, 26, 27, \ldots, \text{etc.}$$

To remember the octal numbers, think of the decimal numbers and cancel any number with a digit greater than 7 (that is, 8 or 9). The numbers that remain are octal numbers. In other words, after thinking of the decimal numbers and canceling out numbers containing 8 or 9, we have

0, 1, 2, 3, 4, 5, 6, 7, 8̸, 9̸, 10, 11, 12, 13, 14, 15, 16, 17, 1̸8̸, 1̸9̸, 20, 21, ..., 75, 76, 77, 7̸8̸, 7̸9̸, 8̸0̸, ..., 100, 101, ...

The numbers left over are the octal numbers.

How do we convert octal numbers to decimal numbers? In the octal number system each digit corresponds to a power of 8. The weights of each digit position in an octal number are as follows:

etc. ← | 8^3 | 8^2 | 8^1 | 8^0 | 8^{-1} | 8^{-2} | 8^{-3} | → etc.

↑
Octal point

Therefore, to convert from octal to decimal we need only multiply each octal digit by its weight and add the resulting products. For instance, in powers of 8, octal number 23 becomes

$$\text{Octal} \quad 23 = 2(8^1) + 3(8^0) = 16 + 3 = 19 \quad \text{decimal}$$

What decimal number does 257_8 represent? Expressed in powers of 8, 257_8 becomes

$$257_8 = 2(8^2) + 5(8^1) + 7(8^0) = 128 + 40 + 7 = 175_{10}$$

How do we convert in the opposite direction, that is, from decimal to octal? A method like double-dabble is useful in octal numbers. Instead of

Binary and Octal Numbers

progressively dividing by 2 (the base of binary numbers), we progressively divide by 8 (the base of octal numbers), writing down the remainders after each division. The remainders taken in reverse order form the octal number. As an example, we convert decimal 19 as follows:

$$19 \div 8 = 2 \text{ with a remainder of } 3$$
$$2 \div 8 = 0 \text{ with a remainder of } 2$$

In reverse order the remainders give 23, the octal equivalent of decimal 19. To convert 175_{10} to an octal number :

$$175 \div 8 = 21 \text{ with a remainder of } 7$$
$$21 \div 8 = 2 \text{ with a remainder of } 5$$
$$2 \div 8 = 0 \text{ with a remainder of } 2$$

We can condense these steps by writing

```
8 | 175
    | 21   7
       | 2   5
          | 0   2
```

With decimal fractions we progressively multiply instead of divide, writing down the carry into the integer position. An example of this is to convert decimal 0.23 into an octal fraction.

$$0.23 \times 8 = 1.84 = 0.84 \text{ with a carry of } 1$$
$$0.84 \times 8 = 6.72 = 0.72 \text{ with a carry of } 6$$
$$0.72 \times 8 = 5.76 = 0.76 \text{ with a carry of } 5$$
$$\text{etc.}$$

The carries taken in forward order give the octal fraction 0.165. We terminated after three places; if more accuracy were required, we would continue multiplying to obtain more octal digits.

2-9 Octal-Binary Conversion

By far, the most important use of octal numbers lies in octal-binary conversions. The relation between octal digits and binary digits is obtained by counting to 7 in each system:

000	001	010	011	100	101	110	111
0	1	2	3	4	5	6	7

With this tabulation we can convert any octal number up to 7 into its binary equivalent.

How do we convert larger octal numbers? Because the base 8 of octal numbers is the third power of 2, the base of binary numbers, the conversion from octal to binary numbers becomes very simple. We merely convert one octal digit at a time to its binary equivalent. For instance, to change octal 23 to its binary equivalent, we proceed as follows:

$$\begin{array}{cc} 2 & 3 \\ \downarrow & \downarrow \\ 010 & 011 \end{array}$$

We have converted each octal digit into its binary equivalent (2 becomes 010, and 3 becomes 011). The binary equivalent of octal 23 is 010 011, or 010011. Sometimes, the space is left between each group of 3 bits; this makes it easier to read the binary number. In other words, we can leave 010011 in the form of 010 011.

As another example, convert octal 3574 to its binary equivalent.

$$\begin{array}{cccc} 3 & 5 & 7 & 4 \\ \downarrow & \downarrow & \downarrow & \downarrow \\ 011 & 101 & 111 & 100 \end{array}$$

Hence, binary 011101111100 is equivalent to octal 3574. Clearly, the binary number is easier to read if we leave a space between each group of 3 bits.

Mixed octal numbers present no problem at all. We simply convert each octal digit to its equivalent binary value. Octal 34.562 becomes 011 100.101 110010 in binary.

Converting from binary to octal is merely a reversal of the foregoing procedures. Simply remember to group the bits in threes, starting at the binary point; then convert each group of three to its octal equivalent (0s are added at each end, if necessary). For instance, binary number 1011.01101 converts as follows:

$$\begin{array}{c} 1011.01101 \rightarrow 001\ 011.011\ 010 \\ \downarrow\ \ \ \downarrow\ \ \ \ \downarrow\ \ \ \downarrow \\ 1\ \ \ \ 3\ \ .\ \ 3\ \ \ \ 2 \end{array}$$

We start at the binary point and, working both ways, separate the bits in groups of three. When necessary, as in this case, we add 0s to complete the outside groups. Then we convert each group of three into its binary equivalent. Hence,

$$1011.01101_2 = 13.32_8$$

The simplicity of converting octal to binary and vice versa has many advantages in digital work. For one thing, getting information into and out

Binary and Octal Numbers

of a digital system is easier and requires less circuitry because it is easier to program, read, record, and print out octal numbers than binary numbers. Another advantage is that large decimal numbers are more easily converted to binary if we first convert to octal, and then to binary; the reason for this is that a direct decimal-to-binary conversion requires many more divisions than a decimal-to-octal conversion when large numbers are involved.

EXAMPLE 2-13

What is the binary equivalent of decimal number 363? Convert to octal, then to binary.

SOLUTION

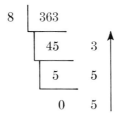

Next, we convert octal 553 to its binary equivalent, which is

101 101 011   (the binary equivalent of 363_{10})

We could have converted 363 directly to binary by progressively dividing by 2, but this would have been tedious, requiring 9 divisions. The decimal-octal-binary method speeds up the conversion process.

SUMMARY

A number system relates quantities and symbols. The base or radix tells us how many digits the number system uses. For instance, the decimal system has a base of 10, binary has a base of 2, and octal uses a base of 8.

In binary numbers $1 + 1 = 10$ because we are combining • with • to get ••. When necessary, a subscript can be used to indicate the base. Thus for clarity we can write

$$1_2 + 1_2 = 10_2$$

To convert from binary to decimal numbers, add the weights of each bit position (1, 2, 4, 8, . . .) when there is a 1 in that position. To convert from decimal to binary numbers, double-dabble is useful.

Octal numbers are most important in making octal-binary conversions. Octal numbers also expedite decimal-binary conversions.

GLOSSARY

base The number of digits or basic symbols in a number system. For instance, the decimal system has a base of 10 because it uses 10 digits. Octal has a base of 8, binary a base of 2.

binary Refers to a number system with a base of 2, that is, with 2 basic symbols or digits.

bit An abbreviated form of *binary digit*. For instance, instead of saying that 10110 has 5 binary digits, we can say that it has 5 bits.

digit A single basic symbol used in a number system. The decimal system has 10 digits, 0 through 9.

end-around carry Used with the 1's complement in subtracting. It means we take the last carry and add it to the least significant digit (Sec. 2-5).

integer A whole number. For instance, 5, 17, and 163 are integers.

octal This refers to a number system with a base of 8, that is, one that uses eight basic symbols or digits. Normally, these are 0, 1, 2, 3, 4, 5, 6, and 7.

quinary This describes a number system with a base of 5. A quinary system would use 0, 1, 2, 3, and 4 (or any other five distinct symbols).

radix A synonym for the word *base*.

vigesimal This describes a number system with a base of 20.

weight Refers to the decimal value of each digit position of a number. For decimal numbers the weights are 1, 10, 100, 1000, . . . , working from the decimal point to the left. For binary numbers the weights are 1, 2, 4, 8, . . . to the left of the binary point. And with octal numbers, the weights become 1, 8, 64, . . . to the left of the octal point.

REVIEW QUESTIONS

1. How many digits are there in the decimal number system? What are they?
2. Count to 7 using binary numbers.
3. What is the base or radix of a number system?
4. What does 110_2 mean? And 110_{10}?
5. What are the decimal weights of each binary position to the left of the binary point? What are the weights to the right of the binary point?
6. Describe the double-dabble method for converting a decimal integer into a binary integer.

Binary and Octal Numbers

7. What is the condensed word for *binary digit?*
8. Define the 1's and 2's complements of a binary number.
9. In the 2's-complement method of subtraction, what is done with a carry in the last position? What is done with a last carry when the 1's complement is used?
10. What does end-around carry mean?
11. What is the radix of the octal number system? What are the digits in this system? The octal number following 17 is what?
12. Why is the octal number system so useful in digital work?

PROBLEMS

2-1 Write the first 27 binary numbers starting with 1.
2-2 A base-3 number system is to use the symbols X, Y, and Z instead of 0, 1, and 2. Write the first 20 numbers in this system, starting with 1.
2-3 Add the following binary numbers:
   (a) 1011 and 1001.
   (b) 1110001 and 1010101.
   (c) 1111010 and 1001101.
2-4 Add the following:
   (a) 1111 and 1011.
   (b) 111, 111, and 111.
   (c) 1110010111 and 1100110011.
2-5 Convert the following decimal numbers to binary numbers: 24, 65, and 106.
2-6 What binary number does decimal 268 stand for?
2-7 Convert decimal 23.45 to a binary number.
2-8 What decimal number does 110101_2 represent?
2-9 What decimal number does 11001.011_2 represent?
2-10 Subtract binary 1101 from 11110.
2-11 Subtract the following binary numbers:
   (a) $111 - 1010$.
   (b) $110011 - 100011$.
   (c) $100011 - 111010$.
2-12 What are the 1's and 2's complements of the following binary numbers: 101010, 111101, 10001000, 00000010?
2-13 What 4-bit number is equal to its 2's complement? Is there any 4-bit number that is equal to its 1's complement?
2-14 In Fig. 2-2b the waveform can be interpreted as a binary number if we let $+5$ volts stand for binary number 1, and 0 volts for binary number 0. What is the 1's complement of the input waveform and how is this complement related to the output waveform?

2-15 Use the 1's and 2's complements to perform the following binary subtractions:
  (a) 1111 − 1011.
  (b) 110011 − 100101.
  (c) 100011 − 111010.

2-16 What binary number does 1100 × 1010 equal?

2-17 Multiply the following binary numbers:
  (a) 111 × 101.
  (b) 11.110 × 100.1.
  (c) −11101 × 100.1.

2-18 What does 101.111 × 1000.101 equal in binary?

2-19 Divide 11011_2 by 100_2.

2-20 Divide these binary numbers:
  (a) 11001 by 110.
  (b) 110.10 by 100.1.

2-21 Convert octal 65 to a decimal number.

2-22 What is the octal equivalent of decimal number 583?

2-23 Perform the following octal-decimal conversions:
  (a) $654_{10} = x_8$.
  (b) $x_{10} = 327_8$.

2-24 Convert the following octal numbers to binary numbers: 34, 567, 4673.

2-25 Convert the following binary numbers to octal numbers:
  (a) 10101111.
  (b) 1101.0110111.
  (c) 1010011.101101.

2-26 The following decimal numbers are to be converted to binary numbers by converting first to octal, then to binary: 352, 850, 7563.

CHAPTER 3

BINARY CODES

Having used decimal numbers for many years, we would like to keep using these numbers. Digital systems, however, force us to use some form of binary numbers. Fortunately, we can strike a compromise. We can use binary-coded decimals (BCD). These are codes that combine some of the features of both decimal and binary numbers. There are an enormous number of BCD codes. In this chapter we will examine some of the common ones.

3-1 The 8421 Code

The 8421 code expresses each decimal digit by its 4-bit binary equivalent. For instance, take a decimal number like 429. Each decimal digit is changed to its binary equivalent as follows:

$$\begin{array}{ccc} 4 & 2 & 9 \\ \downarrow & \downarrow & \downarrow \\ 0100 & 0010 & 1001 \end{array}$$

Thus, in the 8421 code, 0100 0010 1001 stands for decimal number 429. As another example, let us encode 8963.

$$
\begin{array}{cccc}
8 & 9 & 6 & 3 \\
\downarrow & \downarrow & \downarrow & \downarrow \\
1000 & 1001 & 0110 & 0011
\end{array}
$$

Again, we have changed each decimal digit to its natural binary equivalent.

Table 3-1 shows more of the 8421 code. As you can see, each decimal digit is changed to its equivalent 4-bit group. Especially note that 1001 is the largest 4-bit group in the 8421 code. In other words, we use only 10 of the 16 possible 4-bit groups. The 8421 code does not use the numbers 1010, 1011, 1100, 1101, 1110, and 1111. (If any of these forbidden numbers appeared as an answer in a machine using the 8421 code, we would know that an error had occurred.)

Table 3-1

Decimal	8421	Binary
0	0000	0000
1	0001	0001
2	0010	0010
3	0011	0011
4	0100	0100
5	0101	0101
6	0110	0110
7	0111	0111
8	1000	1000
9	1001	1001
10	0001 0000	1010
11	0001 0001	1011
12	0001 0010	1100
13	0001 0011	1101
.	.	.
.	.	.
.	.	.
98	1001 1000	1100010
99	1001 1001	1100011
100	0001 0000 0000	1100100
101	0001 0000 0001	1100101
102	0001 0000 0010	1100110
.	.	.
.	.	.
.	.	.
578	0101 0111 1000	1001000010
.	.	.
.	.	.
.	.	.

(handwritten annotation: "Forbidden states" pointing to binary values 1010–1101)

Binary Codes   **59**

The 8421 code is identical to natural binary through decimal number 9. Because of this, it is called the 8421 code; the weights in each group are 8, 4, 2, 1, reading from left to right — the same as for natural binary numbers.

Above 9, the 8421 code differs sharply from the binary-number code. For instance, the straight binary number for 12 is 1100, but the 8421 number for 12 is 0001 0010. Or, decimal number 24 is simply 11000 in straight binary, but it becomes 0010 0100 in the 8421 code. Thus, above 9 every binary number differs from the corresponding 8421 number.

The main advantage of the 8421 code is the ease of converting to and from decimal numbers; we need only remember the binary numbers for 0 through 9 because we encode only one decimal digit at a time. A disadvantage, however, of the 8421 code is that the rules for binary addition do not apply to the entire 8421 number, but only to the individual 4-bit groups. For instance, to add 12 and 9 in straight binary is easy.

$$\begin{array}{rr} 12 & 1100 \\ +\ 9 & +1001 \\ \hline 21 & 10101 \to 21 \end{array}$$

If we try it in the 8421 code, we get an unacceptable answer.

$$\begin{array}{rr} 12 & 0001\ 0010 \\ +\ 9 & +\ \ \ \ 1001 \\ \hline 21 & 0001\ 1011 \end{array}$$

We are unable to decode 0001 1011 because 1011 does not exist in the 8421 code. Remember the largest 8421 code group is 1001 (9). Therefore, addition of 8421 numbers is not so simple as for straight binary numbers. (We discuss this further in Chap. 5.)

The 8421 code illustrates one of the many codes referred to as binary-coded decimals (BCD). An enormous number of such codes exist.* In general, a BCD code is one in which the digits of a decimal number are encoded — one at a time — into groups of binary digits. For this encoding we can use 4-bit groups, or 5-bit groups, or 6-bit, etc.

The 8421 code is a mixed-base code; it is straight binary within each group of 4 bits, but it is decimal from group to group.

Because the 8421 BCD code is the most natural type of BCD code, it is often simply referred to as BCD without qualifying it. In other words, if we simply refer to *BCD code*, we mean the 8421 code.

EXAMPLE 3-1

Encode the following decimal numbers into 8421 numbers:

(a) 45
(b) 732
(c) 94,685

*For 4-bit BCD codes we use only 10 of 16 possible states, so that there are 16!/6! possible codes, which is approximately 30 billion codes.

Solution

(a)
$$\begin{array}{cc} 4 & 5 \\ \downarrow & \downarrow \\ 0100 & 0101 \end{array}$$

(b)
$$\begin{array}{ccc} 7 & 3 & 2 \\ \downarrow & \downarrow & \downarrow \\ 0111 & 0011 & 0010 \end{array}$$

(c)
$$\begin{array}{ccccc} 9 & 4 & 6 & 8 & 5 \\ \downarrow & \downarrow & \downarrow & \downarrow & \downarrow \\ 1001 & 0100 & 0110 & 1000 & 0101 \end{array}$$

EXAMPLE 3-2

Decode the following 8421 numbers:
(a) 1000 0101 0110 0011
(b) 0011 0101 0001 1001 0111

Solution

(a)
$$\begin{array}{cccc} 1000 & 0101 & 0110 & 0011 \\ \downarrow & \downarrow & \downarrow & \downarrow \\ 8 & 5 & 6 & 3 \end{array}$$

(b)
$$\begin{array}{ccccc} 0011 & 0101 & 0001 & 1001 & 0111 \\ \downarrow & \downarrow & \downarrow & \downarrow & \downarrow \\ 3 & 5 & 1 & 9 & 7 \end{array}$$

3-2 The Excess-3 Code

The excess-3 code is another important BCD code. To encode a decimal number into its excess-3 form, we add 3 to *each* decimal digit before converting to binary. For example, let us convert 12 to an excess-3 number.

$$\begin{array}{cc} 1 & 2 \\ +3 & +3 \\ \hline 4 & 5 \end{array}$$

We have added 3 to each decimal digit. Now we convert each digit to its binary equivalent.

$$\begin{array}{cc} 4 & 5 \\ \downarrow & \downarrow \\ 0100 & 0101 \end{array}$$

So, 0100 0101 in the excess-3 code stands for decimal 12.

Binary Codes

Take another example. To convert 29 to an excess-3 number:

$$\begin{array}{cc} 2 & 9 \\ +3 & +3 \\ \hline 5 & 12 \\ \downarrow & \downarrow \\ 0101 & 1100 \end{array}$$

Especially note that after adding 9 and 3, we do *not* carry the 1 into the next column; instead we leave the result intact as 12; then we convert as shown. Thus 0101 1100 in the excess-3 code stands for decimal 29.

Table 3-2 shows the excess-3 code. We added 3 to each decimal digit before converting each group into binary.

The excess-3 code uses only 10 of the 16 possible 4-bit groups. The six unused groups are 0000, 0001, 0010, 1101, 1110, and 1111. If any of these forbidden combinations turned up as an answer in an excess-3 computer, we would know that an error had occurred.

Table 3-2

Decimal	Excess-3
0	0011
1	0100
2	0101
3	0110
4	0111
5	1000
6	1001
7	1010
8	1011
9	1100
10	0100 0011
11	0100 0100
12	0100 0101
13	0100 0110
⋮	⋮
98	1100 1011
99	1100 1100
100	0100 0011 0011
101	0100 0011 0100
102	0100 0011 0101
⋮	⋮
578	1000 1010 1011
⋮	⋮

The excess-3 code is a *self-complementing* code. This means that the 1's complement of any excess-3 number represents the 9's complement of the decimal number. For instance, 0101 stands for decimal 2 when we use the excess-3 code. The 1's complement of 0101 is 1010, which stands for decimal 7. But decimal 7 is the 9's complement of 2. A check on any other excess-3 number gives the same result — the 1's complement represents the 9's complement of the encoded decimal number. (Not all BCD codes are self-complementing. The 8421 code, for instance, does not have this property.)

The excess-3 code is an unweighted code, whereas the 8421 code is weighted. In a weighted code each bit position has a fixed value. For instance, in 8421 code the number 0101 has weights of 8, 4, 2, and 1, reading from left to right. Whenever a 1 appears, we add the weight of its position. Thus,

$$\begin{array}{cccc} 0 & 1 & 0 & 1 \\ \cancel{8} + & 4 + & \cancel{2} + & 1 \end{array} = 5$$

Similarly, in 8421 code the number 1001 has a decimal equivalent of

$$\begin{array}{cccc} 1 & 0 & 0 & 1 \\ 8 + & \cancel{4} + & \cancel{2} + & 1 \end{array} = 9$$

The excess-3 code, however, is unweighted; there is no way to assign fixed weights or values to the bit positions.

The meaning of the term *excess-3* is clear at this point: each 4-bit group is the binary equivalent of a decimal digit that is 3 larger than the encoded digit. In a similar way, an excess-6 number means that each 4-bit group is the binary equivalent of a decimal number that is 6 larger than the encoded digit. For instance, decimal number 2 would be 0101 (5) in excess-3 code, but would be 1000 (8) in excess-6 code.

As pointed out in the preceding section, an obstacle arises when we try to add 8421 numbers whose decimal sum exceeds 9. The excess-3 code was devised to get rid of this obstacle. To understand why, we must discuss two cases.

Case 1: In excess-3 code whenever we add two decimal digits whose sum is 9 or less, an excess-6 number results. To return to excess-3 form, we must subtract 3. For instance, for 2 + 5 we get

$$\begin{array}{rrl} 2 & 0101 & \text{excess-3 equivalent of 2} \\ +5 & +1000 & \text{excess-3 equivalent of 5} \\ \hline 7 & 1101 & \text{excess-6 equivalent of 7} \\ & -0011 & \text{subtract 3} \\ \hline & 1010 & \text{excess-3 equivalent of 7} \end{array}$$

What we did should make sense. We added two excess-3 numbers and got an excess-6 number. To restore the answer to excess-3 form, we subtracted

Binary Codes

3 from it. The final answer is 1010, which is the excess-3 equivalent of 7. (Note: In digital systems it is common practice to return all answers to whatever code the machine uses.)

As another example, let us add 43 and 36.

```
  43        0111 0110      excess-3 for 43
 +36       +0110 1001      excess-3 for 36
 ---       ---------
  79        1101 1111      excess-6 for 79
           -0011 0011      subtract 3 from each group
           ---------
            1010 1100      excess-3 for 79
```

There was no carry in either group, which means that the decimal sum was 9 or less for each group. As a result, the answer is in excess-6 form and we must subtract 3 from each group to return to excess-3 code.

Case 2: Whenever we add decimal digits whose sum exceeds 9, there will be a carry from one group into the next. When this happens, the group that produced the carry will revert to 8421 form; this occurs because of the excess-6 and the six unused 4-bit groups. To restore the answer to excess-3 code, we must add 3 to the group that produced the carry. For instance, let us add 29 and 39.

```
              A      B
  29        0101   1100    excess-3 for 29
 +39       +0110   1100    excess-3 for 39
 ---       -------------
  68        1100   1000    first results
           -0011  +0011    subtract and add 3
           -------------
            1001   1011    excess-3 for 68
```

Here is what happens. In column B we add 1100 and 1100 to get 1000 *with a carry of 1 into column A*. In column A we add 0101 and 0110 and the carry to get 1100 *with no carry*. The first result in column B is back in 8421 form because this column produced a carry. The first result in column A is still in excess-6 form because this column does not produce a carry. Therefore, we must subtract 3 from column A, and add 3 to column B. The final answer is 1001 1011, which is the excess-3 number for 68.

To summarize addition with excess-3 numbers:

- Add the numbers using the rules for binary addition.
- If any group produces a decimal carry, add 0011 to that group.
- If any group does not produce a decimal carry, subtract 0011 from that group.

The excess-3 code has the advantage that all operations in addition use ordinary binary addition. (The 8421 code is not like this. It requires special operations to handle decimal carries.) Because the excess-3 code is self-complementing, it also has the advantage that the 1's or 2's complements can be used to subtract excess-3 numbers. (This is not true for 8421 numbers.)

EXAMPLE 3-3

What are the excess-3 numbers for the following decimal numbers:
(a) 35
(b) 569
(c) 2468

SOLUTION

(a)
$$\begin{array}{cc} 3 & 5 \\ +3 & +3 \\ \hline 6 & 8 \\ \downarrow & \downarrow \\ 0110 & 1000 \end{array}$$

Thus, 0110 1000 is the excess-3 equivalent for decimal 35.

(b)
$$\begin{array}{ccc} 5 & 6 & 9 \\ +3 & +3 & +3 \\ \hline 8 & 9 & 12 \\ \downarrow & \downarrow & \downarrow \\ 1000 & 1001 & 1100 \end{array}$$

Hence, decimal 569 becomes 1000 1001 1100 in excess-3 code.

(c)
$$\begin{array}{cccc} 2 & 4 & 6 & 8 \\ +3 & +3 & +3 & +3 \\ \hline 5 & 7 & 9 & 11 \\ \downarrow & \downarrow & \downarrow & \downarrow \\ 0101 & 0111 & 1001 & 1011 \end{array}$$

The excess-3 number 0101 0111 1001 1011 stands for 2468_{10}.

EXAMPLE 3-4

Decode the following excess-3 numbers:
(a) 1100 0110
(b) 1010 1011 0011 1001

SOLUTION

(a)
$$\begin{array}{cc} 1100 & 0110 \\ \downarrow & \downarrow \\ 12 & 6 \\ -3 & -3 \\ \hline 9 & 3 \end{array}$$

Binary Codes

Decimal 93 is the decoded value of 1100 0110 (excess-3).

(b)
$$\begin{array}{cccc} 1010 & 1011 & 0011 & 1001 \\ \downarrow & \downarrow & \downarrow & \downarrow \\ 10 & 11 & 3 & 9 \\ -3 & -3 & -3 & -3 \\ \hline 7 & 8 & 0 & 6 \end{array}$$

Decimal 7806 is the decoded value of the excess-3 number 1010 1011 0011 1001.

EXAMPLE 3-5

Add 567 and 295 using excess-3 numbers.

SOLUTION

$$\begin{array}{cccc} 567 & 1000 & 1001 & 1010 \\ +295 & +0101 & 1100 & 1000 \\ \hline 862 & 1110 & 0110 & 0010 \\ & -0011 & +0011 & +0011 \\ \hline & 1011 & 1001 & 0101 \end{array}$$

The two least significant groups produced carries; therefore, we add 3 to each group. The most significant group did not produce a carry; therefore, we subtract 3 from it. The final answer is 1011 1001 0101, which is the excess-3 equivalent of 862.

3-3 Other 4-bit BCD Codes

Many other 4-bit codes exist. Tables 3-3 and 3-4 show some of the common ones. As usual, decimal numbers greater than 9 are encoded a digit at a time. For instance, decimal number 16 becomes the following in 2421 code:

$$\begin{array}{ccc} 1 & 6 & \text{decimal} \\ \downarrow & \downarrow & \downarrow \\ 0001 & 1100 & 2421 \text{ code} \end{array}$$

Or, decimal 75 in 5421 code is

$$\begin{array}{ccc} 7 & 5 & \text{decimal} \\ \downarrow & \downarrow & \downarrow \\ 1010 & 1000 & 5421 \text{ code} \end{array}$$

Or, decimal 693 in 6311 code is

$$\begin{array}{cccc} 6 & 9 & 3 & \text{decimal} \\ \downarrow & \downarrow & \downarrow & \downarrow \\ 1000 & 1100 & 0100 & 6311 \text{ code} \end{array}$$

Table 3-3 4-bit BCD Codes

Decimal	7421	6311	5421	5311	5211
0	0000	0000	0000	0000	0000
1	0001	0001	0001	0001	0001
2	0010	0011	0010	0011	0011
3	0011	0100	0011	0100	0101
4	0100	0101	0100	0101	0111
5	0101	0111	1000	1000	1000
6	0110	1000	1001	1001	1001
7	1000	1001	1010	1011	1011
8	1001	1011	1011	1100	1101
9	1010	1100	1100	1101	1111

Table 3-4 4-bit BCD Codes

Decimal	4221	3321	2421	$84\bar{2}\bar{1}$	$74\bar{2}\bar{1}$
0	0000	0000	0000	0000	0000
1	0001	0001	0001	0111	0111
2	0010	0010	0010	0110	0110
3	0011	0011	0011	0101	0101
4	1000	0101	0100	0100	0100
5	0111	1010	1011	1011	1010
6	1100	1100	1100	1010	1001
7	1101	1101	1101	1001	1000
8	1110	1110	1110	1000	1111
9	1111	1111	1111	1111	1110

All the codes in the tables are weighted codes.* All use positive weights except the last two codes of Table 3-4. These two codes use negative weights, as well as positive. For instance, in the $84\bar{2}\bar{1}$ code the least sig-

*Sometimes a given set of weights may generate a number of different codes. Tables 3-3 and 3-4 show the most popular code forms.

Binary Codes   67

nificant digit has a weight of −1 and the next position has a weight of −2. Therefore, an 842$\bar{1}$ number like 1011 is decoded as follows:

$$\begin{array}{cccc} 1 & 0 & 1 & 1 \\ 8 + 4 & - 2 & - 1 & = 5 \end{array}$$

The 2421 code uses the first five and the last five 4-bit binary numbers (0000, 0001, 0010, 0011, 0100 and 1011, 1100, 1101, 1110, 1111). Electronic counters often use this code.

The 842$\bar{1}$ code can be derived from the excess-3 code. By complementing the 0s and 1s in the last two significant positions of the excess-3 code, we get the 842$\bar{1}$ code. For instance, 5 in excess-3 code is 1000. If we complement the two least significant digits, we obtain 1011, which is the 842$\bar{1}$ number for 5. As another example, 1100 is the excess-3 equivalent of 9. Complementing the two least significant digits, we get 1111, which is the 842$\bar{1}$ number for 9.

Recall that the excess-3 code is a self-complementing code; that is, the 1's complement of any excess-3 number stands for the 9's complement of the encoded decimal number. By inspection of Tables 3-3 and 3-4 we find that the following codes also are self-complementing: the 4221 code, the 3321, the 2421, and the 842$\bar{1}$.

Tables 3-3 and 3-4 will be very useful for future reference.

EXAMPLE 3-6

Encode the following:
(a) Decimal number 842 into 4221 form
(b) Decimal 951 into 6311 form

SOLUTION

(a) Table 3-4 gives the 4221 code. So,

$$\begin{array}{cccc} 8 & 4 & 2 & \text{decimal} \\ \downarrow & \downarrow & \downarrow & \downarrow \\ 1110 & 1000 & 0010 & 4221 \text{ code} \end{array}$$

(b) Table 3-3 shows the 6311 code.

$$\begin{array}{cccc} 9 & 5 & 1 & \text{decimal} \\ \downarrow & \downarrow & \downarrow & \downarrow \\ 1100 & 0111 & 0001 & 6311 \text{ code} \end{array}$$

EXAMPLE 3-7

Decode the following BCD numbers:
(a) 0100 1011 1101, 5311 code
(b) 1100 0101 1111 0001, 3321 code

SOLUTION

(a) Using Table 3-3, we decode the 5311 number as follows:

$$\begin{array}{cccc} 0100 & 1011 & 1101 & \text{5311 code} \\ \downarrow & \downarrow & \downarrow & \downarrow \\ 3 & 7 & 9 & \text{decimal} \end{array}$$

(b) Using Table 3-4, we decode the 3321 number as follows:

$$\begin{array}{ccccc} 1100 & 0101 & 1111 & 0001 & \text{3321 code} \\ \downarrow & \downarrow & \downarrow & \downarrow & \downarrow \\ 6 & 4 & 9 & 1 & \text{decimal} \end{array}$$

3-4 The Parity Bit

In a digital system a *word* is a group of bits that is treated, stored, and moved around as a unit. For instance, suppose that an 8421 coded computer is about to add 0101 1000 0011 and 0010 0100 0110. (This would be 583 and 246.) Each of these BCD numbers is a word. The computer brings each of these words out of memory (the place for storing numbers and instructions) and puts them into the arithmetic unit (the place where arithmetic operations are performed). The sum is a new word, which is then put back into the memory part of the machine.

While words are being moved and stored, errors can get into the words. For instance, one of the 0s in a word might accidentally be changed to a 1 by an intermittent failure, or by noise, or by a transient, and so on. Under normal operating conditions such a change is very unlikely, but any error at all could be disastrous. Therefore, we should have methods for detecting errors that crop up during the storing or moving of a word.

In all the BCD codes discussed so far, we have used only 10 of the 16 possible 4-bit combinations. Therefore, one way to find errors is to look for forbidden combinations. If any of these appear in memory, we immediately know that an error has occurred.

Checking for forbidden combinations is certainly possible, but it does not give a high enough probability for catching errors. The most widely used approach to detecting errors that arise in storing and moving words is to attach a parity bit to the word.

Even parity means that we attach an extra bit to a group of bits so as to produce an even number of 1s. For instance, suppose we have a word like 0111. There are three 1s in this word, so that we have an odd number of 1s. We attach an extra 1 to the word to get 0111 1. Now we have an even number of 1s. This new word can be moved and stored by the computer, and can be checked for even parity at different points to assure that no errors have crept into the word.

Binary Codes 69

As another illustration of even parity, Table 3-5 shows the 8421 code with an even-parity bit. We deliberately select the parity bit to produce an even number of 1s for each code group.

Table 3-5 Even Parity

0000	0
0001	1
0010	1
0011	0
0100	1
0101	0
0110	0
0111	1
1000	1
1001	0

add 1 to make 0001 even

Table 3-6 Odd Parity

0000	1
0001	0
0010	0
0011	1
0100	0
0101	1
0110	1
0111	0
1000	0
1001	1

There is also odd parity. In this case, the added parity bit makes the number of 1s odd. Table 3-6 shows the 8421 code with an odd-parity bit. Note that the odd-parity bit is always the complement of the even-parity bit.

Both types of parity are commonly used; there is no strong argument in favor of either type.

Using a parity bit to detect errors rests upon two assumptions that apply to most digital systems:

- The probability is very small that any errors at all will occur.
- In the small number of cases where an error does get into a word, the error is most likely to be a 1-bit error. The chances of two or more bits changing accidentally are very remote* unless there is a complete failure, which would be readily detected by other means.

In other words, a parity check will catch all 1-bit errors but will not find the double errors that might occur. However, in most parts of a digital system the chances for a double error are extremely small.

Parity checks are especially common in storage devices like magnetic tapes, drums, cores, and paper tapes. Magnetic tape is a very inexpensive way to store large amounts of digital information, but it is more susceptible to error than other techniques. The chances for a double error in magnetic tape are higher, so that a single-parity check is not so reliable as in other parts of a digital system.

To improve matters, we can use a double-parity check with magnetic tape. Figure 3-1 illustrates a double-parity check, using odd parity. The information is put on the tape in what is called a "field" or "block." We have shown a block with seven rows and seven columns. The black dots represent magnetized points, and the light dots stand for unmagnetized points. (Or, black could be one polarity and light the other.) To simplify matters, suppose that straight binary numbers have been encoded into the first 6 bits of each row, and that the seventh bit in each row is a horizontal parity bit. In the first row we read 101001 with a parity bit of 0. In the second row we read 000110 with a parity bit of 1. Each row contains an odd-parity bit, so that when the tape is read electronically, an odd-parity check can be made.

The entire seventh row contains odd-parity bits, one for each column. For instance, the first column is 101111 with a parity bit of 0 (reading from the top down). The second column reads 001010 with a parity bit of 1. In this way, every column has an odd-parity bit. The advantage of this double-parity system is that double errors can be electronically detected during the reading of the tape. For instance, suppose that first row (101001 0) is somehow incorrectly read as 011001 0 (the first two bits have been complemented). A double error has occurred, so that we still have an odd number of 1s horizontally. The horizontal parity still checks. However, the vertical parity will not check because the first and second columns will then have an even number of 1s.

*For example, suppose there are five holes in the ground. The chances of a meteor falling into one of these holes are small. But the chances of two meteors falling into two holes are far smaller.

Binary Codes

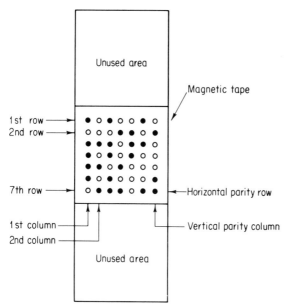

Fig. 3-1 Double-parity check with magnetic tape.

In later chapters we will discuss circuits for checking even and odd parity. As already indicated, checking for parity is the most widespread method of error detection.

EXAMPLE 3-8

Encode the following numbers into 8421 code and attach an odd-parity bit to the end of each word.
(a) 592
(b) 8307

SOLUTION

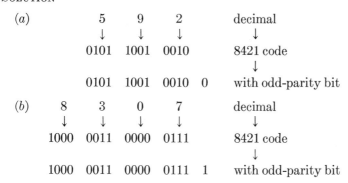

3-5 5-bit Codes

5-bit codes also exist. Although we need only 4 bits to encode any decimal digit from 0 to 9, adding an extra bit will allow us to decode the number more easily, as well as to detect errors more readily. Table 3-7 shows some of these 5-bit codes.

As usual, any decimal number greater than 9 is encoded a digit at a time. For instance, to encode decimal 847 into the shift-counter code,

$$
\begin{array}{cccc}
8 & 4 & 7 & \text{decimal} \\
\downarrow & \downarrow & \downarrow & \downarrow \\
11000 & 01111 & 11100 & \text{shift-counter code}
\end{array}
$$

The 2-out-of-5 code is an unweighted code that has been used in telephone and communication work. Not only does it have even parity, it also has exactly two 1s in each code group. Because of this, errors can be detected more reliably than with just an even-parity check.

Table 3-7 5-bit BCD Codes

Decimal	2-out-of-5	63210	Shift-counter	86421	51111
0	00011	00110	00000	00000	00000
1	00101	00011	00001	00001	00001
2	00110	00101	00011	00010	00011
3	01001	01001	00111	00011	00111
4	01010	01010	01111	00100	01111
5	01100	01100	11111	00101	10000
6	10001	10001	11110	01000	11000
7	10010	10010	11100	01001	11100
8	10100	10100	11000	10000	11110
9	11000	11000	10000	10001	11111

The shift-counter code (sometimes called the "Johnson code") is an unweighted code that is much used in electronic counters. It has the marked advantage of being easy to decode electronically (see Chap. 9).

The 51111 code is weighted and self-complementing. It is similar to the shift-counter code, so that it is easily decoded with electronic circuitry.

The 63210 code is weighted, except for the decimal value of 0. Above decimal number 0, the weights are 6, 3, 2, 1, 0, reading from left to right. It

Binary Codes

has exactly two 1s in each code group, allowing reliable error detection. This code has been used for storage of digital data on magnetic drums.

EXAMPLE 3-9

Encode 529_{10} into 2-out-of-5 code.

SOLUTION

$$\begin{array}{cccc} 5 & 2 & 9 & \text{decimal} \\ \downarrow & \downarrow & \downarrow & \downarrow \\ 01100 & 00110 & 11000 & \text{2-out-of-5 code} \end{array}$$

3-6 Codes with More than 5 Bits

We can decode and detect errors more easily if we add more bits. Table 3-8 shows some BCD codes that use more than 5 bits. These are all weighted codes.

The 9876543210 code, sometimes called the "ring-counter code," uses only a single 1 in each code group; this makes it very easy to decode and to detect errors. The code has the disadvantage that it requires more electronic circuitry than the simpler 4- and 5-bit codes.

Table 3-8 More-than-5-bit Codes

Decimal	50 43210	543210	9876543210
0	01 00001	000001	0000000001
1	01 00010	000010	0000000010
2	01 00100	000100	0000000100
3	01 01000	001000	0000001000
4	01 10000	010000	0000010000
5	10 00001	100001	0000100000
6	10 00010	100010	0001000000
7	10 00100	100100	0010000000
8	10 01000	101000	0100000000
9	10 10000	110000	1000000000

The 50 43210 code, also called the "biquinary code,"* is often used in electronic counters. (Note that biquinary means two-five.) Each member of

---
*The abacus uses the biquinary code. Its beads are arranged in twos and fives.

this code contains a group of 2 bits and a group of 5 bits. The group of 2 bits indicates whether or not the number is less than five. The group of 5 bits denotes the count. Reliable error detection is possible because there is a single 1 in the group of 2 bits, and a single 1 in the group of 5 bits.

The 543210 code is like the biquinary code, except that 6 bits are used instead of 7. The most significant digit (far left) tells us if the number is less than five or not. This code is often used in electronic counters.

3-7 The Gray Code = "UNIT DISTANCE CODE"

The Gray code, sometimes called the "reflected binary system," is an unweighted code not particularly suited to arithmetic operations, but quite useful for input-output devices, analog-to-digital converters, and other peripheral equipment.

Table 3-9 shows part of the Gray code, along with the corresponding binary numbers. The main characteristic of the Gray code is that each Gray number differs from the preceding Gray number by a single bit. For instance, in going from decimal 7 to 8, the Gray-code numbers change from 0100 to 1100. Note that these numbers differ only in the most significant bit. As another example, decimal numbers 13 and 14 are represented by Gray-code numbers 1011 and 1001. Again note that these numbers differ in only one

Table 3-9 Gray Code up to 15_{10}

Decimal	Gray code	Binary
0	0000	0000
1	0001	0001
2	0011	0010
3	0010	0011
4	0110	0100
5	0111	0101
6	0101	0110
7	0100	0111
8	1100	1000
9	1101	1001
10	1111	1010
11	1110	1011
12	1010	1100
13	1011	1101
14	1001	1110
15	1000	1111
⋮		

Binary Codes 75

digit position (the second position from the right). So it is with the entire Gray code — every Gray number differs by only 1 bit from the preceding Gray number.

Sometimes, we have to convert binary numbers into Gray-code numbers and vice versa. Here is how we convert from *binary to Gray:*

- The first Gray digit is the same as the first binary digit.
- Add each pair of adjacent bits to get the next Gray digit. *Disregard* any carries that occur.*

An example is the best way to describe the conversion from binary to Gray. Take binary number 1100. Here is how we find the corresponding Gray-code number:

Step 1: The first Gray digit is the same as the first binary digit. So, we repeat the first digit.

$$\begin{array}{l} 1\ 1\ 0\ 0 \quad \text{binary} \\ \downarrow \\ 1 \quad\quad\quad \text{Gray} \end{array}$$

Step 2: Now, add the first 2 bits of the binary number, disregarding any carries. The sum is the next Gray-code digit.

$$\begin{array}{l} \widehat{1\ 1}\ 0\ 0 \quad \text{binary} \\ \downarrow \\ 1\ 0 \quad\quad \text{Gray} \end{array}$$

In other words, we add the first 2 bits of the binary number to get $1 + 1 = 0$ carry a 1. We write down the 0, but we discard the 1.

Step 3: Add the next 2 binary digits to get the next Gray-code digit.

$$\begin{array}{l} 1\ \widehat{1\ 0}\ 0 \quad \text{binary} \\ \downarrow \\ 1\ 0\ 1 \quad\quad \text{Gray} \end{array}$$

Step 4: Add the last 2 binary digits to get the Gray-code digit.

$$\begin{array}{l} 1\ 1\ \widehat{0\ 0} \quad \text{binary} \\ \downarrow \\ 1\ 0\ 1\ 0 \quad\quad \text{Gray} \end{array}$$

Thus 1010 is the Gray-code equivalent of binary number 1100.

Take another example. To convert binary 110100110 to Gray code, we proceed as follows:

*This is formally called modulo-2 addition, or more simply exclusive-OR addition. The four rules for this kind of addition are: $0 + 0 = 0, 0 + 1 = 1, 1 + 0 = 1, 1 + 1 = 0$.

Step 1: Repeat the most significant digit.

$$\begin{array}{c} 1\ 1\ 0\ 1\ 0\ 0\ 1\ 1\ 0 \quad \text{binary} \\ \downarrow \\ 1 \quad\quad\quad\quad\quad\quad\quad \text{Gray} \end{array}$$

Step 2: Add the first 2 binary digits to get the next Gray digit. (Disregard all carries.)

$$\begin{array}{c} \overset{\frown}{1\ 1}\ 0\ 1\ 0\ 0\ 1\ 1\ 0 \quad \text{binary} \\ \downarrow \\ 1\ 0 \quad\quad\quad\quad\quad\quad \text{Gray} \end{array}$$

Step 3: Continue adding each pair of adjacent bits to get

$$\begin{array}{c} 1\ \overset{\frown}{1\ 0}\overset{\frown}{\ 1}\ 0\ 0\ 1\ 1\ 0 \quad \text{binary} \\ \downarrow\ \downarrow \\ 1\ 0\ 1\ 1\ 1\ 0\ 1\ 0\ 1 \quad \text{Gray} \end{array}$$

To convert *from Gray code back to binary,* we use a method that is similar, but not exactly the same. Again, an example best describes the method. Let us convert Gray-code number 101110101 back to its binary equivalent.

Step 1: Repeat the most significant digit.

$$\begin{array}{c} 1\ 0\ 1\ 1\ 1\ 0\ 1\ 0\ 1 \quad \text{Gray} \\ \downarrow \\ 1 \quad\quad\quad\quad\quad\quad\quad \text{binary} \end{array}$$

Step 2: Add *diagonally* as shown to get the next binary digit.

$$\begin{array}{c} 1\ 0\ 1\ 1\ 1\ 0\ 1\ 0\ 1 \quad \text{Gray} \\ \nearrow \downarrow \\ 1\ 1 \quad\quad\quad\quad\quad\quad \text{binary} \end{array}$$

$(1 + 0 = 1.)$

Step 3: Continue adding *diagonally* to get the remaining Gray-code digits.

$$\begin{array}{c} 1\ 0\ 1\ 1\ 1\ 0\ 1\ 0\ 1 \quad \text{Gray} \\ \nearrow \downarrow \\ 1\ 1\ 0\ 1\ 0\ 0\ 1\ 1\ 0 \quad \text{binary} \end{array}$$

By using these methods, we can convert Gray to binary and vice versa whenever the need arises.

EXAMPLE 3-10

(a) Convert binary 1000110111 to its Gray-code equivalent.
(b) Convert Gray number 11100100011 to its binary equivalent.

Binary Codes   77

SOLUTION

(a) To go from binary to Gray, we add pairs of adjacent bits as follows:

$$1\ 0\ 0\ 0\ 1\ 1\ 0\ 1\ 1\ 1 \quad \text{binary}$$
$$\downarrow$$
$$1\ 1\ 0\ 0\ 1\ 0\ 1\ 1\ 0\ 0 \quad \text{Gray}$$

(b) To convert from Gray to binary, we must add diagonally as shown.

$$1\ 1\ 1\ 0\ 0\ 1\ 0\ 0\ 0\ 1\ 1 \quad \text{Gray}$$
$$\nearrow\downarrow$$
$$1\ 0\ 1\ 1\ 1\ 0\ 0\ 0\ 0\ 1\ 0 \quad \text{binary}$$

SUMMARY

BCD codes are a compromise between the binary and the decimal number systems.

The best-known BCD code is the 8421 code. The weights from left to right are 8, 4, 2, 1 — the same as for straight binary numbers. Above 9 the 8421 code is completely different from the straight binary system. To encode in 8421 form, we change *each* decimal digit to its binary equivalent. The main advantage of the 8421 code is the ease of converting to and from decimal numbers.

To encode a decimal number into excess-3 form, we must add 3 to each decimal digit before converting to binary. The excess-3 code is a self-complementing code, which means that the 1's complement of any excess-3 number represents the 9's complement of the encoded decimal number.

There are many 4-bit BCD codes (about 30 billion). Tables 3-3 and 3-4 show some of the popular ones.

Sometimes, a parity bit is attached to a word or group of bits to make the total number of 1s either even or odd. By doing this, we can more easily detect errors that occasionally get into words as the words are being moved around or stored.

5-bit codes also exist. The extra bit allows us to decode the number more easily, as well as facilitating error detection. The shift-counter code is especially important in electronic counters.

Some BCD codes have more than 5-bits. The 50 43210 code, also called the biquinary code, is quite important and is often used in electronic counters.

The Gray code has the outstanding characteristic that each Gray-code number differs from the preceding Gray-code number by only 1 bit.

GLOSSARY

BCD Stands for binary-coded decimal, and refers to changing each decimal digit into a group of binary digits according to some predetermined code.

biquinary Stands for two-five and refers to the 50 43210 code, whose code groups contain a group of 2 bits and a group of 5 bits.

8421 code The most popular BCD code. Each decimal digit is simply changed into its binary equivalent.

excess-3 A BCD code in which 3 is added to each decimal digit before converting to equivalent binary groups.

Gray code Sometimes called the reflected binary code. This code has the outstanding property that each Gray-code number differs from the preceding one by only 1 bit.

parity bit A bit that is deliberately attached to a group of bits to make the total number of 1s either even or odd.

self-complementing This means that a BCD code has the property that the 1's complement of a code group represents the 9's complement of the encoded decimal digit. Only some of the BCD codes have this property.

weighted code A code whose bit positions carry a fixed value, usually given by the name of the code. For instance, the 8421 code has weights of 8, 4, 2, and 1 in the corresponding bit positions. Some codes like the excess-3 are unweighted; that is, the bit positions have no fixed values.

word In a digital system this is a group of bits that is treated, stored, and moved around as a unit.

REVIEW QUESTIONS

1. What is the 8421 code? What is the 8421 code number for 395?
2. What is a BCD code?
3. How are decimal numbers changed into excess-3 numbers?
4. What does *self-complementing* mean in reference to a BCD code?
5. What is a weighted code? An unweighted code?
6. Name some of the 4-bit codes. Are these weighted or unweighted?
7. What is a parity bit? What are the two kinds of parity?
8. What does *word* refer to?
9. Name some of the 5-bit codes. Are these weighted or unweighted?
10. What are the weights in the biquinary code? What does *biquinary* mean?
11. What is the main characteristic of the Gray code?
12. Describe how a binary number is converted to a Gray-code number.

Binary Codes

PROBLEMS

3-1 Encode the following decimal numbers into 8421 BCD numbers:
 (a) 59.
 (b) 39,584.

3-2 What are the 8421 BCD numbers for these decimal numbers:
 (a) 649.
 (b) 71,465.

3-3 Decode the following 8421 BCD numbers:
 (a) 0011 1000 0111.
 (b) 1001 0110 0111 1000 0111 0011.

3-4 Decode the following 8421 BCD numbers:
 (a) 1001 0111 1000.
 (b) 0101 0110 0111 0001 0000 0100.

3-5 What are the excess-3 numbers for these decimal numbers:
 (a) 467.
 (b) 5839.

3-6 Change the following decimal numbers into excess-3 numbers:
 (a) 5932.
 (b) 386,258.

3-7 Decode these excess-3 numbers:
 (a) 1011 0111 1001.
 (b) 1010 1000 0110 0101 0011.

3-8 What is the excess-3 number for octal 743?

3-9 Use excess-3 numbers to add these decimal numbers: 457 and 293. What is the final answer in excess-3 code?

3-10 What is the excess-3 sum of decimals 45 and 34?

3-11 What are the 5421 BCD numbers for:
 (a) Decimal 435.
 (b) Octal 435.

3-12 Decode the following BCD numbers:
 (a) 1011 1001 0011 0001 (5421 code).
 (b) 1110 1001 0101 0111 ($74\bar{2}\bar{1}$ code).

3-13 An electronic counter uses the 4221 code. Suppose that its output is given in the following form:
 on-on-on-on, off-on-on-on, off-off-on-off
What decimal number does this output represent if *on* stands for 1? What does it represent if *on* stands for 0?

3-14 The 2421 code is used in an electronic counter. What do the following outputs represent?
 (a) 0001 1011 1110 1011 0010.
 (b) 0011 1011 1101 0100 0000.

3-15 Add an even-parity bit to each of the following words:
  (a) 1011001.
  (b) 110001100011.
  (c) 10101110011100.

3-16 Determine whether the parity bit in the following words is even or odd:
  (a) 1001100101 1.
  (b) 1000110011 1.
  (c) 1000011101 0.

3-17 Suppose we want to add an even-parity bit to the 3321 code. What bits should we add?

3-18 To add an odd-parity bit to the 5211 code, what bits should we add?

3-19 Encode the following decimal numbers into shift-counter BCD numbers: 84, 932, and 5482.

3-20 Decode the following 2-out-of-5 BCD numbers:
  (a) 01010 10010 11000.
  (b) 00011 01100 10100 00101 10100.

3-21 Write the following decimal numbers as biquinary BCD numbers: 25, 452.

3-22 Decode the following biquinary BCD numbers:
  (a) 10 00010.
  (b) 01 01000.

3-23 Convert the following binary numbers into Gray numbers:
  (a) 10110.
  (b) 100111011001.
  (c) 10101101110011.

3-24 Convert the following Gray numbers into binary numbers:
  (a) 10101.
  (b) 100110011.
  (c) 00111000010110.

CHAPTER 4

BOOLEAN ALGEBRA

When we hear the word "algebra," we usually think of high school algebra with its positive and negative numbers, simultaneous equations, quadratic equations, and so on. This is not the only kind of algebra, however.

George Boole (1815–1864) invented a new algebra to describe logic and thought. Since the time of the early Greek philosophers, much logic and reasoning has been done using *true* and *false* statements. For many years mathematicians tried to include the laws of true-false logic in the realm of algebra but failed. Boole succeeded with the publication of "An Investigation of the Laws of Thought, ..." in 1854. What Boole did was to symbolize logic by a new kind of algebra. In other words, he showed that some types of thinking and reasoning could be done by manipulating symbols.

Boole's algebra stayed in the domain of pure mathematics until almost a century later. In 1938, Claude Shannon wrote a paper titled "A Symbolic Analysis of Relay and Switching Circuits." This paper applied the new algebra to switching circuits, and since then Boolean algebra has been widely used in telephone and digital systems.

In this chapter we become familiar with Boolean algebra, basic logic circuits, and other topics that are important in digital work.

4-1 The OR Gate

In digital electronics a *gate* is a circuit with one output and two or more input channels; an output signal occurs only for certain combinations of input signals.

The first kind of gate that we study is the OR gate. In the OR gate an output occurs when there is a signal in *any* of the input channels. Figure 4-1a shows a two-input OR gate where A and B are the inputs and x is the output. For the moment, let us analyze this OR gate by restricting the input voltages to either 0 or 1 volt. Also, we will use the ideal-diode approximation (Chap. 1) to simplify the analysis. There are only four possible cases to analyze:

Case 1: $A = 0$ and $B = 0$. With both input voltages at zero the output voltage must be zero because no voltage exists anywhere in the circuit. Therefore, $x = 0$.

Case 2: $A = 0$ and $B = 1$. The B battery forward-biases the lower diode, causing the output to be 1 volt. Since the A battery is 0 volts, it looks like a short. Figure 4-1b shows the circuit for this condition. The upper diode is off, the lower diode is on, and the output $x = 1$.

Case 3: $A = 1$ and $B = 0$. Because of the horizontal symmetry of the circuit the argument is similar to case 2. The upper diode is on, the lower diode is off, and $x = 1$.

Case 4: $A = 1$ and $B = 1$. With both inputs at 1 volt both diodes are forward-biased or shorted. Since the voltages are in parallel, the output voltage is 1 volt. Therefore, $x = 1$.

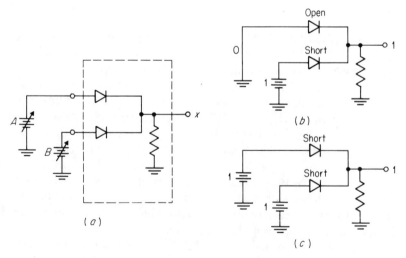

Fig. 4-1 The OR gate.

Boolean Algebra

Table 4-1 lists the input-output conditions of an OR gate. Examine this table carefully and memorize the following: the OR gate has a 1 output when either A or B or both are 1. In other words, the OR gate is an *any-or-all* gate; an output occurs when any or all of the inputs are present.

Table 4-1 The OR-gate Truth Table

A	B	x
0	0	0
0	1	1
1	0	1
1	1	1

Incidentally, a *truth table* (sometimes called a table of combinations) is a table that shows all input-output possibilities for a logic circuit. Table 4-1 is one example of a truth table. All possible AB inputs, 00, 01, 10, and 11, are shown along with the resulting outputs.

As implied in Chaps. 1 and 2, most digital circuits use diodes and transistors as switches to change from one voltage level to another. When we analyze digital circuits, we determine whether a voltage is low or high. The exact magnitude or value is unimportant, as long as the voltage is distinguishable as low or high. In digital circuits low and high voltages are often represented by 0 and 1, respectively. For instance, in the OR gate of Fig. 4-2 the input levels are either 2 or 10 volts — low or high. The operation of this circuit is like that of Fig. 4-1; Table 4-2 shows the truth table. If we let 0 stand for 2 volts and 1 for 10 volts, we can show an equivalent truth table with 0s and 1s (Table 4-3). The point here is simply that we can use the actual voltages, or we can use 0s and 1s to represent the low and high voltages; in either case, the OR gate has a high output when any or all inputs are high.

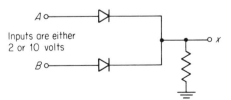

Fig. 4-2 A two-input OR gate.

Table 4-2

A	B	x
2	2	2
2	10	10
10	2	10
10	10	10

Table 4-3

A	B	x
0	0	0
0	1	1
1	0	1
1	1	1

The OR gate can have any number of inputs. For example, Fig 4-3 shows a three-input OR gate. If A or B or C is high, x will be high because the diode associated with the high input will turn on. Letting 0 stand for low and 1 for high, we get the truth table of Table 4-4. In general, no matter how many inputs there are, the OR gate has a 1 output when any or all of the inputs are 1.

Table 4-4 Three-input OR Gate

A	B	C	x
0	0	0	0
0	0	1	1
0	1	0	1
0	1	1	1
1	0	0	1
1	0	1	1
1	1	0	1
1	1	1	1

Incidentally, the number of horizontal rows in a truth table equals 2^n, where n is the number of inputs. For instance, for a two-input gate, the

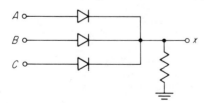

Fig. 4-3 A three-input OR gate.

Boolean Algebra

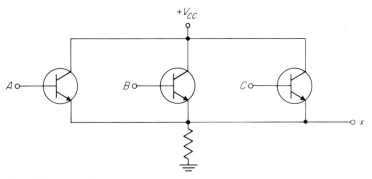

Fig. 4-4 A transistor OR gate.

truth table has 2^2 or 4 rows. A three-input gate will have a truth table with 2^3 or 8 rows, while a four-input gate results in 2^4 or 16 rows, and so on. To include all possible entries it helps to list the entries in a binary-number progression. For instance, in Table 4-4 we deliberately listed the ABC entries as 000, 001, 010, 011, ..., following the binary-number count; this guarantees that we will not forget any possibility.

We can also use transistors to make OR gates. Figure 4-4 shows a three-input OR gate where A, B, and C are the inputs and x is the output. The circuit has three *n-p-n* emitter followers in parallel. With all input voltages in the low state, the output must be low because the emitters follow the inputs. When any or all of the inputs go high, the output follows. Therefore, we have OR-gate action. If we let 0 denote low voltage and 1 denote high voltage, we get a truth table that is the same as Table 4-4.

There are many other ways to build OR gates by using various combinations of resistors, diodes, and transistors. For our purposes the important thing to remember is that the OR gate has a 1 output when any or all inputs are 1.

4-2 The AND Gate

The AND gate is another basic kind of digital circuit — it has an output only when *all* inputs are present. As an example, consider the two-input AND gate of Fig. 4-5a. Again, let us use ideal diodes and restrict all voltages to either 0 or 1 volt. There are four cases to analyze:

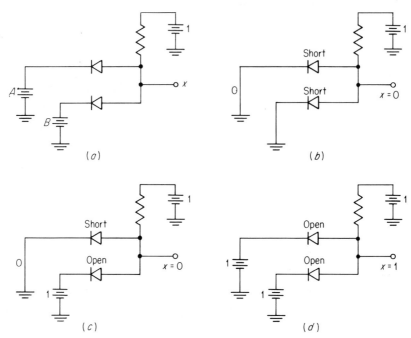

Fig. 4-5 The AND gate.

Case 1: $A = 0$ and $B = 0$. Since both input batteries are at 0 volts, they are like short circuits (Fig. 4-5b). The internal 1-volt battery forces conventional current in the direction of each diode triangle; therefore, both diodes are on or shorted. Carefully examine Fig. 4-5b and realize that the output is shorted to ground through the diodes and the batteries. Therefore, $x = 0$.

Case 2: $A = 0$ and $B = 1$. The upper diode is still forward-biased as shown in Fig. 4-5c; the output is still shorted to ground through the upper diode and the battery. Therefore, $x = 0$.

Case 3: $A = 1$ and $B = 0$. Because the circuit has horizontal symmetry, the argument is similar to case 2, and $x = 0$.

Case 4: $A = 1$ and $B = 1$. No current flows in the circuit (Fig. 4-5d) because all batteries are in parallel. With no current in R there is no voltage drop across R; therefore, x must equal 1 volt.

As usual, we can summarize the circuit action by a truth table (Table 4-5). Examine this table carefully and memorize the following: the AND gate has a 1 output when A *and* B are 1. In other words, the AND gate is an *all-or-nothing* gate; an output occurs only when all inputs are present.

Boolean Algebra

Table 4-5 AND-gate Truth Table

A	B	x
0	0	0
0	1	0
1	0	0
1	1	1

The use of 0 and 1 volt was only a convenience for analysis. We can use any two distinct voltages. For instance, in Fig. 4-5a suppose that the 1-volt battery is changed to a 10-volt battery and that the inputs are either 2 or 10 volts. If we use actual voltages, we get the truth table of Table 4-6.

Table 4-6

A	B	x
2	2	2
2	10	2
10	2	2
10	10	10

The AND gate can have more than two inputs. By adding diodes, we can have any number of input channels that we want. For instance, Fig. 4-6 shows a three-input AND gate. Assume that the two distinct voltage levels are 0 and 10 volts. If any input is at 0 volts (grounded), the diode connected to that input is forward-biased or shorted; therefore, the output will be shorted to ground. Thus $x = 0$ when any input is 0. The only way to make

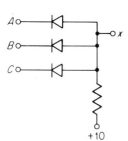

Fig. 4-6 A three-input AND gate.

$x = 10$ volts is to have all inputs simultaneously equal to 10 volts. In this case no current flows through R, and the output voltage rises to the supply voltage. If we let 0 stand for low voltage and 1 for high voltage, we get the truth table of Table 4-7. In general, the AND gate is an *all-or-nothing* gate; no matter how many inputs there are, the AND gate has an output only when all inputs are present.

Table 4-7 Three-input AND Gate

A	B	C	x
0	0	0	0
0	0	1	0
0	1	0	0
0	1	1	0
1	0	0	0
1	0	1	0
1	1	0	0
1	1	1	1

We can make AND gates by using transistors instead of diodes. Figure 4-7 shows one way to make a three-input AND gate. Here is how it works. Suppose that the input voltages (A, B, and C) are either 0 or 10 volts. If any of the inputs is 0 volts (grounded), the base-emitter diode of the associated transistor will be on (ideally, a short); therefore, the output will be held approximately at ground, so that $x = 0$. The only way to make $x = 10$ volts is to raise all inputs to 10 volts. Under this condition none of the

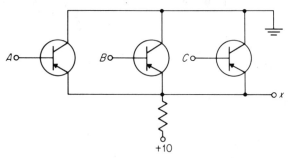

Fig. 4-7 A transistor AND gate.

Boolean Algebra

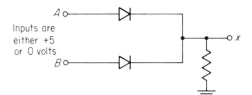

Fig. 4-8

base-emitter diodes is on, and the output is free to rise to 10 volts. By letting 0 stand for low voltage and 1 for high voltage, we get a truth table like Table 4-7.

4-3 Positive and Negative Logic Systems

A positive logic system is a system in which a 1 represents the more positive of the two voltage levels; in a negative logic system the 1 stands for the more negative voltage. For instance, suppose a digital system has voltage levels of +5 and 0 volts. We are free to choose the definitions for a 0 and a 1. If we say that a 1 stands for +5 volts and a 0 for 0 volts, the system becomes a positive logic system. On the other hand, if we let 1 denote 0 volts and let 0 symbolize +5 volts, we have a negative logic system. (Up to now, we have been using positive logic.)

The distinction between positive and negative logic is important because an OR gate in a positive logic system becomes an AND gate in a negative logic system. To understand clearly why this is so, consider the circuit of Fig. 4-8. The truth table for this circuit (Table 4-8) gives actual voltage levels. The output is 5 volts when either A or B is 5 volts. Is this circuit an OR gate or an AND gate? The answer depends on how we define 0 and 1. If we say that 1 stands for +5 volts and 0 for 0 volts, we have a positive logic system, and we can show the truth table as in Table 4-9. Clearly, this is the truth table of an OR gate because there is a 1 output when A or B is 1.

Table 4-8

A	B	x
0	0	0
0	5	5
5	0	5
5	5	5

Table 4-9

A	B	x
0	0	0
0	1	1
1	0	1
1	1	1

Table 4-10

A	B	x
1	1	1
1	0	0
0	1	0
0	0	0

On the other hand, if we say that a 1 stands for 0 volts and a 0 for +5 volts, we have a negative logic system. By looking at Table 4-8 and writing a 1 for each 0-volt entry and a 0 for each 5-volt entry, we get the truth table of Table 4-10. By inspection of this table, a 1 output occurs only when A and B are 1; therefore, the circuit is an AND gate in a negative logic system.

A digital system can be either a positive or a negative logic system; it all depends on how we define 0 and 1. The choice of which system to use is arbitrary — like defining the direction of current from + to − or from − to +. For consistency we will use positive logic throughout this book — 1 always will stand for the more positive of the two voltage levels.

4-4 The NOT Circuit

Another of the basic digital circuits is the NOT circuit, also called a complementary circuit or an inverter. This circuit has one input and one output. All it does is invert the input signal; if the input is high, the output is low, and vice versa.

Figure 4-9 shows one way to build a NOT circuit. When the input voltage is high enough, the transistor saturates; therefore, the output is low. On the other hand, when the input voltage is low enough, the transistor cuts off, and the output voltage is high. The truth table of the NOT circuit is simply

Input	Output
0	1
1	0

We call this circuit a NOT circuit because the output is *not* the same as the input.

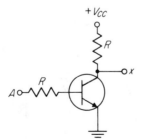

Fig. 4-9 The NOT circuit.

Boolean Algebra

4-5 OR Addition

Boolean algebra differs from ordinary algebra in some ways. In ordinary algebra when we solve an equation for its roots, we can get any real number — positive, negative, fractional, and so forth. In other words, the set of numbers in ordinary algebra is infinite. In Boolean algebra when we solve an equation, we get either a 0 or a 1. No other answers are possible because the set of numbers includes only the binary digits 0 and 1.

Another startling difference about Boolean algebra is the meaning of the plus sign. To bring out this meaning, consider Fig. 4-10 where we have symbolized a two-input OR gate with A and B inputs and an x output. In Boolean algebra the $+$ sign symbolizes the action of an OR gate. In other words, we may think of the OR gate as an adding device that combines A with B to give a result of x. In Boolean algebra when we write

$$x = A + B$$

we mean that A and B are to be combined in the same way that an OR gate combines A and B. To remind us of this, we should read the expression $x = A + B$ as x equals A OR B. To repeat, the $+$ sign does not stand for ordinary addition; it stands for OR addition whose rules are given by the OR truth table (Table 4-1).

To get used to OR addition, let us work out the value of $x = A + B$ for the four possible input conditions.

Case 1: $A = 0$ and $B = 0$. We have

$$x = A + B = 0 + 0 = 0$$

because an OR gate combines 0 with 0 to give 0.

Case 2: $A = 0$ and $B = 1$. This gives

$$x = A + B = 0 + 1 = 1$$

because an OR gate combines 0 with 1 to give 1.

Case 3: $A = 1$ and $B = 0$. This is like case 2.

$$x = A + B = 1 + 0 = 1$$

Case 4: $A = 1$ and $B = 1$. We get

$$x = A + B = 1 + 1 = 1$$

because an OR gate combines 1 with 1 to give 1.

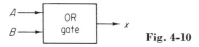

Fig. 4-10

Fig. 4-11 Schematic symbol for OR gate.

It may take a while to get used to the last result because of our built-in understanding of the + sign. We simply must remember that in digital work the + sign has several meanings. With decimal numbers it means ordinary addition — the first kind that we learned about. With binary numbers, however, it refers to binary addition. In Boolean algebra the + sign stands for OR addition — the kind of addition that an OR gate does. To display the three different meanings:

$$1 + 1 = 2 \quad \text{decimal addition}$$
$$1 + 1 = 10 \quad \text{binary addition}$$
$$1 + 1 = 1 \quad \text{OR addition}$$

Which of these meanings the + sign has is usually clear from the context. In other words, in solving decimal-arithmetic problems, we use the ordinary meaning of the + sign. In solving a Boolean equation, however, we will use the new meaning of OR addition. In case of doubt, we will indicate what type of addition the + sign refers to.

Incidentally, Fig. 4-11 shows the symbol that we will use for the OR gate. It should be memorized.

4-6 AND Multiplication

The multiplication sign (either × or ·) has a new meaning in Boolean algebra. To define this meaning, consider the AND gate of Fig. 4-12. We can think of an AND gate as a device that combines A and B to give a result of x. Thus, in Boolean algebra when we write

$$x = A \cdot B$$

or simply

$$x = AB$$

we mean that A and B are to be combined in the same way that an AND gate combines A with B to give an x output.

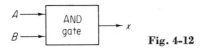

Fig. 4-12

Boolean Algebra

For practice, let us solve $x = AB$ for the four possible cases.
Case 1: $A = 0$ and $B = 0$. This gives

$$x = AB = 0 \cdot 0 = 0$$

because an AND gate combines 0 with 0 to give 0.
Case 2: $A = 0$ and $B = 1$. We get

$$x = AB = 0 \cdot 1 = 0$$

because an AND gate combines 0 with 1 to give 0.
Case 3: $A = 1$ and $B = 0$. This is like case 2.

$$x = AB = 1 \cdot 0 = 0$$

because an AND gate combines 1 with 0 to give 0.
Case 4: $A = 1$ and $B = 1$. This gives

$$x = AB = 1 \cdot 1 = 1$$

because an AND gate combines 1 with 1 to give 1.

These four results are easy to remember. Even though the dot does not mean multiplication in the ordinary sense, the results of AND multiplication are exactly the same as for ordinary multiplication.

Figure 4-13 shows the block-diagram symbol that we will use for the AND gate. Memorize it.

EXAMPLE 4-1

What is the Boolean expression for the output of the system shown in Fig. 4-14a?

SOLUTION

The output of the AND gate is AB. This goes into the OR gate along with B. Therefore, the final output is

$$x = AB + B$$

EXAMPLE 4-2

Evaluate the Boolean expression in the preceding example for
(a) $A = 0$ and $B = 1$
(b) $A = 1$ and $B = 0$

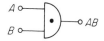
Fig. 4-13 Schematic symbol for AND gate.

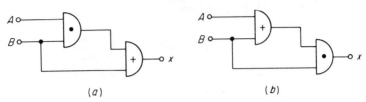

Fig. 4-14 (a) Example 4-1. (b) Example 4-3.

SOLUTION

(a) To evaluate the expression $x = AB + B$, we merely substitute the values of A and B into the expression and work out an answer.

$$x = AB + B = 0 \cdot 1 + 1 = 0 + 1 = 1$$

(Note that we multiply before we add, just as we would in ordinary algebra.)

(b) For $A = 1$ and $B = 0$, we get

$$x = AB + B = 1 \cdot 0 + 0 = 0 + 0 = 0$$

EXAMPLE 4-3

Find the Boolean expression for the output in Fig. 4-14b, and evaluate the expression for $A = 0$ and $B = 1$.

SOLUTION

The output of the OR gate is $A + B$. This goes into the AND gate along with the other input, which is B. Therefore, the final output is

$$x = (A + B)B$$

(Note that the parentheses are also used to denote AND multiplication.)

To evaluate this expression for $A = 0$ and $B = 1$, we substitute to get

$$x = (0 + 1)1 = (1)1 = 1$$

4-7 The NOT Operation

In Fig. 4-15 the A input to the NOT circuit is inverted; that is, if a 0 goes in, a 1 comes out; if a 1 goes in, a 0 comes out. In Boolean algebra the expression

$$x = \bar{A}$$

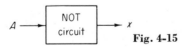

Fig. 4-15

Boolean Algebra

A ○─▷○─ $\bar{A}$ **Fig. 4-16** Schematic symbol for NOT circuit.

means that we are to change A in the same way that a NOT circuit changes A. We read the expression $x = \bar{A}$ as x equals NOT A. The bar over A simply means that we change or complement the quantity to the alternate digit (same as the 1's complement of Chap. 2). In other words, when $A = 0$, we get

$$x = \bar{A} = \bar{0} = 1 \quad \text{because NOT 0 is 1}$$

When $A = 1$, we get

$$x = \bar{A} = \bar{1} = 0 \quad \text{because NOT 1 is 0}$$

The NOT operation is sometimes called "negation" or "inversion." Also, the prime is sometimes used instead of a bar to signify the NOT operation. That is,

$$x = A'$$

is sometimes used instead of $x = \bar{A}$.

Figure 4-16 shows the symbol we will use for a NOT circuit. It should be memorized.

The OR, AND, and NOT operations of Boolean algebra may seem strange and unnatural. Why make up new operations like these? Very simply, these operations describe OR, AND, and NOT circuits — the building blocks of complex digital systems. With Boolean algebra we can analyze and design digital systems more easily. This will become clearer in the remainder of the book.

One more point. Because Boolean algebra was invented to solve true-false logic problems, we find many logic terms being used in digital work. For instance, digital circuits are often called "logic" circuits, output voltages are sometimes said to be "true" or "false" instead of 1 or 0, the NOT operation is sometimes called "negation," and so on.

Let us summarize the OR, AND, and NOT rules:

OR	AND	NOT
$0 + 0 = 0$	$0 \cdot 0 = 0$	$\bar{0} = 1$
$0 + 1 = 1$	$0 \cdot 1 = 0$	$\bar{1} = 0$
$1 + 0 = 1$	$1 \cdot 0 = 0$	
$1 + 1 = 1$	$1 \cdot 1 = 1$	

EXAMPLE 4-4

Write the Boolean equation for the output of Fig. 4-17.

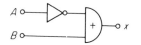

Fig. 4-17 Example 4-4.

SOLUTION

The OR gate has two inputs. One of them is $\bar{A}$ and the other is B. Therefore, the output of the system is

$$x = \bar{A} + B$$

EXAMPLE 4-5

Solve the Boolean equation of the preceding example for all possible input conditions.

SOLUTION

1. $A = 0$ and $B = 0$:
$$x = \bar{A} + B = \bar{0} + 0 = 1 + 0 = 1$$

2. $A = 0$ and $B = 1$:
$$x = \bar{A} + B = \bar{0} + 1 = 1 + 1 = 1$$

3. $A = 1$ and $B = 0$:
$$x = \bar{A} + B = \bar{1} + 0 = 0 + 0 = 0$$

4. $A = 1$ and $B = 1$:
$$x = \bar{A} + B = \bar{1} + 1 = 0 + 1 = 1$$

EXAMPLE 4-6

Write the Boolean expression for the output of Fig. 4-18a.

SOLUTION

One of the inputs to the OR gate is $\bar{A}$; the other input is $\bar{B}$. Therefore, the final output must be

$$x = \bar{A} + \bar{B}$$

EXAMPLE 4-7

Evaluate the Boolean expression of the preceding example for the four possible input combinations.

SOLUTION

When $A = 0$ and $B = 0$, we have
$$x = \bar{A} + \bar{B} = \bar{0} + \bar{0} = 1 + 1 = 1$$

When $A = 0$ and $B = 1$, we get
$$x = \bar{A} + \bar{B} = \bar{0} + \bar{1} = 1 + 0 = 1$$

When $A = 1$ and $B = 0$,
$$x = \bar{A} + \bar{B} = \bar{1} + \bar{0} = 0 + 1 = 1$$

Boolean Algebra

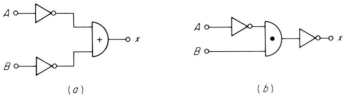

Fig. 4-18 (a) Example 4-6. (b) Example 4-8.

Finally, when $A = 1$ and $B = 1$,

$$x = \bar{A} + \bar{B} = \bar{1} + \bar{1} = 0 + 0 = 0$$

EXAMPLE 4-8

(a) Find the Boolean expression for the output of Fig. 4-18b.
(b) Evaluate this Boolean expression for all possible input combinations.

SOLUTION

(a) Working from input to output, we see that the two inputs to the AND gate are $\bar{A}$ and B. Therefore, the input to the NOT circuit is $\bar{A}B$. The final output is

$$x = \overline{\bar{A}B}$$

(b) When $A = 0$ and $B = 0$, we get

$$x = \overline{\bar{A}B} = \overline{\bar{0}\cdot 0} = \overline{1\cdot 0} = \bar{0} = 1$$

When $A = 0$ and $B = 1$,

$$x = \overline{\bar{A}B} = \overline{\bar{0}\cdot 1} = \overline{1\cdot 1} = \bar{1} = 0$$

When $A = 1$ and $B = 0$,

$$x = \overline{\bar{A}B} = \overline{\bar{1}\cdot 0} = \overline{0\cdot 0} = \bar{0} = 1$$

When $A = 1$ and $B = 1$,

$$x = \overline{\bar{A}B} = \overline{\bar{1}\cdot 1} = \overline{0\cdot 1} = \bar{0} = 1$$

As a general rule, the following order helps when evaluating expressions like the foregoing ones:

1. Take the NOT of all individual terms first.
2. When a NOT is applied to more than one term (like $\overline{0\cdot 1}$), work out the AND or OR operation first, and then take the NOT of the result.

4-8 De Morgan's Theorems

De Morgan was a great logician and mathematician, as well as a friend of Boole's. Among De Morgan's important contributions to logic are these two theorems:

$$\overline{A + B} = \bar{A} \cdot \bar{B} \tag{4-1}$$

$$\overline{A \cdot B} = \bar{A} + \bar{B} \tag{4-2}$$

In words, the first equation says that *the complement of a sum equals the product of the complements*. The second equation says that *the complement of a product equals the sum of the complements*.

These two theorems can easily be proved. For instance, let us start with Eq. (4-1). To prove that

$$\overline{A + B} = \bar{A} \cdot \bar{B}$$

we need to show that the left side equals the right side for all possible values of A and B. Here are the four possible cases:

Case 1: $A = 0$ and $B = 0$.

Left $\quad \overline{A + B} = \overline{0 + 0} = \bar{0} = 1$
Right $\quad \bar{A} \cdot \bar{B} = \bar{0} \cdot \bar{0} = 1 \cdot 1 = 1$

Case 2: $A = 0$ and $B = 1$.

$\overline{A + B} = \overline{0 + 1} = \bar{1} = 0$
$\bar{A} \cdot \bar{B} = \bar{0} \cdot \bar{1} = 1 \cdot 0 = 0$

Case 3: $A = 1$ and $B = 0$.

$\overline{A + B} = \overline{1 + 0} = \bar{1} = 0$
$\bar{A} \cdot \bar{B} = \bar{1} \cdot \bar{0} = 0 \cdot 1 = 0$

Case 4: $A = 1$ and $B = 1$.

$\overline{A + B} = \overline{1 + 1} = \bar{1} = 0$
$\bar{A} \cdot \bar{B} = \bar{1} \cdot \bar{1} = 0 \cdot 0 = 0$

Since no other combinations of A and B exist, we have proved De Morgan's first theorem: $\overline{A + B} = \bar{A} \cdot \bar{B}$. To make the proof really obvious we can show the truth table for each expression (Tables 4-11 and 4-12). From these tables we can see that $\overline{A + B}$ does equal $\bar{A} \cdot \bar{B}$ for each and every case; therefore, the expressions are identical.

Boolean Algebra

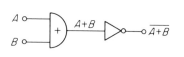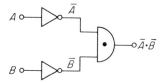

Fig. 4-19 De Morgan's first theorem.

Table 4-11

A	B	$\overline{A+B}$
0	0	1
0	1	0
1	0	0
1	1	0

Table 4-12

A	B	$\overline{A} \cdot \overline{B}$
0	0	1
0	1	0
1	0	0
1	1	0

The physical meaning of the first theorem is important. $\overline{A+B}$ represents a logic system in which a NOT circuit *follows* an OR gate (Fig. 4-19a). Also, $\overline{A} \cdot \overline{B}$ describes a logic system in which the outputs of two NOT circuits are used as the inputs to an AND gate (Fig. 4-19b). De Morgan's theorem tells us that these two systems are interchangeable. Incidentally, in Fig. 4-19a a NOT follows an OR gate; we call this particular combination a NOT-OR, or simply a NOR gate.

The second De Morgan theorem is

$$\overline{A \cdot B} = \overline{A} + \overline{B}$$

(The complement of a product equals the sum of the complements.) This theorem is easily proved (see Example 4-9). Tables 4-13 and 4-14 give the truth tables for each expression. Note that the truth tables are identical.

Table 4-13

A	B	$\overline{A \cdot B}$
0	0	1
0	1	1
1	0	1
1	1	0

Table 4-14

A	B	$\overline{A} + \overline{B}$
0	0	1
0	1	1
1	0	1
1	1	0

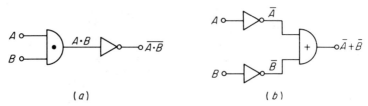

Fig. 4-20 De Morgan's second theorem.

Therefore, the expressions are equivalent and the digital systems represented by $\overline{A \cdot B}$ and $\bar{A} + \bar{B}$ are interchangeable. Figure 4-20 shows these digital circuits.

In Fig. 4-20a a NOT follows an AND gate; we call this particular combination a NOT-AND gate, or simply a NAND gate. In this book we will use the NAND gate a great deal. Because of this, we will abbreviate the NAND-gate symbol as shown in Fig. 4-21. Note that the NAND-gate symbol looks like a D connected to a small circle.

De Morgan's theorems are very important and should be memorized. We repeat them again in words:

- The complement of a sum equals the product of the complements.
- The complement of a product equals the sum of the complements.

These theorems are useful in changing Boolean expressions to alternate forms.

As already indicated, to apply the De Morgan theorems we change plus signs to multiplication signs or vice versa, and take the complement of the individual terms rather than the entire expression. For instance, for $\overline{A + B}$ we

1. Change the $+$ sign to a $\cdot$ sign to get $A \cdot B$
2. Take the complement of each term to get $\bar{A} \cdot \bar{B}$

The A and B can represent complicated expressions; we can still apply De Morgan's theorem. As an example, suppose we have

$$x = \overline{(C + DE)(CE + DF)}$$

Recognize that this is the complement of a product, $(C + DE)$ times $(CE + DF)$. Therefore, we can use the second De Morgan theorem and rewrite the expression as

$$x = \overline{(C + DE)} + \overline{(CE + DF)}$$

Fig. 4-21 Schematic symbol for a NAND gate.

Boolean Algebra

(Geometrically, this is like cutting the original expression down the middle, and replacing the multiplication by addition.)

EXAMPLE 4-9

Prove the second De Morgan theorem: $\overline{A \cdot B} = \bar{A} + \bar{B}$

SOLUTION

We have to show that the left member equals the right member for every possible input combination.

When $A = 0$ and $B = 0$,
$$\overline{A \cdot B} = \overline{0 \cdot 0} = \bar{0} = 1$$
$$\bar{A} + \bar{B} = \bar{0} + \bar{0} = 1 + 1 = 1$$

When $A = 0$ and $B = 1$,
$$\overline{A \cdot B} = \overline{0 \cdot 1} = \bar{0} = 1$$
$$\bar{A} + \bar{B} = \bar{0} + \bar{1} = 1 + 0 = 1$$

When $A = 1$ and $B = 0$,
$$\overline{A \cdot B} = \overline{1 \cdot 0} = \bar{0} = 1$$
$$\bar{A} + \bar{B} = \bar{1} + \bar{0} = 0 + 1 = 1$$

When $A = 1$ and $B = 1$,
$$\overline{A \cdot B} = \overline{1 \cdot 1} = \bar{1} = 0$$
$$\bar{A} + \bar{B} = \bar{1} + \bar{1} = 0 + 0 = 0$$

Since there are no other input combinations, we have proved the second De Morgan theorem. Tables 4-13 and 4-14 give the truth tables for these expressions.

EXAMPLE 4-10

Prove the following identity:
$$\overline{\bar{A} \cdot B} = A + \bar{B}$$

SOLUTION

We use the second De Morgan theorem: the complement of a product equals the sum of the complements. First, note that
$$\overline{\bar{A} \cdot B}$$
means the complement of $\bar{A} \cdot B$. In turn $\bar{A} \cdot B$ is the product of $\bar{A}$ and B. We can rewrite $\overline{\bar{A} \cdot B}$ as the sum of the complements of $\bar{A}$ and B; that is
$$\overline{\bar{A} \cdot B} = \bar{\bar{A}} + \bar{B}$$

(We changed the · sign to a + sign, and took the complements of $\bar{A}$ and $\bar{B}$ individually. If you like the geometrical viewpoint, we cut the expression down the middle and changed the · sign to a + sign.)

The final step in the proof is to recognize that the double complement of a variable is always equal to the variable itself. In other words, $\bar{\bar{A}} = A$. This should be obvious. If $A = 0$, $\bar{\bar{0}} = \bar{1} = 0$. On the other hand, if $A = 1$, $\bar{\bar{1}} = \bar{0} = 1$. Thus, we see that

$$\overline{\bar{A} \cdot \bar{B}} = \bar{\bar{A}} + \bar{\bar{B}} = A + B$$

This proves what we set out to prove.

EXAMPLE 4-11

Prove that the three-input NAND gate of Fig. 4-22a is interchangeable with the logic circuit of Fig. 4-22b.

SOLUTION

The output of the NAND gate of Fig. 4-22a is the NOT-AND of the inputs. The AND of the inputs is ABC. The NOT of this is $\overline{ABC}$. Thus the output of the NAND gate is $\overline{ABC}$.

The output of the system of Fig. 4-22b is easily found. The three-input OR gate has inputs of $\bar{A}$, $\bar{B}$, and $\bar{C}$. Therefore, the final output is $\bar{A} + \bar{B} + \bar{C}$.

To show that the two systems are equivalent or interchangeable we must prove that

$$\overline{ABC} = \bar{A} + \bar{B} + \bar{C}$$

We can do this with the second De Morgan theorem as follows: Think of ABC as the product of AB and C. Now, apply the second De Morgan theorem to get

$$\overline{ABC} = \overline{AB \cdot C} = \overline{AB} + \bar{C}$$

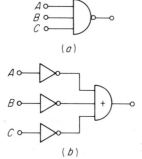

Fig. 4-22 Example 4-11.

Boolean Algebra

Applying De Morgan's theorem again to $\overline{AB}$, we get

$$\overline{AB} + \bar{C} = \bar{A} + \bar{B} + \bar{C}$$

So you see, we have proved that the two systems of Fig. 4-22 are equivalent and interchangeable. Another way to prove the equivalence is to show that the truth table for $\overline{ABC}$ is identical to the one for $\bar{A} + \bar{B} + \bar{C}$ (see Prob. 4-19).

4-9 The Universal Building Block

Given any Boolean expression, we can build a logic circuit for it. Conversely, given a logic circuit, we can write a Boolean expression. As an example, the expression

$$(A + B)\bar{C}$$

suggests a digital circuit in which A and B are first ORed, and then ANDed with the complement of C. Figure 4-23 shows the digital system for this expression.

In general, to build the logic circuit associated with *any* Boolean expression, we can use OR gates for the $+$ signs, AND gates for $\cdot$ signs, and NOT circuits for the overhead bars. In other words, OR, AND, and NOT circuits are the basic building blocks of all logic systems.

The NAND gate has an interesting property: it can be used to build an OR gate or an AND gate or a NOT circuit. Because of this, NAND gates are all that we need to build any logic system. In other words, the NAND gate is a universal building block. (A similar statement can be made for the NOR gate.)

We can make a NOT circuit out of a NAND gate by connecting all inputs together as shown in Fig. 4-24a. Clearly, if A is 0, the output of the NAND gate is

$$\overline{A \cdot A} = \overline{0 \cdot 0} = \bar{0} = 1 = \bar{A}$$

And if the A is 1, the output is

$$\overline{A \cdot A} = \overline{1 \cdot 1} = \bar{1} = 0 = \bar{A}$$

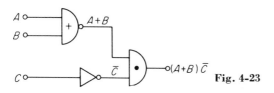

Fig. 4-23

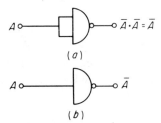

Fig. 4-24 Using a NAND gate as a NOT circuit.

From now on, when we use a NAND gate as a NOT circuit, we will simply show the symbol of Fig. 4-24b — it will be understood that all inputs are connected together.

We can also use NAND gates to make an AND gate. Figure 4-25a shows how. The action here is straightforward. The output of the first NAND gate is $\overline{AB}$. The second NAND gate then complements this, producing $\overline{\overline{AB}}$. As we already know, the double complement of a quantity is the quantity itself. Thus $\overline{\overline{AB}} = AB$. Since this final output is the AND of A and B, we have shown that NAND gates can be connected to perform the AND function.

We can also make an OR gate using only NAND gates. Figure 4-25b shows how to connect the NAND gates to do this. The first two NAND gates invert A and B to produce $\bar{A}$ and $\bar{B}$. The two-input NAND gate then produces an output of

$$\overline{\bar{A} \cdot \bar{B}}$$

Recall De Morgan's second theorem. We can rewrite this expression as

$$\overline{\bar{A} \cdot \bar{B}} = \bar{\bar{A}} + \bar{\bar{B}} = A + B$$

(Cut down the middle and change the · sign to a + sign.) Since the final output is A OR B, we have shown that NAND gates can be used to get the OR function.

Again, we repeat. The point of all this is simply that we can use NAND gates to build OR, AND, and NOT circuits. This, in turn, means that we can

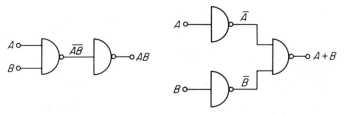

Fig. 4-25 (a) Using NAND gates to get an AND gate. (b) Using NAND gates to get an OR gate.

Boolean Algebra

build any logic system using only NAND gates. (A similar statement applies to the NOR gate.)

4-10 Laws and Theorems of Boolean Algebra

Boolean algebra lets us analyze and simplify logic systems. By manipulating algebraic expressions we rearrange the logic circuits corresponding to these expressions. Thus, when we simplify a complicated Boolean expression, we are changing a complicated digital circuit into a simpler one. For instance, suppose we have an expression like

$$x = ABC + A\bar{B}C + AB\bar{C}$$

To build the logic circuit for this, we need the following:

- A three-input OR gate to add ABC, $A\bar{B}C$, and $AB\bar{C}$
- Three three-input AND gates to produce ABC, $A\bar{B}C$, and $AB\bar{C}$
- Two NOT circuits to produce $\bar{B}$ and $\bar{C}$

Figure 4-26a shows the proper connection of these elements. If diodes are used in the AND and OR gates, and transistors in the NOT circuits, we would need a total of 12 diodes and 2 transistors.

By using Boolean laws and theorems, we can simplify the original expression (this is done in Example 4-14), and get

$$x = A(B + C)$$

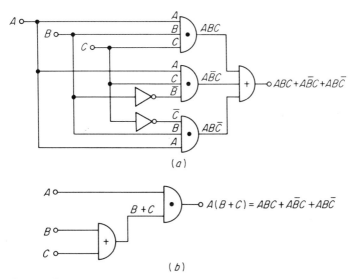

Fig. 4-26

Clearly, this leads to a much simpler logic circuit. Only one two-input AND gate and one two-input OR gate are needed, as shown in Fig. 4-26b. As far as hardware is concerned, we use only four diodes and no transistors.

The systems of Fig. 4-26 are identical as far as input-output operation is concerned; in other words, they are interchangeable. Obviously, we would prefer to build the circuit of Fig. 4-26b since it uses much less hardware and is easier to construct.

For our purposes, we should be familiar enough with the laws and theorems of Boolean algebra to make obvious simplifications. Therefore, we will list some basic laws and theorems of Boolean algebra.

The first group of laws in Boolean algebra is the same as in ordinary algebra.

Commutative laws:

$$A + B = B + A \tag{4-3}$$

$$A \cdot B = B \cdot A \tag{4-4}$$

Associative laws:

$$A + (B + C) = (A + B) + C \tag{4-5}$$

$$A(BC) = (AB)C \tag{4-6}$$

Distributive law:

$$A(B + C) = AB + AC \tag{4-7}$$

The commutative laws tell us that the order of adding or multiplying is unimportant. In other words, we get the same answer by adding A to B as we do when we add B to A. Likewise, the same answer results whether we multiply A by B or B by A.

The associative laws say that we can group any two terms of a sum or any two factors of a product. In other words, given $A + B + C$, we can first add B and C, and then add the result to A; or, we can first add A and B, and then add the result to C. A similar statement applies to multiplication.

The distributive law, Eq. (4-7), says that we can expand expressions by multiplying term by term just as we do in ordinary algebra. This law also implies that we can factor expressions. In other words, when we have a sum of two terms, each containing a common variable, we can factor this common variable. [This is like $AB + AC$. Each term contains A, so that we can factor the A to get $A(B + C)$.]

These first five laws present no difficulties at all. They are identical to those of ordinary algebra and therefore are already ingrained in our minds.

Boolean Algebra

The next group of laws is the backbone of Boolean algebra. First, we have the operations with 0.

$$A + 0 = A \tag{4-8}$$
$$A \cdot 0 = 0 \tag{4-9}$$

These should be thought of in terms of OR and AND gates. In other words, $A + 0 = A$ should remind us of an OR gate with inputs of A and 0 as shown in Fig. 4-27a. Clearly, the output of the OR gate depends on A only; if A is 0, the output must be 0; if A is 1, the output must be 1. Similarly, $A \cdot 0 = 0$ suggests the AND gate of Fig. 4-27a. Since the AND gate is an *all-or-nothing* gate, the output must be 0. The remaining identities in the group are

$$A + 1 = 1 \qquad new \tag{4-10}$$
$$A \cdot 1 = A \tag{4-11}$$
$$A + A = A \qquad new \tag{4-12}$$
$$A \cdot A = A \qquad new \tag{4-13}$$
$$A + \bar{A} = 1 \qquad new \tag{4-14}$$
$$A \cdot \bar{A} = 0 \qquad new \tag{4-15}$$

Again, we should visualize these formulas as logic circuits. Figure 4-27 shows the circuits for each of these. A moment of reflection about each formula and its circuit convinces us of the validity of these identities. (If

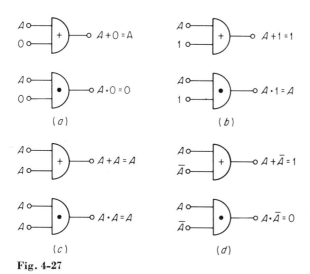

Fig. 4-27

there is any doubt, you can prove each formula by substituting the two possible values of A: 0 and 1. In each case, the left side of the formula equals the right side.) All these formulas are very basic and should be known.

The next three formulas are already familiar to us: the double-complement theorem and De Morgan's theorems.

$$\bar{\bar{A}} = A \qquad \text{new} \qquad (4\text{-}16)$$

$$\overline{A + B} = \bar{A} \cdot \bar{B} \qquad \text{new} \qquad (4\text{-}17)$$

$$\overline{A \cdot B} = \bar{A} + \bar{B} \qquad \text{new} \qquad (4\text{-}18)$$

Finally, there is a group of miscellaneous theorems, all of which can be easily derived from the more basic identities already listed. Some of these theorems are

$$A + AB = A \qquad (4\text{-}19)$$
$$A(A + B) = A \qquad (4\text{-}20)$$
$$(A + B)(A + C) = A + BC \qquad (4\text{-}21)$$
$$A + \bar{A}B = A + B \qquad (4\text{-}22)$$
$$A(\bar{A} + B) = AB \qquad (4\text{-}23)$$
$$(A + B)(\bar{A} + C) = AC + \bar{A}B \qquad (4\text{-}24)$$
$$AB + \bar{A}C = (A + C)(\bar{A} + B) \qquad (4\text{-}25)$$

We can add many more theorems to this list. For someone really interested in all aspects of Boolean algebra, all the foregoing formulas are important.

For our purposes, we should know Eqs. (4-3) through (4-18). Many of these are the same as in ordinary algebra. The ones that are different from ordinary algebra are marked *new*. Almost all of these are easily remembered if we think of the corresponding logic circuits (Fig. 4-27).

Often, we can simplify a complicated Boolean expression by using the foregoing laws and theorems. This is important; it means we can build a simpler circuit instead of a more complicated one. For instance, suppose we need a digital circuit whose output x is given by

$$x = AC + ABC$$

This equation says that the output x is obtained by

1. ANDing A and C
2. ANDing A, B, and C
3. ORing AC and ABC

Boolean Algebra

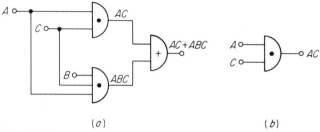

Fig. 4-28

Figure 4-28a shows the digital circuit for $x = AC + ABC$; it uses two AND gates and an OR gate. This circuit is actually more complicated than it has to be. We can simplify the original expression as follows:

$$x = AC + ABC$$
$$= AC(1 + B)$$
$$= AC$$

Figure 4-28b illustrates the digital system for $x = AC$; it uses only one AND gate. The point is clear. Whenever we can simplify a Boolean expression, we should do so; this results in less hardware when the circuit is built.

We will refer back to Eqs. (4-3) through (4-25) as the need arises. As already indicated, the basic group of equations (4-3) through (4-18) should be known.

One more point. Positive and negative logic gives rise to a basic duality in all the identities. We already know that when we change from one logic system to another, the 0 becomes 1, and vice versa. Furthermore, AND gates become OR gates, and OR gates become AND gates. What this means is that, given any Boolean identity, we can produce a dual identity by

- changing + signs to · signs, and vice versa
- complementing all 0s and 1s

For instance, given Eq. (4-8)

$$A + 0 = A$$

we need only complement the + sign and the 0 to get the dual identity, Eq. (4-11):

$$A \cdot 1 = A$$

As another example, Eq. (4-14) is

$$A + \bar{A} = 1$$

To get the dual identity, we complement the + sign and the 1 to get

$$A \cdot \bar{A} = 0$$

This property of duality is sometimes helpful. The Boolean identities are listed at the end of the chapter in dual pairs.

EXAMPLE 4-12

Prove the following:
(a) $A + A = A$
(b) $A \cdot A = A$

SOLUTION

(a) We will prove this by substituting the possible values of A which are 0 and 1. When $A = 0$,

$$\begin{aligned} A + A &= A \\ 0 + 0 &= 0 \\ 0 &= 0 \quad \text{check} \end{aligned}$$

When $A = 1$,

$$\begin{aligned} A + A &= A \\ 1 + 1 &= 1 \\ 1 &= 1 \quad \text{check} \end{aligned}$$

(b) In the same way, we can prove that $A \cdot A = A$. When $A = 0$,

$$\begin{aligned} A \cdot A &= A \\ 0 \cdot 0 &= 0 \\ 0 &= 0 \quad \text{check} \end{aligned}$$

When $A = 1$,

$$\begin{aligned} A \cdot A &= A \\ 1 \cdot 1 &= 1 \\ 1 &= 1 \quad \text{check} \end{aligned}$$

EXAMPLE 4-13

Prove Eq. (4-19), which says that $A + AB = A$.

SOLUTION

One way to prove this is to check the four possible cases.

Case 1: When $A = 0$ and $B = 0$,

$$\begin{aligned} A + AB &= A \\ 0 + 0 \cdot 0 &= 0 \\ 0 + 0 &= 0 \\ 0 &= 0 \quad \text{check} \end{aligned}$$

Boolean Algebra

Case 2: When $A = 0$ and $B = 1$,
$$A + AB = A$$
$$0 + 0 \cdot 1 = 0$$
$$0 + 0 = 0$$
$$0 = 0 \quad \text{check}$$

Case 3: When $A = 1$ and $B = 0$,
$$A + AB = A$$
$$1 + 1 \cdot 0 = 1$$
$$1 + 0 = 1$$
$$1 = 1 \quad \text{check}$$

Case 4: When $A = 1$ and $B = 1$,
$$A + AB = A$$
$$1 + 1 \cdot 1 = 1$$
$$1 + 1 = 1$$
$$1 = 1 \quad \text{check}$$

A different way to prove $A + AB = A$ is the following:

$$A + AB = A(1 + B) = A(1) = A$$

This approach assumes that Eqs. (4-7) and (4-10) are true, that is, that we can factor and that $1 + B = 1$.

EXAMPLE 4-14

Show that $x = ABC + A\bar{B}C + AB\bar{C}$ can be simplified to $x = A(B + C)$.

SOLUTION

$$\begin{aligned}
x &= ABC + A\bar{B}C + AB\bar{C} \\
&= AC(B + \bar{B}) + AB\bar{C} & \text{by using Eq. (4-7)} \\
&= AC + AB\bar{C} & \text{by using Eq. (4-14)} \\
&= A(C + B\bar{C}) & \text{by using Eq. (4-7)} \\
&= A(C + B) & \text{by using Eq. (4-22)} \\
&= A(B + C) & \text{by using Eq. (4-3)}
\end{aligned}$$

SUMMARY

The OR gate is an *any-or-all* gate; it has a 1 output when any or all of the inputs are 1. The AND gate is an *all-or-nothing* gate; it has 1 output only when all inputs are 1. The NOT circuit is an inverter or complementer; it changes a 0 to a 1, and vice versa.

A truth table is a list showing all input-output combinations of a logic circuit. The number of horizontal rows in a truth table equals 2^n, where n is the number of inputs to the logic circuit. To include all possible input combinations, we normally list the combinations in a binary-number progression.

A positive logic system is one in which 1 represents the more positive voltage; a negative logic system uses 1 to represent the more negative voltage. A positive OR gate becomes a negative AND gate, the vice versa, when we switch from one type of logic to the other. For consistency we will use positive logic throughout this book.

Boolean algebra is the algebra of logic circuits. The + sign stands for OR addition, that is, the way that an OR gate combines its inputs to produce its output. The · sign denotes AND multiplication, and the − sign denotes the operation of a NOT circuit.

De Morgan's theorems are quite useful in changing Boolean expressions into alternate forms. These two theorems suggest the NOT-OR (NOR) gate and the NOT-AND (NAND) gate. The NAND gate has a 0 output only when all inputs are 1. The NAND gate is a universal building block and can be used to build any logic system.

Boolean algebra has many laws and theorems. Some of them are identical to those of ordinary algebra; others are quite different. By using Boolean algebra we can often simplify the expressions for logic systems; as a result, we can use less hardware in building these systems.

We can list the laws and theorems of Boolean algebra in dual pairs as follows:

1. $A + B = B + A$ $AB = BA$
2. $A + (B + C) = (A + B) + C$ $A(BC) = (AB)C$
3. $A(B + C) = AB + AC$ $A + BC = (A + B)(A + C)$
4. $A + 0 = A$ $A \cdot 1 = A$
5. $A + 1 = 1$ $A \cdot 0 = 0$
6. $A + A = A$ $A \cdot A = A$
7. $A + \bar{A} = 1$ $A \cdot \bar{A} = 0$
8. $\bar{\bar{A}} = A$ $\bar{\bar{A}} = A$
9. $\overline{A + B} = \bar{A} \cdot \bar{B}$ $\overline{A \cdot B} = \bar{A} + \bar{B}$
10. $A + AB = A$ $A(A + B) = A$
11. $A + \bar{A}B = A + B$ $A(\bar{A} + B) = AB$

GLOSSARY

AND *gate*   An all-or-nothing gate. The output is a 1 only when all the inputs are 1s.

complement   This refers to the NOT operation. The complement of 0 is 1, and the complement of 1 is 0.

gate   In logic circuits this is a device with one output and two or more inputs, designed in such a way that there is an output only for certain combinations of the input signals.

Boolean Algebra

inverter A NOT circuit; also called a complementing circuit; also called a negating circuit.

NAND *gate* An AND gate followed by a NOT circuit, sometimes called a NOT-AND gate.

negative logic In discussing logic circuits this simply means that a 1 stands for the more negative of the two voltage levels.

NOR *gate* An OR gate followed by a NOT circuit. Sometimes called a NOT-OR gate.

NOT *circuit* A circuit that inverts or complements the input signal; that is, it makes 1s out of 0s, and vice versa.

OR *gate* An any-or-all gate. It has a 1 output if any or all of its inputs are 1s.

positive logic In discussing logic circuits this simply means that we define 1 as the more positive of the two voltage levels.

truth table A list of all the input-output possibilities of a logic circuit.

REVIEW QUESTIONS

1. When does an OR gate have a 1 output?
2. Describe the truth table for a two-input OR gate. For a three-input OR gate.
3. How many horizontal rows are there in a truth table if there are n inputs?
4. When does an AND gate have a 1 output? What does the truth table of an AND gate look like?
5. Define a positive logic system; and a negative logic system.
6. In Boolean algebra why does $1 + 1 = 1$?
7. What logic circuits do the $+$ sign, $\cdot$ sign, and $-$ sign represent?
8. What are the two De Morgan theorems (in words)?
9. What is a NOR gate and a NAND gate? Under what input conditions is the output of a NOR gate equal to 1? For what input conditions is the NAND gate output equal to 0?
10. Why is the NAND gate called a universal building block?
11. How do you find the dual identity for any given identity?

PROBLEMS

4-1 In Fig. 4-1a show the truth table for the OR gate if the batteries can have values of 3 or 12 volts. Use actual voltages in the table and ideal diodes.

4-2 In Fig. 4-3 suppose that the A, B, C inputs are at either 2 or 10 volts, and that the diodes are ideal. Show the truth table using actual voltages.

4-3 Show the truth table for Fig. 4-6 using actual voltages. Treat the diodes as ideal and use voltage levels of 1 and 10 volts.

4-4 In Fig. 4-7 the two voltage levels are 2 and 10 volts. Construct the truth table assuming a negative logic system.

4-5 A system has three inputs: A, B, and C. Show the truth table if the output x of this system is

$$x = (A + B)(A + C)$$

4-6 A logic circuit has inputs of A, B, and C. Show its truth table if its output x is

$$x = A + BC$$

Also, draw the logic circuit that has this output.

4-7 Write the Boolean expression for the output x of Fig. 4-29a. Work out the value of x for all possible input conditions, and show the truth table.

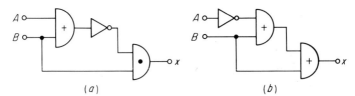

Fig. 4-29

4-8 In Fig. 4-29b write the expression for x. What is the truth table for this system?

4-9 Work out the value of $x = A\bar{B} + \bar{A}B$ for the four possible input conditions.

4-10 Draw the logic circuit for $x = A\bar{B} + \bar{A}B$.

4-11 Prove that $AB + B = B$ by showing that the left side equals the right side for all four input conditions.

4-12 Given that $x = A + \bar{B}C(A + \bar{C}) + B$, what is the value of x when
 (a) $A = 0, B = 1, C = 0$.
 (b) $A = 1, B = 0, C = 1$.

4-13 Show that $\overline{ABCD} = \bar{A} + \bar{B} + \bar{C} + \bar{D}$.

4-14 Prove that $\overline{A + B + C + D} = \bar{A} \cdot \bar{B} \cdot \bar{C} \cdot \bar{D}$.

4-15 Rewrite $\overline{\bar{A} + B}$ using the first De Morgan theorem.

4-16 Use De Morgan's second theorem to rewrite

$$x = \overline{(AB + CD)(ABC + D)}$$

4-17 Show how NAND gates can be used to build the logic circuit for $x = A + B\bar{C}$.

Boolean Algebra

4-18 Show how NAND gates can be used to build the logic circuit for $x = AB + CD$.
4-19 Use truth tables to show that $\overline{ABC} = \bar{A} + \bar{B} + \bar{C}$.
4-20 Prove Eq. (4-22) by constructing the truth table of each side of the equation.
4-21 Prove Eq. (4-25) by showing that the truth table of the left side is identical to that of the right side.
4-22 Prove Eqs. (4-7) and (4-10) using the truth-table approach.
4-23 Reduce $x = (A + B)(A + \bar{B})(\bar{A} + B)$ by using the laws and theorems of Boolean algebra.
4-24 Simplify $x = AB + ABC + \bar{A}B + A\bar{B}C$.

CHAPTER 5

ARITHMETIC CIRCUITS

Digital electronics is exciting; it enables us to build circuits that duplicate some of the processes of our minds — the best known being that of computing. By combining AND, OR, and NOT circuits in the right way, we can build circuits that add and subtract. Since these circuits are electronic, they are fast. Typically, an addition problem is done in microseconds.

In this chapter we will discuss some basic arithmetic circuits like the exclusive-OR gate, the half- and full-adders, the half- and full-subtractors, 8421 adders, and excess-3 adders. Besides giving us a first idea of how a computer works, these circuits lay the base upon which to build our later discussions of digital systems.

5-1 The Exclusive-OR Gate

Figure 5-1 shows an *exclusive*-OR *gate*. It has two inputs and one output. Each input goes into a NAND-gate inverter; the outputs of the NAND gates are $\bar{A}$ and $\bar{B}$. As shown in Fig. 5-1, $\bar{A}$ and B go into the upper AND gate, so

Arithmetic Circuits

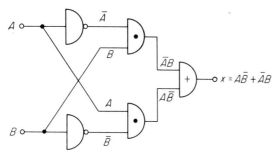

Fig. 5-1 The exclusive-OR gate.

that its output is $\bar{A}B$. Likewise, $A\bar{B}$ comes out of the lower AND gate. The OR gate has inputs of $A\bar{B}$ and $\bar{A}B$, so that the final output is

$$x = A\bar{B} + \bar{A}B \tag{5-1}$$

Why is the circuit of Fig. 5-1 called an exclusive-OR gate? To answer this, we will find the value of x for the four input conditions.

1. When $A = 0$ and $B = 0$,
$$x = 0 \cdot \bar{0} + \bar{0} \cdot 0 = 0 \cdot 1 + 1 \cdot 0 = 0 + 0 = 0$$

2. When $A = 0$ and $B = 1$,
$$x = 0 \cdot \bar{1} + \bar{0} \cdot 1 = 0 \cdot 0 + 1 \cdot 1 = 0 + 1 = 1$$

3. When $A = 1$ and $B = 0$,
$$x = 1 \cdot \bar{0} + \bar{1} \cdot 0 = 1 \cdot 1 + 0 \cdot 0 = 1 + 0 = 1$$

4. When $A = 1$ and $B = 1$,
$$x = 1 \cdot \bar{1} + \bar{1} \cdot 1 = 1 \cdot 0 + 0 \cdot 1 = 0 + 0 = 0$$

These results are summarized in the truth table of Table 5-1.

Table 5-1 Exclusive-OR Truth Table

A	B	x
0	0	0
0	1	1
1	0	1
1	1	0

The reason for the name exclusive-OR is this: a 1 output occurs when A or B is 1, *but not both*. Stated another way, the exclusive-OR gate has a 1 output only when both inputs are different; the output is 0 when the inputs are the same.

The exclusive-OR gate gives us a new kind of function to work with. We will use the symbol $\oplus$ to stand for this function. That is, when we want to describe an exclusive-OR gate, we can write

$$x = A \oplus B \tag{5-2}$$

Read this as x equals A OR B, but not both. Whenever we see $x = A \oplus B$, we will know that the output x is given by the truth table of Table 5-1.

The exclusive-OR operation is sometimes called "modulo-2 addition." Here are the rules for this addition:

$$0 \oplus 0 = 0$$
$$0 \oplus 1 = 1$$
$$1 \oplus 0 = 1$$
$$1 \oplus 1 = 0$$

Modulo-2 addition is really the same as binary addition, provided that we disregard carries.

The circuit of Fig. 5-1 is only one of many ways to build an exclusive-OR gate. Figure 5-2 shows another way. As you can see, $\overline{AB}$ comes out of the NAND gate. The final output is

$$x = (A + B)\overline{AB} \tag{5-3}$$

At first, this may not seem equal to $A\bar{B} + \bar{A}B$ (the output of an exclusive-OR gate). However, we can easily show that Eqs. (5-1) and (5-3) are equal. By De Morgan's second theorem,

$$x = (A + B)\overline{AB} = (A + B)(\bar{A} + \bar{B})$$

Next, we can multiply the binomials to get

$$x = A\bar{A} + A\bar{B} + \bar{A}B + B\bar{B}$$

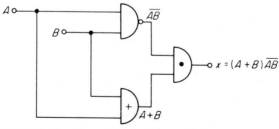

Fig. 5-2 Another way to build an exclusive-OR gate.

Arithmetic Circuits

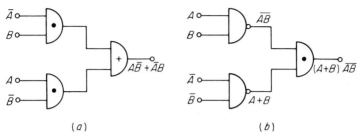

Fig. 5-3 Exclusive-OR gates.

Clearly, the AND of a variable and its complement must always be 0 because either the variable or its complement is 0. Therefore, $A\bar{A} = 0$ and $B\bar{B} = 0$, so that the expression becomes

$$x = A\bar{B} + \bar{A}B$$

which is the exclusive-OR function. Thus the circuits of Figs. 5-1 and 5-2 are equivalent.

Sometimes both a variable and its complement are stored in a digital system (as in flip-flops). We can then use the simpler circuits of Fig. 5-3 for modulo-2 addition.

To condense the block diagram of the exclusive-OR gate, we will use the symbol of Fig. 5-4. Whenever we see this logic symbol, we will know that the output x is a 1 when A or B is 1, but not both.

EXAMPLE 5-1

Show one way of building a parity checker for a 4-bit word.

SOLUTION

Figure 5-5a shows a way to check the parity of the word $ABCD$. The circuit adds the digits of $ABCD$; a final sum of 0 implies even parity; a sum of 1 odd parity. For instance, suppose $ABCD = 1001$. Figure 5-5b shows the circuit for this condition. Note that the final sum is 0, which means even parity.

As another example, the circuit for $ABCD = 1110$ is shown in Fig. 5-5c. The final output is 1, denoting odd parity.

EXAMPLE 5-2

Show a way of building a 6-bit binary-to-Gray converter.

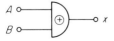

Fig. 5-4 Schematic symbol for exclusive-OR gate.

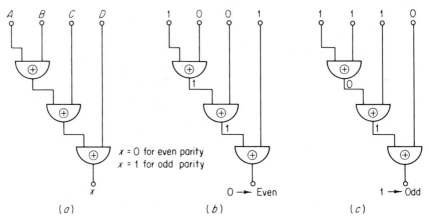

Fig. 5-5 A parity checker.

SOLUTION

Figure 5-6a shows one way to do it. The binary number is $ABCDEF$. The circuit adds the adjacent digits using modulo-2 addition and produces the Gray equivalent.

As a numerical example, suppose $ABCDEF = 100110$. Figure 5-6b shows the circuit for this input. The circuit adds the bits, two at a time, to get the Gray equivalent, which is 110101.

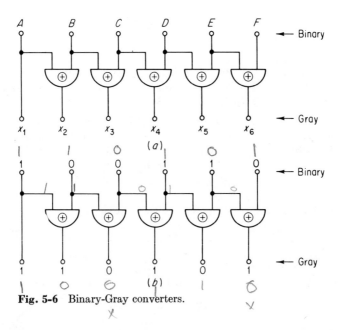

Fig. 5-6 Binary-Gray converters.

5-2 The Half-adder

The *half-adder* adds two binary digits at a time. Figure 5-7 shows how to make a half-adder. The output of the exclusive-OR gate is the sum, and the output of the AND gate is the carry.

Here is how the half-adder works. As with any two-input circuit, there are four distinct cases. These are:

1. When $A = 0$ and $B = 0$,
$$\text{Sum} = A \oplus B = 0 \oplus 0 = 0$$
$$\text{Carry} = AB = 0 \cdot 0 = 0$$

2. When $A = 0$ and $B = 1$,
$$\text{Sum} = 0 \oplus 1 = 1$$
$$\text{Carry} = 0 \cdot 1 = 0$$

3. When $A = 1$ and $B = 0$,
$$\text{Sum} = 1 \oplus 0 = 1$$
$$\text{Carry} = 1 \cdot 0 = 0$$

4. When $A = 1$ and $B = 1$,
$$\text{Sum} = 1 \oplus 1 = 0$$
$$\text{Carry} = 1 \cdot 1 = 1$$

We can summarize these results in the truth table of Table 5-2.

Table 5-2 Half-adder Truth Table

A	B	Carry	Sum
0	0	0	0
0	1	0	1
1	0	0	1
1	1	1	0

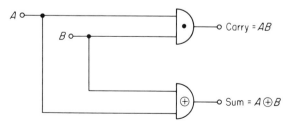

Fig. 5-7 The half-adder.

The half-adder of Fig. 5-7 performs binary addition. It does what we do mentally when we add 2 binary digits. For instance, in the first row of the truth table, 0 combined with 0 gives 0 carry a 0. In the last row, 1 combined with 1 gives 0 carry a 1.

Figure 5-8a shows another way to build a half-adder. The upper AND gate produces AB, which is used for the carry output, and the NAND gate produces $\overline{AB}$. As you can see, the sum output is

$$\text{Sum} = (A + B)\overline{AB} = A\bar{B} + \bar{A}B = A \oplus B$$

and the carry is simply

$$\text{Carry} = AB$$

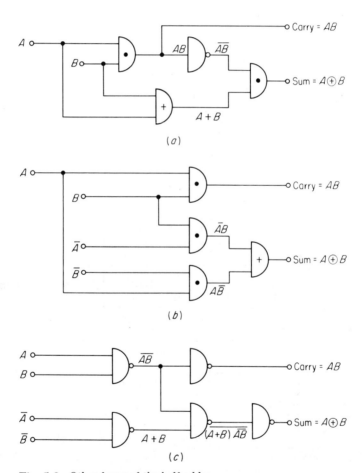

Fig. 5-8 Other forms of the half-adder.

Arithmetic Circuits

Very often, digital systems contain a variable and its complement (as in flip-flops). This leads to simpler forms of the half-adder. For instance, Fig. 5-8b shows a half-adder that can be used when the inputs are A and B, along with their complements $\bar{A}$ and $\bar{B}$.

Figure 5-8c shows another way to make a half-adder. This time only NAND gates are used.

There are many ways to build half-adders. The important thing to remember is that the half-adder adds 2 binary digits. It is an elementary circuit, but it does take the first step toward circuits capable of more difficult arithmetic.

5-3 The Full-adder

When we add two binary numbers, we sometimes have a carry from one column to the next. For instance,

$$\begin{array}{r} 111 \\ +\ 101 \\ \hline 1100 \end{array}$$

In the least significant column we add two digits:

$$1 + 1 = 0 \quad \text{carry a 1}$$

In adding the next column, we must add three digits because of the carry.

$$1 + 0 + 1 = 0 \quad \text{carry a 1}$$

In adding the last column, we again must add three digits because of the carry.

$$1 + 1 + 1 = 1 \quad \text{carry a 1}$$

To add binary numbers electronically, we need a circuit that can handle three digits at a time. By connecting two half-adders and an OR gate, we get a *full-adder*, which is a circuit that can add three digits at a time.

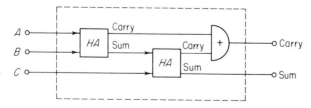

Fig. 5-9 The full-adder.

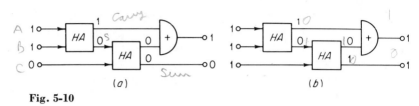

Fig. 5-10

Figure 5-9 shows a full-adder. The boxes labeled HA are half-adders. Since we know how a half-adder and an OR gate work, we can easily determine the final output.

For instance, suppose $A = 1$, $B = 1$, and $C = 0$. Figure 5-10a shows the full-adder with these inputs. The first half-adder has a sum of 0 with a carry of 1. The second half-adder has a sum of 0 with a carry of 0. Thus the final output is a sum of 0 with a carry of 1.

If we change the inputs to $A = 1$, $B = 1$, and $C = 1$, we ought to get a sum of 1 with a carry of 1. Figure 5-10b shows the half-adder with these inputs. Each half-adder has the outputs shown, so that the final output is a sum of 1 with a carry of 1. Thus the full-adder has again given us the correct binary sum of the inputs.

If we work out the sum and carry outputs in the same way for the other input combinations, we get the truth table of Table 5-3. (The reader should work out enough of these to satisfy himself that the full-adder does work.)

Table 5-3 Full-adder Truth Table

A	B	C	Carry	Sum
0	0	0	0	0
0	0	1	0	1
0	1	0	0	1
0	1	1	1	0
1	0	0	0	1
1	0	1	1	0
1	1	0	1	0
1	1	1	1	1

Remember the key idea of the full-adder: it adds 3 binary digits at a time. The next section shows how we can use full-adders to give us the sum of binary numbers with more than 1 bit.

Arithmetic Circuits

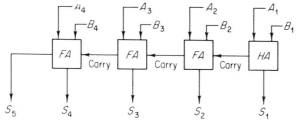

Fig. 5-11 A parallel binary adder.

5-4 A Parallel Binary Adder

We can connect full-adders and half-adders as shown in Fig. 5-11 to add two binary numbers. (The boxes labeled FA are full-adders.) The binary numbers being added are $A_4A_3A_2A_1$ and $B_4B_3B_2B_1$. The answer is

$$\begin{array}{r} A_4A_3A_2A_1 \\ + B_4B_3B_2B_1 \\ \hline S_5S_4S_3S_2S_1 \end{array}$$

In the first column we are adding only two digits A_1 and B_1; to handle this column we need only a half-adder. For any column above the first, there may be a carry from the preceding column; therefore, we must use a full-adder for each column above the first.

As an example of how the system of Fig. 5-11 works, suppose we want to add decimal numbers 11 and 7. The binary equivalent of decimal 11 is 1011, and the binary equivalent of decimal 7 is 0111. Figure 5-12 shows the binary adder for these inputs. Starting with the half-adder, we know the sum must be 0 with a carry of 1 as shown. The carry goes into the first full-adder, which adds $1 + 1 + 1$ to get a sum of 1 with a carry of 1. This carry goes into the next full-adder (third adder from right), which adds $0 + 1 + 1$ to get a sum of 0 with a carry of 1. The last full-adder adds

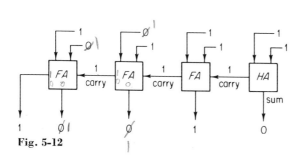

Fig. 5-12

$1 + 0 + 1$ to get a sum of 0 with a carry of 1. Thus, the final output of the system is 10010. The decimal equivalent of this is

$$\begin{array}{c} 1\ 0\ 0\ 1\ 0 \\ 16\ \cancel{8}\ \cancel{4}\ 2\ \cancel{1} = 18 \end{array}$$

which is the correct decimal sum of 11 and 7. Hence, the parallel binary adder of Fig. 5-11 gives us the binary sum of two 4-bit numbers. (The reader should work out a few other cases to convince himself that the system does add binary numbers.)

The system of Fig. 5-11 has limited capacity. The largest binary numbers that can be added are 1111 and 1111. Thus, the maximum capacity is

$$\begin{array}{r} 15 \\ +\ 15 \\ \hline 30 \end{array} \qquad \begin{array}{r} 1111 \\ +\ 1111 \\ \hline 11110 \end{array}$$

To increase the capacity of the system, we can connect more full-adders to the left end of the system. For instance, to add 6-bit numbers, we can connect two more full-adders.

How about subtraction? Figure 5-13 illustrates a system that subtracts $B_4B_3B_2B_1$ from $A_4A_3A_2A_1$. Here is what it does. First, the four NAND gates complement each B bit to get $\bar{B}_4\bar{B}_3\bar{B}_2\bar{B}_1$, which is the 1's complement of $B_4B_3B_2B_1$. The full-adders now add $A_4A_3A_2A_1$, $\bar{B}_4\bar{B}_3\bar{B}_2\bar{B}_1$, and the end-around carry. In other words, the system does what we do in our minds when we use the 1's complement to subtract. (Sometimes the NAND gates are unnecessary because the 1's complement of the number is already stored in the system.)

At last we are beginning to remove some of the mystery about digital computers. We have just seen how simple circuits can be connected to perform binary addition and subtraction. There are other aspects of computer operation like storing numbers, bringing the numbers out of storage into the arithmetic section, and putting the answer back into storage. These processes are covered in later chapters.

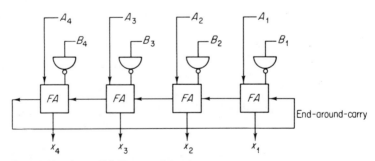

Fig. 5-13 A parallel binary subtractor.

5-5 An 8421 Adder

Some digital systems use straight binary numbers throughout. But there are many machines that use BCD numbers. As we said earlier, the 8421 code is the most popular BCD code, so popular, in fact, that it is sometimes called "the BCD code."

In the 8421 code each decimal digit is converted into its 4-bit equivalent. For instance, decimal 539 becomes

$$\begin{array}{cccc} 5 & 3 & 9 & \text{decimal} \\ \downarrow & \downarrow & \downarrow & \downarrow \\ 0101 & 0011 & 1001 & \text{8421 equivalent} \end{array}$$

In the 8421 code there is no 4-bit group above 1001. In other words, combinations like 1010, 1011, 1100, 1101, 1110, and 1111 do not exist in the 8421 code. Because of this, we run into trouble when we try to add 8421 numbers whose sum exceeds 9. For instance, if we add 8 and 5 using binary addition, we get

$$\begin{array}{rr} 8 & 1000 \\ +\ 5 & +\ 0101 \\ \hline 13 & 1101 \end{array}$$

The answer 1101 is fine in straight binary code, but it is meaningless in the 8421 code. The 8421 answer should be 0001 0011.

The basic problem with 8421 addition lies in the fact that 8421 code uses only 10 of 16 possible 4-bit groups. When the decimal sum exceeds 9, we must somehow skip the six forbidden groups to restore the answer to 8421 form. To skip these six forbidden combinations, we need to add 6 (0110) to the sum. For instance, let us add 8 and 5 using 8421 numbers.

$$\begin{array}{ll} 1000 & \\ +\ 0101 & \\ \hline 1101 & \text{binary equivalent of 13} \\ +\ 0110 & \text{add 6 to return to 8421 form} \\ \hline 0001\ 0011 & \text{8421 equivalent of 13} \end{array}$$

Note that the final answer is the 8421 equivalent of 13. As another example, suppose we add 9 and 3.

$$\begin{array}{rrl} 9 & 1001 & \\ +\ 3 & +\ 0011 & \\ \hline 12 & 1100 & \text{binary answer} \\ & +\ 0110 & \text{add 6 to return to 8421} \\ \hline & 0001\ 0010 & \text{8421 answer} \end{array}$$

128 **Digital Principles and Applications**

We can summarize the rules for 8421 addition as follows:

- Add the 8421 numbers using the rules of binary addition.
- If the sum of the 4-bit groups is greater than decimal 9, add 0110 to that sum to return to 8421 form.
- If the sum is 9 or less, leave it alone because it is in 8421 form.

How can we build a logic system that will do 8421 addition? Quite simply, we need a system that can do the following:

1. Add the 4-bit groups using binary addition.
2. Detect when the decimal sum of 4-bit groups is greater than 9.
3. Add 0110 (6) to the 4-bit answer when the sum is greater than 9.

With enough patience we can all arrive at the design of Fig. 5-14 (or something like it). This system adds 8421 digits using binary addition, and adds the correction of 0110 when the sum exceeds 9.

As an example of how the 8421 adder of Fig. 5-14 works, let us trace the addition for 8 and 5. At the top of the figure the two input BCD digits are $A_4A_3A_2A_1$ (1000) and $B_4B_3B_2B_1$ (0101). The upper row of adders combines these two inputs to get

$$\begin{array}{rr} A_4A_3A_2A_1 & 1000 \\ + B_4B_3B_2B_1 & + 0101 \\ \hline S_5S_4S_3S_2S_1 & 01101 \end{array}$$

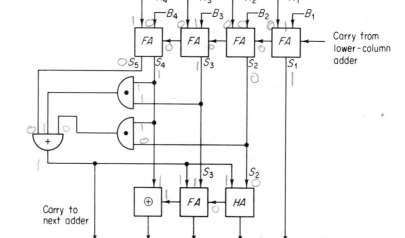

Fig. 5-14 An 8421 adder.

Arithmetic Circuits

The sum is greater than 9, and we detect this by feeding S_4 and S_3 into the upper AND gate. Both the inputs are 1s, so that the AND gate produces a 1 output; this 1 output is fed into the OR gate, so that there is a 1 coming out of the OR gate.

Next, we must restore $S_4S_3S_2S_1$ (1101) to 8421 form by adding 0110 to it.

$$\begin{array}{r} S_4S_3S_2S_1 \\ +\ 0\ 1\ 1\ 0 \\ \hline x_5x_4x_3x_2x_1 \end{array}$$

In the least significant column x_1 must equal S_1, so that we can bring S_1 down to the final output as shown in Fig. 5-14. The next row of adders now adds the 1 out of the OR gate to S_3 and S_2 as shown. This restores the answer to 8421 form.

The system of Fig. 5-14 will correctly add any two 8421 BCD digits. In general, it detects when the decimal sum is greater than 9 — this is the purpose of the two AND gates and the OR gate — and it adds 0110 to the answer. When the sum is 9 or less, 0000 is added.

Remember: The system of Fig. 5-14 takes care of only one column of decimal digits. To add decimal numbers with several digits, we must use several 8421 adders, one for each decimal column. For instance, to add 531 and 326, we need three 8421 adders like that of Fig. 5-14, one adder for each decimal column. Thus, by cascading 8421 adders we can add numbers of any size.

5-6 An Excess-3 Adder

We discussed the excess-3 code in Chap. 3. Recall that a decimal number is expressed as an excess-3 number by adding 3 to each decimal digit before converting to its 4-bit equivalent. Also recall the rules for excess-3 addition. Whenever a 4-bit group does not produce a carry, we must subtract 0011 (3) from it to return to excess-3 form. But when a group does produce a carry, we must add 3 to it. (See Sec. 3-2.)

Figure 5-15 shows how to build an excess-3 adder. The two excess-3 inputs go into the upper row of adders. The uncorrected sum is $S_5S_4S_3S_2S_1$. We either add or subtract 0011 to this answer, depending upon whether there is a carry or not. S_5 is a 1 whenever there is a carry, and is 0 when there is no carry. Note that $\bar{S}_5\bar{S}_5S_5S_5$ is added to the uncorrected sum $S_4S_3S_2S_1$. When there is a carry of 1, S_5 is 1, so that $\bar{S}_5\bar{S}_5S_5S_5$ is 0011. In other words, we are adding 3 to return the answer to excess-3 form. When there is no carry, we subtract 3 from the answer by using the 1's complement. Thus, when $S_5 = 0$, $\bar{S}_5\bar{S}_5S_5S_5 = 1100$ (the 1's complement of 0011). The full-

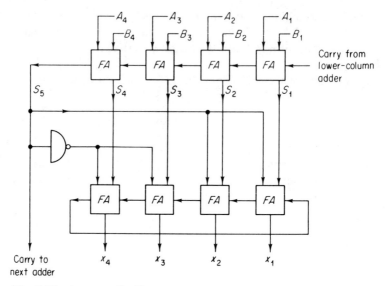

Fig. 5-15 An excess-3 adder.

adder on the left produces the end-around carry, so that the final answer is the correct excess-3 sum.

As a numerical example, let us trace the addition of 5 and 2. In the upper adders we get

$$
\begin{array}{ll}
A_4A_3A_2A_1 & 1000 \quad \text{excess-3 for 5} \\
+\,B_4B_3B_2B_1 & +\,0101 \quad \text{excess-3 for 2} \\
\hline
S_5S_4S_3S_2S_1 & 1101 \quad \text{excess-6 for 7}
\end{array}
$$

Since the decimal sum is less than 10, there is no final carry. In other words, S_5 is 0. The second row of adders now subtracts 0011 by using the 1's-complement approach.

$$
\begin{array}{l}
1101 \quad \text{uncorrected sum} \\
+\,1100 \quad \text{1's complement of 0011} \\
\hline
1\,1001 \\
\llcorner\!\!\!\rightarrow 1 \quad \text{end-around carry} \\
\hline
1010 \quad \text{excess-3 for 7}
\end{array}
$$

The excess-3 adder of Fig. 5-15 handles one column of decimal digits. By cascading adders of this type we can add larger excess-3 numbers. For instance, to add 3946 and 2679 we need four excess-3 adders, one for each decimal column.

Arithmetic Circuits

5-7 Half- and Full-subtractors

Instead of using complements to subtract, we can design circuits that subtract binary numbers in a more direct way. Recall the rules for binary subtraction:

$$0 - 0 = 0 \quad \text{with a borrow of } 0$$
$$0 - 1 = 1 \quad \text{with a borrow of } 1$$
$$1 - 0 = 1 \quad \text{with a borrow of } 0$$
$$1 - 1 = 0 \quad \text{with a borrow of } 0$$

We can summarize these results by Table 5-4, which shows the subtraction rules for $A - B$.

Table 5-4 Half- and Full-subtractors

A	B	Borrow	Difference
0	0	0	0
0	1	1	1
1	0	0	1
1	1	0	0

What kind of logic system has a truth table like Table 5-4? First, note that the difference output is 0 whenever the inputs A and B are the same; the difference output is 1 whenever A and B are different. So, we can use an exclusive-OR gate to produce the difference output. Second, note that the borrow output is 1 only when A is 0 and B is 1. We can get this borrow output by ANDing $\bar{A}$ and B.

Figure 5-16 shows one way to build a half-subtractor, which is a circuit that subtracts one binary digit from another. The circuit of Fig. 5-16 has a truth

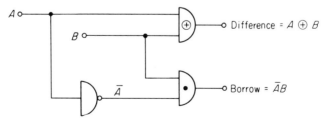

Fig. 5-16 A half-subtractor.

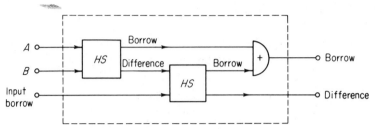

Fig. 5-17 A full-subtractor.

table that is identical to Table 5-4. You can see that there will be a borrow only when $A = 0$ and $B = 1$. Further, the difference output will be correct for each of the four possible $A - B$ combinations. Therefore, we have built a half-subtractor.

The half-subtractor handles only 2 bits at a time and can be used for the least significant column of a subtraction problem. To take care of a higher-order column, we need a full-subtractor. Figure 5-17 shows a full-subtractor; it uses two half-subtractors and an OR gate.

Half- and full-subtractors are analogous to the half- and full-adders; by cascading half- and full-subtractors as shown in Fig. 5-18, we have a system that directly subtracts $B_4B_3B_2B_1$ from $A_4A_3A_2A_1$.

The adders and subtractors give us the basic circuits we need for binary arithmetic; multiplication and division can be done by repeated additions and subtractions (this is discussed in later chapters after we have studied registers).

5-8 Classifying Logic Systems

We can make logic systems using a variety of parts — like diodes and resistors, or diodes and transistors, or resistors and transistors. Because of these different combinations, logic systems are often classified by the parts used. In this section we discuss some of these classifications.

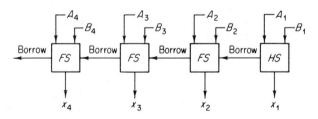

Fig. 5-18 A parallel binary subtractor.

Arithmetic Circuits 133

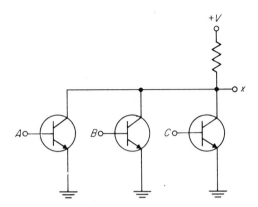

Fig. 5-19 A three-input DCTL NOR gate.

Direct-coupled transistor logic (DCTL) uses circuits in which input signals are coupled directly into bases and output signals are taken directly from collectors or emitters. For instance, Fig. 5-19 shows a three-input DCTL NOR gate. The inputs are A, B, and C; the output is x. When all inputs are 0s, all transistors are cut off, so that the output rises to a 1. When any input is a 1, the transistor for that input will saturate, causing the output to drop to a 0. Hence, we have a NOR gate. This is only one example of a DCTL; there are many others.

The fan-in of a logic circuit is the number of inputs that the logic circuit can handle. The fan-out is the maximum number of loads that can be connected to the output of the logic circuit. For example, suppose that the NOR gate of Fig. 5-19 drives several circuits as shown in Fig. 5-20. The NOR

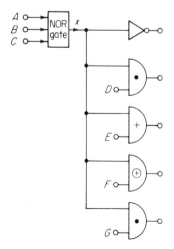

Fig. 5-20 Current hogging.

gate must supply current to the five circuits it drives. There is a limit to how much current it can supply. Suppose the NOR gate can drive a maximum of five circuits. Then, it has a fan-out of five. Also, since the NOR gate of Fig. 5-19 has three inputs, it has a fan-in of three.

When a DCTL circuit drives two or more loads, a serious problem called "current hogging" can occur. For instance, in Fig. 5-20 if the loads happen to be the bases of five transistors, there is no guarantee that the i-v characteristics of these bases are identical; one of the bases may turn on before the others, and literally hog most of the current, thereby preventing the other loads from turning on.

One way to eliminate current hogging is to use base resistors like those of Fig. 5-21a; the input impedances can be made almost equal, so that no one base predominates. Adding base resistors also increases the fan-out of these units.

Figure 5-21a is an example of resistor-transistor logic (RTL).

The NOR gate of Fig. 5-21b is another example of RTL. When all inputs are 0s, the voltage divider back-biases the base. The transistor cuts off, and the output goes to 1. When any of the inputs is a 1, the base turns on and the output drops to a 0.

Although resistors in the base circuit reduce current hogging and increase input impedance, they degrade the switching time because the capacitance at each base must now charge through the base resistor. To reduce this charging time, speed-up capacitors can be added as shown in Fig. 5-22. Circuits like these, which use resistors and capacitors in the base, are called resistor-capacitor-transistor logic (RCTL).

Diode-transistor logic (DTL) has high fan-in and fan-out capabilities. The gates of Fig. 5-23 are examples of DTL circuits. In Fig. 5-23a the diodes and resistors in the base circuit form an OR gate; the transistor is a NOT

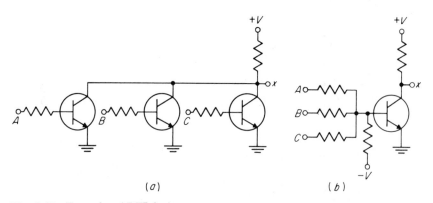

Fig. 5-21 Examples of RTL logic.

Arithmetic Circuits

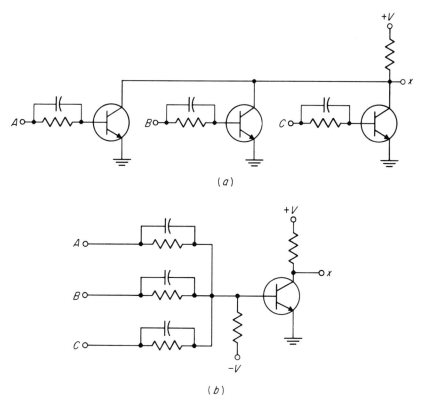

Fig. 5-22 Examples of RCTL logic.

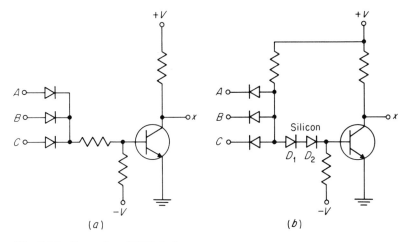

Fig. 5-23 Examples of DTL logic.

circuit; therefore, we have a NOT-OR, or simply a NOR gate. When all the inputs are low, the base shuts off, so that x goes high. When any input is high, the base turns on, so that x goes low.

The circuit of Fig. 5-23b has an AND gate in the base; the transistor provides the NOT function; therefore, the circuit is a NOT-AND, or a NAND gate. When all inputs are 0, the junction of all diodes is about $+0.7$ volt with respect to ground. The drop across the two diodes D_1 and D_2 is about 1.4 volts, so that the base is around -0.7 volt with respect to ground. So, the base is back-biased. D_1 and D_2 are in the circuit to guarantee that the voltage is around -0.7 volt when A, B, and C are 0s. (If D_1 and D_2 were replaced by a short, $+0.7$ volt would appear at the base, thereby incorrectly turning on the transistor.) Only when all inputs go high does the transistor turn on; under this condition the output x drops to a low voltage.

Transistor-transistor logic (TTL or T²L) is still another classification. Figure 5-24 shows a TTL NAND gate. The input transistor has three emitters, so that the emitters-base part of the transistor acts like a three-input AND gate. The output transistor is simply a NOT circuit, so that the overall circuit is a NAND gate.

Finally, there is current-mode logic (CML). In this type of logic the transistors do not saturate in order to get fast switching action. (Recall from Chap. 1 that avoiding saturation eliminates the storage time t_s of a transistor.) In CML circuits a fixed amount of current is switched from a transistor to another part of the circuit. This current is less than the value needed to saturate, so that the transistor never goes into saturation.

To bring out the current-switching idea, consider the circuit of Fig. 5-25. The current through the 10-kilohm resistor is fixed at about 1 ma because the right base is referenced at ground. Q1 and Q2 share this 1 ma, the split being determined by the A input. If A is slightly positive, all of the 1 ma flows through Q1, and none through Q2. If A is negative, Q1 is off and the

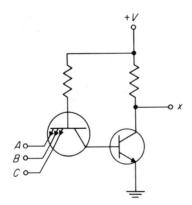

Fig. 5-24 A TTL NAND gate.

Arithmetic Circuits 137

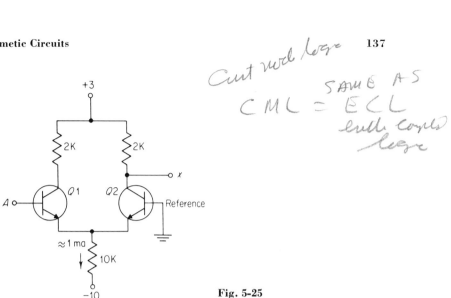

Fig. 5-25

entire 1 ma flows through Q2. Note that Q2 cannot saturate because even when it gets the entire 1 ma, the maximum voltage drop across its load resistor is only 2 volts, which is not enough to saturate. This is the basic idea behind current-mode logic. A fixed amount of current (less than the saturation value) is switched into or out of a transistor.

Figure 5-26 shows a NOR gate using current-mode logic. The three transistors on the left act like Q1 in Fig. 5-25. The reference voltage is between the low and high levels of the inputs. This reference voltage establishes a threshold; when the inputs are less than the reference value,

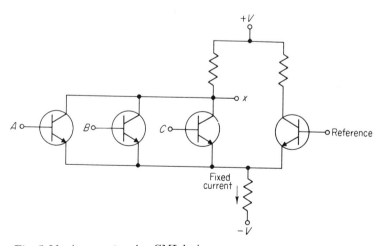

Fig. 5-26 A NOR gate using CML logic.

the circuit is in one state, but when the inputs are greater than this reference value, the circuit goes into the other state. For instance, when all inputs are 0s (less than the reference), the output is a 1. When any of the inputs is a 1 (greater than the reference), the output is a 0.

The different kinds of logic classes that we have discussed are used in building OR, AND, NOT, NOR, NAND, half-adders and full-adders, half- and full-subtractors, as well as many other circuits that we will study in later chapters. Many of these circuits are commercially available in integrated-circuit packages.

SUMMARY

The exclusive-OR gate has two inputs and one output. The output is a 1 when either input, but not both, is a 1. Stated another way, the output is a 1 when the inputs are different.

The exclusive-OR gate does modulo-2 addition. This kind of addition is the same as binary addition, provided that we disregard the carries. The symbol for modulo-2 addition is $\oplus$.

The half-adder has two inputs and two outputs; it adds 2 bits at a time. The full-adder has three inputs and two outputs; it adds 3 bits at a time.

By connecting half-adders and full-adders, we can make a parallel binary adder that is capable of adding numbers with many bits. With a few modifications this binary adder can subtract using the 1's complement.

To make an 8421 adder we must add 0110 (6) whenever the decimal sum of the 4-bit groups exceeds 9. By cascading 8421 adders we can add numbers of any size.

An excess-3 adder is also easy to make. First, we use a straight binary adder to give us the uncorrected sum. Next, we add or subtract 0011 (3), depending on whether the uncorrected sum is greater or less than 9. By cascading excess-3 adders, we can handle numbers of any size.

The use of complements to subtract numbers is not necessary. Half- and full-subtractors give us a more direct method. These units are analogous to half- and full-adders. By cascading half- and full-subtractors, we get a parallel binary subtractor.

All the basic logic circuits and many advanced ones are available in integrated form.

The classifications of logic circuits are:

- Direct-coupled transistor logic (DCTL)
- Resistor-transistor logic (RTL)
- Resistor-capacitor-transistor logic (RCTL)
- Diode-transistor logic (DTL)

Arithmetic Circuits

- Transistor-transistor logic (TTL)
- Current-mode logic (CML)

GLOSSARY

current hogging When the output of a logic circuit drives several loads, it is possible that one of the loads may turn on before the others, and take most of the current. If this happens, the other loads may not be turned on at all. This problem arises in DCTL circuits.

*exclusive-*OR *gate* A logic circuit with two inputs and one output. The output is a 1 when the two inputs are different, and is a 0 when the inputs are the same.

fan-in This is the number of inputs to a logic circuit. For instance, a three-input AND gate has a fan-in of three.

fan-out This is the maximum number of loads that a logic circuit can drive.

full-adder A logic circuit with three inputs and two outputs. This circuit adds 3 bits at a time, giving a sum and a carry output.

full-subtractor A three-input two-output circuit. It is used to subtract one binary digit from another, as well as subtracting a borrow produced by a lower-order column.

half-adder A logic circuit with two inputs and two outputs. The circuit adds 2 bits at a time, producing a sum and a carry output.

half-subtractor A two-input two-output logic circuit. Used to subtract one binary digit from another.

modulo-2 addition A binary-type addition whose rules are the same as ordinary binary addition, provided that we disregard the carries. It is the same as exclusive-OR addition.

REVIEW QUESTIONS

1. What is the output of an exclusive-OR gate for the four input cases?
2. What are the rules for modulo-2 addition? What symbol do we use for this addition?
3. Describe the truth table of a half-adder. What kind of logic circuit can we use to produce the carry output? The sum output?
4. What does a full-adder do? How many inputs and outputs does it have?
5. To make a parallel binary adder that can handle 10-bit inputs, how many half-adders and how many full-adders do we need?
6. In an 8421 adder how much is added when the decimal sum is less than 10? What do we add if the decimal sum is greater than 9?

7. In an excess-3 adder it is necessary to correct the answer to return it to excess-3 form. Describe what number is used to correct the answer, and when it is added or subtracted.
8. What is the truth table for a half-subtractor? How is the borrow output produced?
9. What is a full-subtractor? How many half-adders and OR gates make a full-adder?
10. Define the fan-in and fan-out of a logic circuit.
11. DCTL means direct-coupled transistor logic. Name some of the other classifications.

PROBLEMS

5-1 If we let 1 stand for a minus sign and let 0 stand for a plus sign, numbers like -110 and $+101$ would be written as 1 110 and 0 101. In multiplying binary numbers we may get plus or minus answers. Show how an exclusive-OR gate can be used to produce the correct sign in a multiplication problem.
5-2 To check the parity of a 10-bit word using exclusive-OR gates, how many of these gates do we need?
5-3 Draw a logic circuit that converts 10-bit binary numbers into Gray-code numbers. Use exclusive-OR gates.
5-4 Show a logic system that converts 10-bit Gray numbers into binary numbers. Use exclusive-OR gates.
5-5 Instead of using De Morgan's second theorem to show that

$$A\bar{B} + \bar{A}B = (A + B)\overline{AB}$$

use the truth-table method of proof. In other words, show that the truth table of the left member equals the truth table of the right member.
5-6 Work out the truth table of the circuit shown in Fig. 5-27. What kind of circuit is this?
5-7 Redraw the 8421-adder of Fig. 5-14, showing the inputs and outputs of all logic circuits for input numbers of $A_4A_3A_2A_1 = 0110$, and $B_4B_3B_2B_1 = 1001$.
5-8 Redraw the excess-3 adder of Fig. 5-15, showing the inputs and outputs of all logic circuits for decimal-number inputs of 4 and 7.
5-9 Show a logic system that converts a 4-bit 8421 number into a 4-bit excess-3 number.
5-10 Draw a logic circuit that converts a 4-bit excess-3 number into the equivalent 4-bit 8421 number.
5-11 To subtract 35-bit numbers, how many half-subtractors and how many full-subtractors are needed?

Arithmetic Circuits

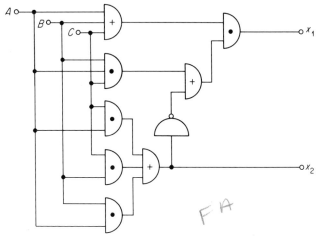

Fig. 5-27

CHAPTER 6

MULTIVIBRATORS

A multivibrator is a regenerative circuit with two active devices, designed so that one device conducts while the other cuts off. There are three basic kinds of multivibrators — the bistable, the astable, and the monostable.

Multivibrators can store binary numbers, count electrical pulses, control digital circuits, synchronize arithmetic operations, produce rectangular pulses, and do many other things that are vital to modern digital systems.

In this chapter we will discuss the types of multivibrators needed for our later work.

6-1 The T Flip-flop

A flip-flop is a multivibrator whose output can be either a low voltage or a high voltage, a 0 or a 1. This output stays low or high until the circuit is driven by an input called a *trigger*.

The T flip-flop is one type of flip-flop. It will change from low to high voltage, or vice versa, each time that a trigger drives it. For instance, in

Multivibrators

Fig. 6-1 The T flip-flop.

Fig. 6-1 negative triggers occur at points A, B, C, and D. Each time one of these arrives, the T flip-flop changes to the opposite state. Thus, at point A in time the output changes from a 0 to a 1. At point B it changes from a 1 to a 0, and so on. After the last pulse arrives at point D, the T flip-flop changes from a 1 to a 0, and stays in this state.

There are many ways to build T flip-flops. We will discuss one of these to show how it is done. But first, consider Fig. 6-2a. In this circuit one transistor saturates while the other cuts off. For instance, suppose that Q1 is saturated. The collector of Q1 will therefore be about 0 volts with respect to ground. With 0 volts applied to the top of the voltage divider, the base voltage Q2 becomes about -2.5 volts with respect to ground. Therefore, the base diode of Q2 back-biases and Q2 cuts off. With Q2 cut off, the collector of Q2 rises to almost $+10$ volts with respect to ground. With this 10 volts at the top of the voltage divider, the base of Q1 forward-biases enough to saturate Q1.

We assumed that Q1 was on and showed that Q2 would be off. Suppose we assume that Q2 is on. Because of the vertical symmetry of the circuit it is clear that Q1 will be cut off.

Since there are two distinct states (Q1-on, Q2-off; or Q1-off, Q2-on), the circuit is bistable; it can stay indefinitely in either of these two states.

To make the circuit change states, we can couple a negative trigger into the "on" transistor. For instance, in Fig. 6-2b suppose Q1 is on and Q2 is off. A narrow pulse or trigger goes into the base of Q1, the "on" transistor. This back-biases the base of Q1, forcing Q1 to go from saturation to cutoff. When this happens, the collector voltage of Q1 rises from about 0 to almost 10 volts. This rise drives the base of Q2 hard enough to saturate Q2. Therefore, the collector of Q2 drops to 0 volts. Thus, the trigger flips the circuit into the opposite state. If we now couple a negative trigger into the base of Q2, the circuit will flop back to the original state.

We do not have to use a negative trigger to shut off the "on" transistor; we can use a positive trigger to turn on the "off" transistor. For example, in Fig. 6-2c suppose that Q1 is off and Q2 is on. When the positive trigger hits the base, Q1 will conduct and its collector voltage will drop. This drop makes Q2 conduct less, which forces the collector voltage of Q2 to rise. This rise goes back to the base of Q1, reinforcing the positive trigger, which means we have regeneration. The circuit now drives itself into a condition where Q1 saturates and Q2 cuts off.

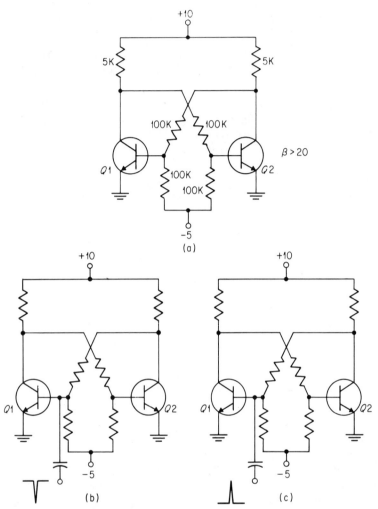

Fig. 6-2 Triggering a flip-flop.

To build a T flip-flop, we need a circuit that changes states each time a trigger pulse arrives. Figure 6-3 shows one way to build a T flip-flop — it uses negative triggering. The diodes in this circuit are called "steering diodes" because they steer the negative trigger into the base of the "on" transistor via the voltage divider of 100-kilohm resistors. The capacitors across the 100-kilohm resistors are called "speed-up" capacitors. These compensate for the stray and internal capacitance from each base to ground.

Multivibrators

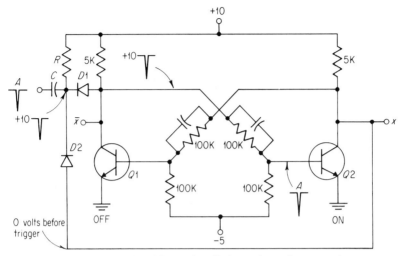

Fig. 6-3 A flip-flop circuit with steering diodes and speed-up capacitors.

To understand how the circuit works, assume that the output x of the flip-flop is a 0. This means that Q1 is off and Q2 is on. At point A in time the negative trigger appears. The capacitor couples the trigger into the junction of the diodes as shown in Fig. 6-3. Note that the trigger goes from 10 volts downward. Diode D2 does not conduct because the collector of Q2 is at 0 volts and this keeps D2 back-biased. Diode D1 does conduct because the collector of Q1 is initially at 10 volts; since the trigger applied to D1 goes from 10 volts downward, D1 becomes forward-biased. With D1 on, the negative trigger appears at the top of the voltage divider. The speed-up capacitor couples the negative trigger into the base of Q2, and so Q2 comes out of saturation. The flip-flop then regenerates, driving Q1 into saturation and Q2 into cutoff. The output x is now in the 1 state and remains in this state until another trigger appears. When the next trigger comes in, D2 will conduct, but D1 will not. Thus, the steering diodes automatically couple the trigger into the base of the "on" transistor.

The flip-flop has two outputs: the collector of Q2 produces the output x; the collector of Q1 produces the complementary output $\bar{x}$. Hence, whenever x is a 0, $\bar{x}$ is a 1, and vice versa.

Very often, we want to drive a flip-flop with a square-wave input. One way to do this is by using a small input capacitor to differentiate the square wave. Figure 6-4 illustrates this idea. By using an RC time constant that is small compared with the period of the input square wave, we get the positive and negative triggers shown. The T flip-flop responds only to the

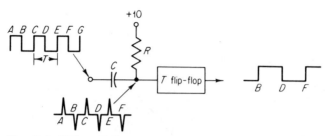

Fig. 6-4 Differentiating a square wave to get narrow triggers.

negative triggers. Because of this, the output of the flip-flop is a square wave with one-half the frequency of the input square wave. In other words, a T flip-flop *divides the input frequency by two.*

The circuit of Fig. 6-3 is one way to build a T flip-flop. Several variations are possible. T flip-flops can be designed to trigger on positive pulses instead of negative. Also, to increase the switching speed, we can design the circuit to prevent saturation. (Remember from Chap. 1 that avoiding saturation eliminates the storage time t_s.) Another modification is direct coupling of trigger instead of capacitive coupling. In this case, we do not need a narrow pulse; we simply need a trigger that changes from one voltage level to another.

In general, to indicate any T flip-flop, we will use the block-diagram symbol of Fig. 6-5. The x output is sometimes called the 1 output or the true output, while the $\bar{x}$ output is the 0 output or the false output.

The flip-flop has acquired many names from its wide use in electronics — like bistable multivibrator, binary multivibrator, Eccles-Jordan multivibrator, toggle, trigger, scale-of-two, divide-by-two, and binary counter.

To summarize, the T flip-flop has one input and two outputs. Each time a trigger arrives at the T input, the flip-flop changes states; its x output toggles (switches) from a 1 to a 0, or vice versa.

EXAMPLE 6-1

The waveform of Fig. 6-6a drives a T flip-flop that responds to negative-going transitions. Initially, $x = 0$. Sketch the x and $\bar{x}$ outputs.

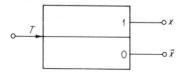

Fig. 6-5 Schematic symbol for a T flip-flop.

Multivibrators

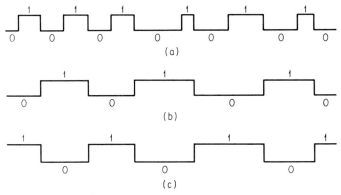

Fig. 6-6 Example 6-1.

SOLUTION

Each time the input waveform goes from a 1 to a 0, the flip-flop toggles. Hence, we can draw the true or 1 output as shown in Fig. 6-6b. The false or 0 output is shown in Fig. 6-6c.

EXAMPLE 6-2

Suppose we connect three T flip-flops as shown in Fig. 6-7. If the input to the first flip-flop is a 1000-Hz square wave, what is the final output? Assume negative triggering.

SOLUTION

Each time the input square wave goes from a 1 to a 0, the first flip-flop changes states; therefore, the output of the first flip-flop is a 500-Hz square wave.

This 500-Hz square wave drives the second flip-flop, and on each negative-going change the second flip-flop changes states. So, the output of the second flip-flop is a 250-Hz square wave.

The 250-Hz square wave drives the last flip-flop; the final output is a 125-Hz square wave.

We have seen that the three flip-flops of Fig. 6-7 divide the input frequency by 8. This is an example of a *scaler*, which is a circuit that delivers

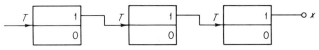

Fig. 6-7 Example 6-2.

an output pulse when a specified number of input pulses are received. In this case, the scaler gives us one output pulse for every eight input pulses.

6-2 The RS and RST Flip-flops

A reset-set flip-flop or simply an RS flip-flop has two inputs and two outputs. We can modify a T flip-flop to get an RS flip-flop as shown in Fig. 6-8. The difference between a T flip-flop and an RS flip-flop lies in the triggering.

The flip-flop of Fig. 6-8 responds to positive triggers. It works as follows: When a positive trigger arrives at the set input, this trigger goes through the capacitor and diode into the base of Q1. If Q1 is cut off, this positive trigger will turn on the base, causing Q1 to conduct. The circuit then regenerates, driving Q1 into saturation and Q2 into cutoff. With Q2 off, the x output will be a 1. Thus, applying a positive trigger to the set input has *set* the x output to a 1.

If Q1 were already saturated, a positive trigger on the set input would do nothing because the base is already forward-biased.

Thus, we conclude that when a positive trigger hits the set input, the x output becomes a 1 if it is not already a 1.

The reset input is similar. Since it drives the opposite side of the flip-flop, a positive trigger on the reset input will cause the x output to become a 0.

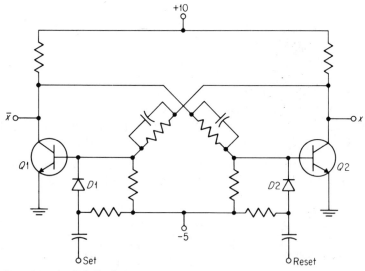

Fig. 6-8 An RS flip-flop.

No matter what state the flip-flop is in, a positive trigger on the set input ensures that the x output will be a 1. Likewise, a positive trigger on the reset input ensures that the x output will be a 0.

Square-wave inputs can be used. The capacitors on the set and reset inputs will differentiate the square waves to get positive and negative triggers. The diodes then clip off the negative triggers, so that only the positive triggers pass into the bases.

Figure 6-9 shows a direct-coupled RS flip-flop. It can respond directly to square waves without differentiation. When x is a 0, a positive voltage applied to the set input will turn on the left side of the flip-flop. This causes the right side of the flip-flop to turn off, so that x goes to a 1. Similarly, a positive voltage on the reset input causes the right side of the flip-flop to turn on, thereby resetting x to a 0.

The flip-flop of Fig. 6-8 responds a-c-type triggers, that is, sharp, narrow triggers. If square waves are used with this flip-flop, they must be differentiated to get narrow pulses. The flip-flop of Fig. 6-9 is different. It can respond to narrow triggers or to d-c voltages.

An RST flip-flop combines an RS flip-flop and a T flip-flop. The RST flip-flop can set, reset, or trigger. Figure 6-10 shows an RST flip-flop — the trigger input responds to negative-going voltages, whereas the set and reset inputs respond to positive-going voltages. When we apply negative pulses to the trigger input, the flip-flop toggles back and forth between its two stable states. When a positive pulse arrives at the set input, the x output becomes a 1 if it is not already a 1; when a positive pulse hits the reset input, the x output becomes a 0 if it is not already a 0.

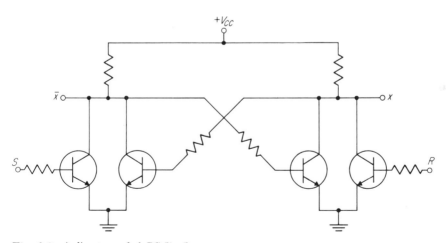

Fig. 6-9 A direct-coupled RS flip-flop.

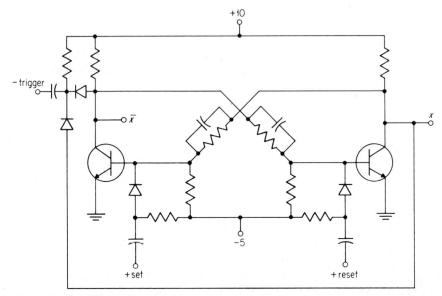

Fig. 6-10 An *RST* flip-flop.

There are many ways to build *RST* flip-flops. All three inputs could be designed to respond to positive triggers, or perhaps to negative triggers, or to a combination of positive and negative triggers. For consistency, we will use positive triggering on the *R* and *S* inputs, and negative triggering on the *T* input, unless otherwise specified. (The use of the master/slave flip-flop in Chap. 8 is one reason for doing this.)

Figure 6-11 shows the symbols we will use for the *RS* and *RST* flip-flops.

EXAMPLE 6-3

Show one way of making a *T* flip-flop using an *RS* flip-flop and some AND gates.

SOLUTION

We want the *RS* flip-flop to toggle each time a trigger comes in. Stated another way, when the *x* output is a 0, we want the next trigger to hit the set input; when *x* is a 1, the next trigger should go to the reset input.

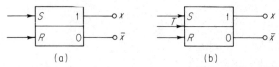

Fig. 6-11 Schematic symbols for the *RS* and the *RST* flip-flop.

Multivibrators

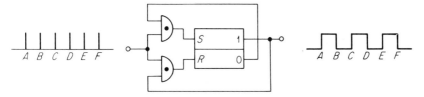

Fig. 6-12 Example 6-3.

Thus, by steering the triggers into the correct inputs, we can make the *RS* flip-flop toggle.

Figure 6-12 shows one way to do this. The x and $\bar{x}$ outputs cross-couple back to the input AND gates. Let us start with the assumption that x is 0 before point A in time. Then $\bar{x}$ is a 1, and it goes back to the upper AND gate. When the trigger arrives at point A in time, it passes through the upper AND gate into set input because both upper inputs are 1s. The lower AND gate does nothing because one of its inputs is a 0. Thus, the positive trigger is steered into the set input. This causes the flip-flop to change states.

The x output now is a 1. This 1 *enables* the lower AND gate, that is, it makes the lower AND gate ready to respond to the next trigger. At the same time, the $\bar{x}$ output, which is a 0, disables the upper AND gate. When the narrow trigger arrives, it passes through the lower AND gate into the reset input. The x output then goes back to a 0.

Each time a trigger arrives, it is alternately steered into the set and reset inputs. This causes the flip-flop to toggle. In effect, we have made a T flip-flop.

6-3 The JK Flip-flop

The JK flip-flop is another important and widely used flip-flop. Figure 6-13 shows the symbol we will use for the JK flip-flop. It has three inputs and two outputs. The middle input is called the trigger or "clock" input; the two other inputs are the J and K inputs.* The flip-flop response is deter-

*Throughout this book JK flip-flop refers to a clocked JK flip-flop.

Fig. 6-13 The JK flip-flop.

mined by the values of J and K at the instant that the trigger or clock pulse arrives. There are four cases to describe:

1. When $J = 1$ and $K = 1$, the flip-flop toggles each time a trigger or clock pulse arrives. In other words, when J and K are 1s, the JK flip-flop acts like a T flip-flop.
2. When $J = 1$ and $K = 0$, the flip-flop will set on the next clock pulse. On succeeding triggers the flip-flop stays set. In other words, it is like setting an RS flip-flop.
3. When $J = 0$ and $K = 1$, the flip-flop will reset on the next trigger, and then stay reset on succeeding clock pulses (like resetting an RS flip-flop).
4. When $J = 0$ and $K = 0$, the flip-flop remains in whatever state it is in.

Table 6-1 summarizes the JK flip-flop action. Since b is the value just before the trigger arrives, the first entry of the truth table tells us that the output is the same before and after the trigger. On the other hand, the last entry says that the output complements after the trigger arrives.

Table 6-1 The JK Flip-flop Truth Table

J	K	Output after trigger
0	0	b (same)
0	1	0 (reset)
1	0	1 (set)
1	1	$\bar{b}$ (toggle)

b = value of the output just before the trigger or clock pulse arrives.

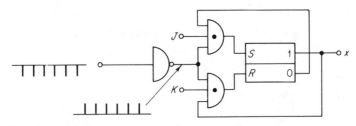

Fig. 6-14 Modifying an RS flip-flop to get a JK flip-flop.

Multivibrators

One easy way to build a JK flip-flop is to modify an RS flip-flop as shown in Fig. 6-14. Here is how it works:

- When $J = 0$ and $K = 0$, both AND gates are disabled; therefore, none of the clock pulses can get through to the set or reset inputs. So the flip-flop stays in whatever state it is in.
- When $J = 0$ and $K = 1$, the upper AND gate is disabled, but the lower one is enabled if $x = 1$. On the next clock pulse, the flip-flop will reset if it is not already in this state. The succeeding clock pulses do nothing.
- When $J = 1$ and $K = 0$, the lower AND gate is disabled, but the upper one is enabled if $\bar{x}$ is a 1. Therefore, the next clock pulse will pass through the upper AND gate and set the x output to a 1, unless this output is already a 1.
- When $J = 1$ and $K = 1$, both AND gates are enabled as far as the J and K inputs are concerned. Since the outputs cross-couple back to the inputs, the set AND gate is enabled whenever x is a 0. When the trigger comes in, the x output will be set to a 1. This now causes the reset AND gate to be enabled, and on the next clock pulse the x output will return to 0. Thus, the flip-flop toggles when each clock pulse comes in.

Remember how the JK flip-flop works; it is very important in our later discussions.

One more point. For convenience, we often will show the toggle condition ($J = 1$ and $K = 1$) by leaving the J and K inputs unconnected. When we do this, simply remember that even though not shown, 1s are going into the J and K inputs (see Example 6-5).

EXAMPLE 6-4

The JK flip-flop of Fig. 6-15 responds to the trailing edges of the input square wave. What will the x output be like?

SOLUTION

When x is a 0, J is a 1 and K is a 0; therefore, the flip-flop will set on the next trailing edge of the input.

When x becomes a 1, J becomes a 0 and K becomes a 1. For this condition the next trailing edge will reset x to a 0.

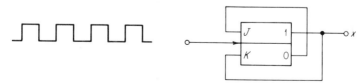

Fig. 6-15 Example 6-4.

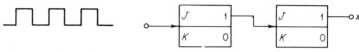

Fig. 6-16 Example 6-5.

So, the flip-flop toggles as each trailing edge arrives, and as a result, the output is a square wave of one-half the input frequency.

EXAMPLE 6-5

A 1000-Hz square wave drives the first JK flip-flop of Fig. 6-16. Both flip-flops respond to negative-going changes. What is the frequency of the final output?

SOLUTION

As indicated earlier, when no inputs are connected to J and K, by definition we understand that 1s are going into J and K. As a result, each flip-flop toggles on negative-going changes. Since 1000 Hz drives the first flip-flop, 500 Hz comes out of this flip-flop and drives the second flip-flop. Therefore, the output of the second flip-flop is a 250-Hz square wave.

6-4 The Schmitt Trigger

The Schmitt trigger is stable in either of two states, and therefore falls into the category of bistable multivibrators. However, instead of coupling from the collector of one transistor into the base of another, it couples by way of a common-emitter resistor.

Figure 6-17a shows one way of building a Schmitt trigger. When Q1 is on, Q2 is off, and vice versa. To understand how the circuit works, assume that the input voltage v_{in} is initially set to 0 volts. Under this condition no base current flows in Q1; therefore, there is no collector current in Q1. The base of Q2 is connected to a voltage divider formed by the 100-kilohm resistors. The voltage at the top of the voltage divider is enough to turn on the base and saturate Q2. With Q2 saturated, the collector and emitter are effectively shorted together, as shown in Fig. 6-17c. As a result, the voltage from the emitter to ground is 2 volts.

With the battery at 0 volts, we have Q1 cut off and Q2 saturated. Under this condition the output voltage of the Schmitt trigger is about 2 volts, as seen in Fig. 6-17c. Since the emitter voltage is 2 volts, no base current can flow in Q1 for any value of v_{in} less than 2 volts (the base diode remains

Multivibrators

back-biased). Therefore, for any input less than 2 volts, the output voltage of the Schmitt trigger stays at 2 volts because Q2 remains saturated.

If we raise the input voltage above 2 volts, base current flows in Q1. Once this happens, the collector current begins to flow, and the voltage at the Q1 collector decreases. The speed-up capacitor couples this drop into the base of Q2, bringing Q2 out of saturation. The drop in Q2 current reduces the voltage from the emitter to ground. Since v_e has been reduced, Q1 is turned on harder. The increase in the Q1 current drops the collector voltage of Q1 even further, and we have regeneration. This regeneration continues until Q1 saturates and Q2 cuts off. With Q2 cut off, the output voltage goes to 10 volts because there is no current through the 4-kilohm load resistor. The regenerative switching action takes place very quickly, so that we can think of v_{out} as suddenly having gone from 2 to 10 volts.

Any further increase in the input voltage will increase the Q1 base current. However, since Q2 is already cut off, the output voltage must stay at 10 volts.

Note that with Q1 saturated the emitter voltage to ground is only 1 volt, as shown in Fig. 6-17b.

With the output voltage at 10 volts, the only way of bringing the output voltage back to 2 volts is by reducing the input voltage driving Q1. Since the emitter voltage with Q1 saturated is only 1 volt, the input voltage must be reduced to about 1 volt to bring Q1 out of saturation. At this value the collector of Q1 comes out of saturation. The collector voltage will increase,

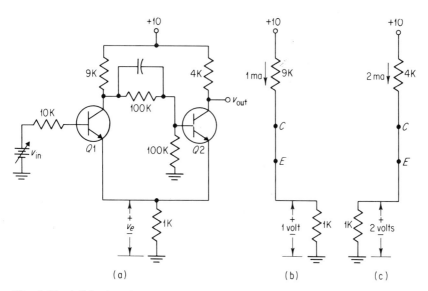

Fig. 6-17 A Schmitt trigger.

and this in turn causes Q2 to conduct. With Q2 conducting, the emitter voltage increases, causing the current in Q1 to decrease further. This raises the collector voltage of Q1, and again we have regenerative action. This regeneration continues until Q1 cuts off and Q2 saturates. The circuit is then back in the original state with an output voltage of 2 volts.

Figure 6-18a summarizes the action of a Schmitt trigger. Note that the output voltage is either 2 or 10 volts. When v_{out} is in the low state, it is necessary to raise v_{in} to slightly more than 2 volts to cause v_{out} to jump into the high state. Once in the high state, v_{out} stays at 10 volts until v_{in} drops to slightly more than 1 volt. The output then jumps back to the low state of 2 volts. The rapid switching action is indicated by the dashed lines.

Figure 6-18b shows the general characteristic of any Schmitt trigger. The value of v_{in} that causes the output to jump from a low to a high state is called the upper trip point (UTP); the value of v_{in} that causes the output to jump from the high state to the low state is called the lower trip point (LTP).

The Schmitt trigger is an amplitude-sensitive circuit. Once the input voltage exceeds the upper trip point, the output voltage goes from a low value to a high value, from a 0 to a 1. When the input voltage drops below the lower trip point, the output voltage drops from a 1 to a 0. Because of this, we can use the Schmitt trigger to detect when the input voltage crosses certain voltage levels.

In Fig. 6-18b the difference between the UTP and the LTP produces a hysteresis or backlash in the switching action. It is possible to eliminate this hysteresis by making the LTP and UTP equal. But a small amount of hysteresis is good because it ensures fast switching from one state to another.

The Schmitt trigger is an example of a category of circuits known as "voltage comparators." These circuits detect when a voltage has crossed a specified level. Comparators are useful in analog-to-digital converters, which are discussed in Chap. 11.

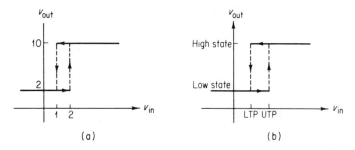

Fig. 6-18 Schmitt-trigger characteristics.

Multivibrators

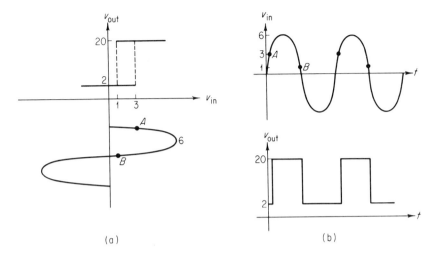

Fig. 6-19 Example 6-6.

EXAMPLE 6-6

A Schmitt trigger has the following characteristics: UTP = 3 volts, LTP = 1 volt, high state = 20 volts, low state = 2 volts. Sketch the output voltage for a sine-wave input with a peak value of 6 volts.

SOLUTION

First we can sketch the Schmitt-trigger characteristic and the input sinusoid as shown in Fig. 6-19a. When the sinusoid increases above 3 volts, the Schmitt trigger goes from the low to the high state. This occurs at point A in time. The output stays in the high state until the input sinusoid drops below 1 volt. Then the output drops back to the low state at point B in time. Hence, we can sketch the input and output as shown in Fig. 6-19b. This shows one major use of the Schmitt trigger — to change sine waves into square waves.

EXAMPLE 6-7

The Schmitt trigger of the preceding example is driven by a sawtooth voltage with a peak value of 10 volts. Sketch the Schmitt-trigger output.

SOLUTION

The Schmitt-trigger characteristic and the sawtooth are shown in Fig. 6-20a. It is clear that the Schmitt trigger trips as the sawtooth rises through the 3-volt level. On the sawtooth flyback, the Schmitt trigger trips at 1 volt. Figure 6-20b shows the output voltage.

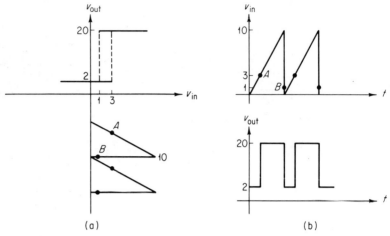

Fig. 6-20 Example 6-7,

In general, note that whenever the Schmitt trigger is driven by a periodic signal whose peak value exceeds the UTP, the output will be a rectangular waveform. Because of this, the Schmitt trigger is sometimes known as a "squaring circuit." Also note that the output frequency will equal the input frequency.

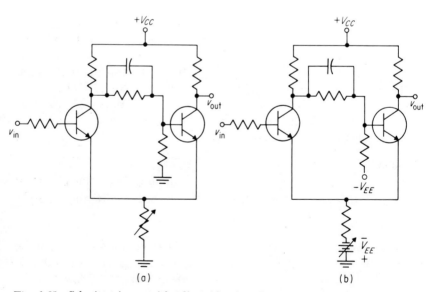

Fig. 6-21 Schmitt triggers with adjustable trip points.

Multivibrators

The trip points of a Schmitt trigger can be moved. Figure 6-21a shows an adjustable emitter resistor. This allows us to control the trip points over a range. Figure 6-21b shows the emitter resistor returned to a negative supply; this permits us to use negative trip points if we want.

6-5 The Astable Multivibrator

The astable multivibrator has two states, but is stable in neither; it oscillates back and forth between these two states, producing a square-wave output. In other words, we can think of an astable multivibrator as being a square-wave oscillator. A "free-running" multivibrator is another name for the astable multivibrator.

Often in digital systems a single astable multivibrator keeps all the different circuit operations in step with one another. In a sense, this multivibrator is like a master clock that synchronizes all parts of the digital system. Because of this, the astable multivibrator is often called a "clock."

Figure 6-22 shows a transistorized astable multivibrator. The transistors in this circuit alternately saturate and cut off. A complete analysis of the circuit shows that the period of the output square wave is

$$T \cong 1.4RC \tag{6-1}$$

where T = period of the square wave
R = value of the base resistor
C = value of the coupling capacitor

Or, the frequency of the output square wave is

$$f \cong \frac{0.7}{RC} \tag{6-2}$$

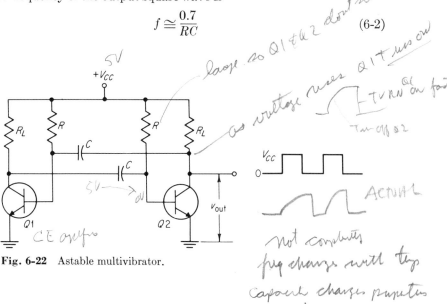

Fig. 6-22 Astable multivibrator.

To ensure that the transistor does saturate, the beta of each transistor must be greater than the ratio of the base to load resistance. That is,

$$\beta > \frac{R}{R_L} \qquad (6\text{-}3)$$

For instance, if the base resistor R is 400 kilohms, and if the load resistor R_L is 20 kilohms, then the beta of each transistor must be greater than 400/20, or 20.

The circuit of Fig. 6-22 is symmetrical about the vertical centerline; that is, both base resistors, both load resistors, and both capacitors are equal. Because of this, each transistor stays off for the same amount of time, and the output waveform is square. Occasionally, an unsymmetrical output is needed, and can be obtained by using either two different base resistances or two different capacitor values.

If we need a variable-frequency clock, we can gang variable capacitors. In this way, the frequency changes when we tune the capacitors.

Figure 6-23 shows another way to control the frequency. A separate supply drives the base resistors. When the base supply voltage changes, the frequency changes. In terms of natural logarithms the period of the square wave is

$$T \cong 2RC \ln\left(1 + \frac{V_{cc}}{V_{bb}}\right) \qquad (6\text{-}4)$$

EXAMPLE 6-8

In the multivibrator of Fig. 6-22 the base resistors have a value of 100 kilohms and the capacitors have a value of 1000 pf. Find the approximate frequency of the square wave.

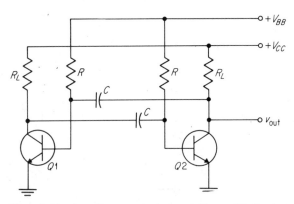

Fig. 6-23 A voltage-controlled astable multivibrator.

Solution

Using Eq. (6-2), we get

$$f \cong \frac{0.7}{RC} = \frac{0.7}{100(10^3)1000(10^{-12})} = 7 \text{ kHz}$$

The astable multivibrator delivers an output square wave with a fundamental frequency of 7 kHz.

EXAMPLE 6-9

Suppose the betas of the transistors in the preceding example are in the range of 50 to 150. To ensure saturation, what is the largest permissible value of R_L?

Solution

The worst case is $\beta = 50$. Using Eq. (6-3), we get

$$50 = \frac{100(10^3)}{R_L}$$

or

$$R_L = 2(10^3) = 2 \text{ kilohms}$$

stay ONE STABLE STATE   BISTABLE = F/F

6-6 The Monostable Multivibrator — ONE SHOT

In this section we study the last basic type of multivibrator — the monostable multivibrator. This kind of multivibrator is stable in one state, but unstable in the other. When it is triggered, it goes from the stable state into the unstable state. It remains in the unstable state *temporarily* and then returns to the stable state.

Figure 6-24 shows the general idea of a monostable multivibrator. At point A in time a trigger hits the input. This causes the output voltage to go from a low to a high value. The high state is an unstable state, so that after a while the output voltage returns to the low state. The output remains in the low state until the next trigger arrives at point B in time. Again the output jumps to the high state, and after a while returns to the low state, where it stays until the trigger comes in at point C in time.

INDEPENDENT of INPUT WIDTH

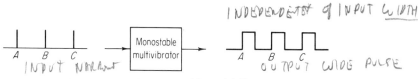

INPUT NARROW   OUTPUT WIDE PULSE

Fig. 6-24 The concept of the monostable multivibrator.

PULSE STRETCHER → Can't see pulse on scope

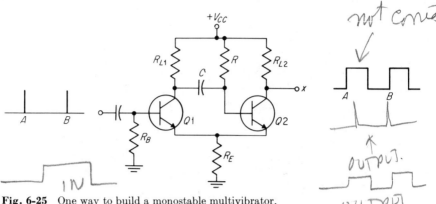

Fig. 6-25 One way to build a monostable multivibrator.

The general idea of a monostable multivibrator is now clear. Each time a trigger arrives, the output temporarily changes to the high state, but then returns to the low state. For each input trigger there is one rectangular output pulse. Because of this, the monostable multivibrator is often called a "one-shot" multivibrator, or simply a one-shot.

There are many ways to build a one-shot. Figure 6-25 shows one of these. The low state has Q1 off and Q2 on. When a positive trigger arrives, it turns Q1 on, which drops the collector voltage of Q1. This drop couples into the base of Q2, shutting this transistor off. But this condition, Q1 on and Q2 off, is only temporary, because as the charge on the capacitor changes, the back bias on the base of Q2 disappears. After an amount of time determined by the RC time constant of the base circuit, Q2 again turns on and Q1 turns off. Each time that a positive trigger hits the base of Q1, the output voltage x temporarily goes from low to high, but then returns to a low voltage. There is one rectangular output pulse for each input trigger.

A square wave can drive a one-shot multivibrator, provided we differentiate the square wave with a small input capacitor.

The one-shot multivibrator is useful for reshaping ragged pulses. It also can generate gating signals, introduce time delays, and change the width of a pulse. We will discuss it again in later chapters.

SUMMARY

A multivibrator is a regenerative circuit with two active devices, designed so that one device conducts while the other cuts off. The three basic types of multivibrators are the bistable, the astable, and the monostable.

A T flip-flop is a bistable multivibrator that changes to the opposite state each time a trigger arrives. A T flip-flop divides the input frequency by 2.

An RS flip-flop has two inputs — a reset and a set input. When a trigger hits the reset input, the x output becomes a 0 if it is not already a 0. When a trigger hits the set input, the x output becomes a 1 if it is not already a 1.

An RST flip-flop combines an RS flip-flop with a T flip-flop; it can reset, set, or trigger. For convenience, we will use negative triggering for the T input, and positive triggering for the R and S inputs.

The JK flip-flop has three inputs: a trigger input, a J input, and a K input. The values of J and K determine what the flip-flop does when a pulse arrives at the trigger input. If J and K are 1s, the flip-flop toggles on the next trigger. If J is a 1 and K is a 0, the flip-flop sets on the next trigger. When J is a 0 and K is a 1, the next trigger will reset the flip-flop. If J and K are 0s, the next trigger does nothing.

The Schmitt trigger is a bistable multivibrator. It changes states when the input voltage crosses certain levels called the "trip points." The Schmitt trigger falls into the category of voltage comparators.

The astable multivibrator free-runs, producing a rectangular-wave output. It often synchronizes the operations of circuits throughout digital systems and because of this is sometimes called a "clock."

The monostable multivibrator, also known as a one-shot multivibrator, delivers one rectangular output pulse for each input trigger.

GLOSSARY

astable Without a stable state. In a multivibrator it means that the circuit oscillates back and forth between the two unstable states.

bistable Two stable states. In a bistable multivibrator the circuit can stay in either state indefinitely.

clock An oscillator that synchronizes all the different operations in a digital system.

differentiate In reference to a square wave, this means that narrow positive and negative triggers are generated by the leading and trailing edges of the square wave.

monostable One stable state. In a monostable multivibrator the output can remain indefinitely in its stable state. When a trigger comes in, the output goes into the unstable state, but will return to the stable state.

multivibrator A regenerative circuit with two active devices. It is designed so that one device conducts while the other cuts off.

scaler A circuit that delivers one output pulse when a specified number of input pulses have arrived.

toggle To switch into the opposite state, as in a T flip-flop.

trigger The input signal to a multivibrator that causes it to change states. Triggers can be sharp pulses, or they can be changes from one voltage level to another.

voltage comparator A circuit that delivers an output when the input voltage has crossed certain levels.

REVIEW QUESTIONS

1. Describe the input-output action of a T flip-flop.
2. What are steering diodes and speed-up capacitors?
3. How is the output frequency of a T flip-flop related to the input frequency?
4. What is a scaler?
5. Describe the input-output operation of the RS and the RST flip-flops.
6. The JK flip-flop operates in four possible ways. What are these?
7. In general, what does a Schmitt trigger do? What does UTP mean? LTP?
8. What is a voltage comparator?
9. How many stable states does an astable multivibrator have? What are some other names for the astable multivibrator? What determines the frequency of its output?
10. How does a monostable multivibrator respond to an input trigger? How many stable states does it have? What is another name for it?

PROBLEMS

6-1 A 500-kHz square wave drives a T flip-flop. The flip-flop responds to the negative-going changes. What frequency does the output of the flip-flop have?

6-2 A square wave with a period $T = 10$ μsec drives the flip-flop of Fig. 6-3. We can differentiate this square wave by using an RC time constant (on the input) that is less than one-tenth of T. If $R = 100$ kilohms, what size C should we use?

6-3 Suppose we cascade seven T flip-flops (similar to Fig. 6-7). If the input frequency is 512 kHz, what is the final output frequency?

6-4 We want to build a scaler that delivers one output pulse for every 1024 input pulses received. How many T flip-flops do we need to cascade to do this?

6-5 In Fig. 6-26, if the right side of the flip-flop is cut off, x will be a high voltage. Approximately how much base current flows in Q2? How much voltage is needed at the R input to just saturate Q4? Assume a beta of 100.

6-6 The circuit of Fig. 6-10 has T and S inputs as shown in Fig. 6-27. The T input responds to negative-going changes, but the S input responds to positive-going changes. Sketch the output waveform. Initially, x is a 0.

Multivibrators

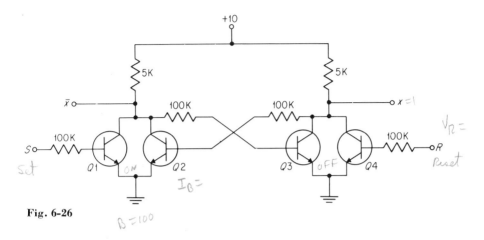

Fig. 6-26

6-7 An RST flip-flop has the T and R inputs shown in Fig. 6-27. The T input responds to negative triggering, but the R input responds to positive triggering. If x is initially 0, what is the x waveform?

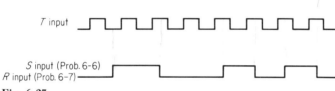

Fig. 6-27

6-8 In Fig. 6-28, four narrow triggers drive the circuit as shown. Sketch the output x_2 for each of the following initial conditions:
  (a) $x_1 = 1$ and $x_2 = 0$.
  (b) $x_1 = 0$ and $x_2 = 1$.
  (c) $x_1 = 0$ and $x_2 = 0$.
  (d) $x_1 = 1$ and $x_2 = 1$.

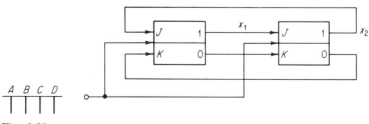

Fig. 6-28

6-9 In Fig. 6-29 list the pairs of x_1, x_2 values for each of the following initial conditions:
(a) $x_1 = 1$ and $x_2 = 1$.
(b) $x_1 = 0$ and $x_2 = 0$.

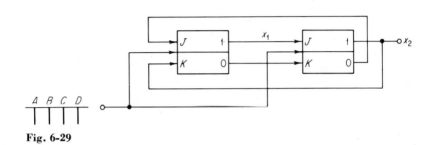

Fig. 6-29

6-10 A Schmitt trigger has the following characteristics: UTP = 5 volts, LTP = 4 volts, high state = 12 volts, low state = 3 volts. A 10-volt-peak sine wave drives the Schmitt trigger. Sketch the output.

6-11 A Schmitt trigger has a UTP of 5 volts and an LTP of 3 volts. A 5-volt-peak sine wave drives the Schmitt trigger, producing a rectangular output pulse. If the input frequency is 1000 Hz, what is the time duration of the output pulse?

6-12 In Fig. 6-22 the value of R is 50 kilohms, and the value of C is 0.005 μf. What is the approximate frequency of the output waveform?

6-13 In Fig. 6-22, if R equals 20 kilohms and if the capacitors have an adjustable range from 50 to 500 pf, what is the minimum and maximum frequency of the free-running multivibrator?

6-14 The triggers of Fig. 6-30 drive a monostable multivibrator. The monostable multivibrator delivers a 3-μsec pulse for each input trigger. Sketch the output waveform.

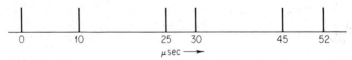

Fig. 6-30

CHAPTER 7

BASIC ELECTRONIC COUNTERS

In this chapter we will show how flip-flops are used to count pulses. We examine binary counters first and, after understanding their operation, discuss how they can be modified to obtain decimal counters. Also, we discuss some important applications — like frequency counters and digital voltmeters.

7-1 Using Flip-flops to Count

We can connect flip-flops to get an electronic counter. For instance, in Fig. 7-1 there are three flip-flops in cascade. A square wave drives the A flip-flop. From now on, we will call this square wave the "clock" signal. Note that

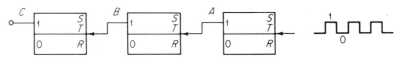

Fig. 7-1 Counting with flip-flops.

167

the output of the A flip-flop drives the B flip-flop. The B flip-flop in turn drives the C flip-flop.

What happens as each clock pulse comes in?

We begin our discussion by assuming that all flip-flops are initially in the 0 state. (We can produce this condition by feeding a positive pulse into all R inputs.) When a clock pulse comes in, the A flip-flop changes on the negative-going part of the pulse, as shown in Fig. 7-2. Therefore, the A flip-flop changes states *each* time the clock changes from a 1 to a 0.

The B flip-flop toggles on negative-going changes; therefore, it changes states each time that A goes from a 1 to a 0, as shown in Fig. 7-2.

The C flip-flop toggles each time that B goes from a 1 to a 0, again shown in Fig. 7-2.

Note carefully that the system of Fig. 7-1 actually can count the number of clock pulses that arrive. To bring this counting ability out clearly, look at the truth table of Table 7-1. We made this table by starting with all flip-flops in the 0 state, that is, $CBA = 000$. On the negative-going part of the first clock pulse, A changes from a 0 to a 1. The change in A is positive, so that the B flip-flop does not trigger. Thus, at the end of the first pulse, $CBA = 001$. When the second clock pulse arrives, A again toggles on the negative transition. Now A changes from a 1 to a 0. This negative change in A will trigger the B flip-flop, making B go from a 0 to a 1. Therefore, at the end of the second clock pulse, $CBA = 010$. When the third clock pulse comes in, A changes from a 0 to a 1. This positive change does not trigger the B flip-flop, so that $CBA = 011$ after the third clock pulse.

We can summarize the operation of these flip-flops as follows:

- The A flip-flop toggles on the negative-going part of each clock pulse.
- Any succeeding flip-flop toggles only if the flip-flop on the right has changed from a 1 to a 0.

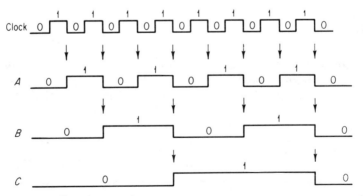

Fig. 7-2 Counting waveforms.

Basic Electronic Counters

With these summarizing rules we can easily verify the remaining entries in Table 7-1.

Here is a very important point. You can see that the value of CBA in Table 7-1 follows a straight binary count. Each time a clock pulse comes in, the count advances to the next binary number. Thus, the system of Fig. 7-1 is a binary counter.

Table 7-1 Binary Counter

C	B	A	Count
0	0	0	0
0	0	1	1
0	1	0	2
0	1	1	3
1	0	0	4
1	0	1	5
1	1	0	6
1	1	1	7
0	0	0	8

Table 7-1 shows that there are eight distinct states or words formed by the CBA values — 000 through 111. If we let 000 stand for 0, the binary counter of Fig. 7-1 can count up to 111, which is the equivalent of decimal 7. On the eighth pulse CBA returns to 000 because all flip-flops reset to 0.

To increase the capacity of a binary counter we need more flip-flops. When we use four flip-flops, we have a binary counter that can count from 0000 through 1111 before it resets. There are 16 distinct states or words in this progression when we include 0000.

In general, if we cascade n flip-flops together, we get 2^n states. For instance, for five flip-flops there are 2^5 or 32 states. For six flip-flops there are 2^6 or 64 states, and so forth.

When the output of one flip-flop drives the input of the next flip-flop as in Fig. 7-1, we call this type of counter a "ripple" counter; the A flip-flop must change states before the B flip-flop can change states, B must change before C can change, and so on. The triggers move through the counter like a ripple in water.

In Fig. 7-1, we can connect the output of each flip-flop to a lamp. When the output of a flip-flop is a 1, it will light its lamp; when the output is a 0,

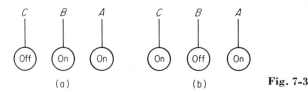

Fig. 7-3

the lamp will be off. By looking at the lamps, we can see at a glance what the binary count is. For instance, after three clock pulses we see the display of Fig. 7-3a, which is 011. After five clock pulses the lamps show 101. In this way, we can read the binary value of CBA.

EXAMPLE 7-1

A binary ripple counter uses seven flip-flops. How many distinct states can it be in? What is the largest binary number that can be stored in this ripple counter?

SOLUTION

With seven flip-flops the number of distinct states is

$$2^n = 2^7 = 128$$

Thus, the binary counter can be in any of these 128 states, depending on the number of pulses received.

The distinct states go from 0000000 through 1111111. Clearly, the largest binary number that can be stored is 1111111. On the next clock pulse the counter returns to 0000000.

Incidentally, the decimal equivalent of 1111111 is 127, one less than the number of states. In general, the decimal equivalent for the largest binary number in a counter is $2^n - 1$. (We subtract the 1 because the reset state stands for 0.)

EXAMPLE 7-2

How many states does a 10-flip-flop binary counter have? What is the largest number we can store in this counter?

SOLUTION

The number of states is

$$2^n = 2^{10} = 1024$$

The largest binary number we can store is 1111111111, which has a decimal equivalent of

$$2^n - 1 = 1024 - 1 = 1023$$

Basic Electronic Counters   171

7-2 A Decade Counter

Because we know the decimal number system so well, we prefer to have a decimal or decade counter instead of a binary counter. A decade counter uses a base of 10; that is, it has 10 distinct states. To get 10 distinct states, we can cause a binary counter with more than 10 states to skip some of its states.

There are many different ways to advance a counter, that is, make it skip some of its natural states. In this chapter we will discuss one way to show how it can be done. (The next chapter shows many ways of advancing a counter.) Consider the counter of Fig. 7-4. It has four flip-flops, which means that there are 16 natural states (2^4). However, the feedback from the D flip-flop to the B and C flip-flops causes the counter to skip some of these natural states.

Here is how the system of Fig. 7-4 works. Starting with $DCBA = 0000$ (the reset condition), the counter advances to the next binary number when each clock pulse comes in. The feedback does nothing for the first seven clock pulses because D stays at 0 for these first seven counts. That is, $DCBA$ changes as follows during the first seven clock pulses:

$$0000, 0001, 0010, 0011, 0100, 0101, 0110, 0111$$

This is a natural binary count up to seven. Throughout this count, the value of D stays at 0, so that the feedback does nothing. The counter acts like a straight binary counter for these first seven counts.

However, on the eighth clock pulse, $DCBA$ changes from 0111 to 1000. This 1000 state is only temporary because the D output has changed from a 0 to a 1, a positive change. This positive change goes back to the set inputs of the B and C flip-flops, forcing each flip-flop to change from a 0 to a 1. Thus, the value of $DCBA$ becomes 1110. We have made the counter skip through some of its natural states because it has gone from 1000 to 1110. On the ninth clock pulse $DCBA$ becomes 1111, and on the tenth clock pulse $DCBA$ resets to 0000.

Let us summarize what has happened.

1. Up to 0111 or seven the counter acts like a straight binary ripple counter.

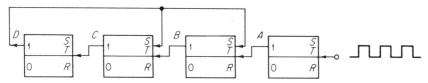

Fig. 7-4 Using feedback to get a decade counter.

2. On the eighth clock pulse, we get

   1000   before feedback
   1110   after feedback

The feedback action is fast, so that for practical purposes we can think of $DCBA$ as having gone from 0111 to 1110.

3. On the tenth clock pulse, $DCBA$ resets to 0000.

Table 7-2 shows the truth table for the modified counter. The count is natural binary through seven. On the eighth clock pulse the feedback advances the counter to 1110. On the ninth clock pulse $DCBA$ becomes 1111; on the tenth clock pulse $DCBA$ resets to 0000. The important fact to note here is that there are 10 distinct states: 0 through 9. Thus, we have a counter with a base or radix of 10.

Table 7-2   Decade Counter

D	C	B	A	Count
0	0	0	0	0
0	0	0	1	1
0	0	1	0	2
0	0	1	1	3
0	1	0	0	4
0	1	0	1	5
0	1	1	0	6
0	1	1	1	7
1	1	1	0	8
1	1	1	1	9

In Chap. 3 we discussed BCD codes. The code of Table 7-2 is a modified form of the 2421 BCD code. By the correct use of feedback and logic circuits we can make a counter with any BCD code that we wish. Also, we can make counters with any number of distinct states. More is said of this in the next chapter.

EXAMPLE 7-3

Show one way to make an 8421 decade counter.

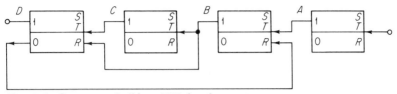

Fig. 7-5 One way to build an 8421 decade counter.

SOLUTION

In the 8421 BCD code we have a natural binary progression up to nine, and then the count resets. In other words, we have

0000, 0001, 0010, 0011, 0100, 0101, 0110, 0111, 1000, 1001, 0000

Figure 7-5 shows one way to get this sequence. It works as follows: The system acts like a straight binary counter up through decimal number 9. At this point in time $DCBA = 1001$. When the tenth clock pulse comes in, $DCBA$ becomes 1010, but this condition is only temporary because the following things happen in rapid order:

- The positive change out of B is fed to the reset input of the D flip-flop, causing D to change from a 1 to a 0. (1010 — 0010.)
- The $\bar{D}$ output (which has changed from a 0 to a 1) goes back to the B flip-flop, causing B to change from a 1 to a 0. (0010 — 0000.) Thus, on the tenth pulse the counter quickly resets to 0000.

Thus, we have shown how to make an 8421 BCD counter.

7-3 Decoding a Counter

In the preceding section we saw how it was possible to modify a counter with feedback to get a decade-counter unit (DCU). To read the contents of a counter we can attach lamps to each flip-flop output. What we read is a BCD number which we must mentally decode. For instance, if we have the decade counter of Fig. 7-4, we would decode the contents by using the truth table of Table 7-2. A lamp display of 0101 would mean decimal number 5; if the lamps show 1110, the decimal number would be 8. Thus, whatever the contents of the counter, we would decode according to Table 7-2.

The decoding of the binary-coded-decimal number can be done electronically. In other words, it is possible to convert the BCD number stored in the counter into its decimal equivalent using logic circuits. When we read the final display, we read decimal numbers.

There are many ways to decode a counter. We will show the most straightforward of all methods, namely, the use of AND gates to drive decimal-numbered lamps. Consider Fig. 7-6. There are 10 AND gates, each driving a numbered lamp. Each AND gate has four inputs which are connected to the flip-flops. For instance, the AND gate for the 0 lamp has its four inputs connected to $\bar{A}$, $\bar{B}$, $\bar{C}$, and $\bar{D}$. The AND gate for the 1 lamp has its inputs connected to A, $\bar{B}$, $\bar{C}$, and $\bar{D}$. The AND gate for the 2 lamp has inputs of $\bar{A}$, B, $\bar{C}$, and $\bar{D}$. Each of the remaining AND gates has the inputs shown in Fig. 7-6.

The connections of Fig. 7-6 ensure that one and only one lamp is lit at a given time. Further, the lamp that is lit shows the decimal equivalent of the BCD number stored in the counter (we are using the BCD code of Table 7-2). To prove this, consider some of the following cases:

1. When the number stored in the counter is $DCBA = 0000$, note that only the 0-lamp AND gate has all four inputs at 1. *All* other AND gates have at least one 0 going into them. Therefore, the only lamp that is lit when $DCBA = 0000$ is the 0 lamp.

2. When the counter stores a $DCBA$ value of 0001, you will note that only the AND gate for the 1 lamp has all four inputs at 1. Therefore, only the 1 lamp is lit when $DCBA = 0001$.

3. If the BCD number in the counter is $DCBA = 0010$, all four inputs to the 2-lamp AND gate are 1s. So, the 2 lamp will light for this condition.

4. By checking each of the remaining values of $DCBA$, we can verify that the correct lamp will light (remember we are using the BCD code of Table 7-2). Thus, the AND gates and lamps of Fig. 7-6 decode the BCD contents of the counter, permitting us to read out the decimal count.

Very often, instead of using 10 separate lamps, a single lamp known as a "nixie tube" is used. It contains the numbers 0 through 9, stacked on top of each other. There are 10 input leads to the nixie tube, and when a voltage is applied to one of these inputs, the decimal number associated with that input is lit.

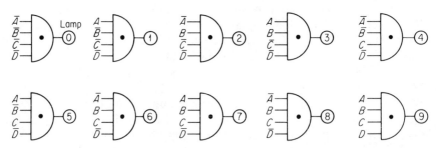

Fig. 7-6 Decoding the counter.

Basic Electronic Counters

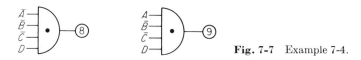

Fig. 7-7 Example 7-4.

EXAMPLE 7-4

If a counter uses the 8421 BCD code instead of the modified 2421 code of Table 7-2, the decoding for lamps 8 and 9 must be different. Show the AND-gates connections for these lamps.

SOLUTION

In the 8421 code, 1000 and 1001 stand for 8 and 9. Clearly, to light number 8 lamp when $DCBA = 1000$, we must use D, $\bar{C}$, $\bar{B}$, and $\bar{A}$ as the inputs to number 8 AND gate. Also, lamp 9 must light when $DCBA = 1001$; therefore, the inputs to number 9 AND gate must be D, $\bar{C}$, $\bar{B}$, and A. Figure 7-7 summarizes the decoding for 8 and 9.

7-4 Cascading Decade-counter Units

The decade-counter unit (DCU) discussed in the previous sections counts from 0 through 9, and then resets to 0. We can cascade as many DCUs as needed to increase the capacity of a counter. For instance, if we want to count from 0 through 999, we need only cascade three DCUs as shown in Fig. 7-8. The counting action of this system is simple. At the beginning of the count, all DCUs are reset to 0. (This is easily done by applying a positive reset pulse to all flip-flops.) After nine clock pulses have arrived, the DCUs will read 009. On the tenth clock pulse the D flip-flop in the *units* DCU will change from a 1 to a 0. This negative change goes into the *tens* DCU, causing it to advance one count. Thus, after ten clock pulses, the DCUs read 010.

As additional clock pulses arrive, the *units* DCU advances one count for each clock pulse. Every time the *units* DCU resets to 0, it produces a negative change that advances the *tens* DCU. After 99 clock pulses have arrived, the DCUs read 099. On the 100th clock pulse the *tens* DCU resets to 0.

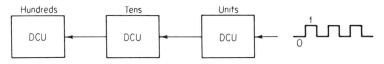

Fig. 7-8 A counter with a capacity of 999.

This produces a negative change that triggers the *hundreds* DCU, advancing it one count. Hence, after 100 clock pulses the DCUs read 100. In this way, the system can handle up to 999 clock pulses.

In general, we need one DCU for each decimal digit. If we want to count to 99,999,999, we need eight DCUs.

7-5 Gating a Counter

Gating a counter means that we turn it on for a specified period of time during which it counts the number of pulses that arrive at its input. Figure 7-9 shows a simple way to gate a counter. A train of clock pulses goes into an AND gate. To count the number of clock pulses during a given time, we feed a rectangular pulse into the AND gate. This rectangular pulse enables the AND gate, so that all clock pulses between points A and B in time pass through the AND gate. Note that before point A in time the AND gate is disabled, so that no clock pulses pass through to the counter. At point A in time the positive-going gate does two things: it resets the counter to zero, and it enables the AND gate, letting the clock pulses pass through to the counter. During the gating time the counter counts each new clock pulse that comes in. At point B in time the AND gate is disabled and the counter stops counting. The final count holds until the next gate comes in. Thus, with the system of Fig. 7-9 we can count the number of pulses that occur in a given time.

We can get the gating pulse in many ways. For instance, the gating pulse could be the output of an astable multivibrator, or perhaps the output of a one-shot multivibrator. The accuracy of the count depends upon how accurate the width of the gating pulse is. For instance, suppose that the gating time is *exactly* one second. The count we read is the number of pulses that occur in one second. Clearly, any error in the gating time produces an error in the count.

Figure 7-10 shows a way to get a very accurate gating pulse. A 1-MHz frequency standard is divided by 10^6 to get an accurate 1-Hz signal. This 1-Hz signal drives a Schmitt trigger, which in turn drives a flip-flop. The

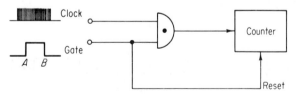

Fig. 7-9 Gating a counter.

Basic Electronic Counters

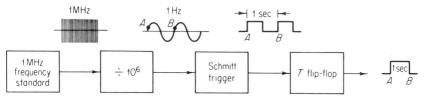

Fig. 7-10 Getting an accurate time base or gate.

gating pulse that comes out of the flip-flop has a very accurate duration of 1 sec. To get other gating times, we can divide by some other multiple of 10. If we divide by 10^7, we get a gating time of 10 sec.

EXAMPLE 7-5

In Fig. 7-9 the gating time of the pulse is 10 sec. If the clock has a frequency of 575 Hz, what does the counter read at the end of the gating pulse?

SOLUTION

There are 575 clock pulses during 1 sec. Therefore, during 10 sec, there will be 5750 clock pulses. The counter reads the number of pulses during a 10-sec period; so it will read 5750.

EXAMPLE 7-6

In Fig. 7-10 what values of frequency division give gating times of 0.01, 0.1, 1, and 10 sec?

SOLUTION

For a 1-sec gating time we already know that the frequency division must be 10^6. Hence, the frequency-division values are 10^4, 10^5, 10^6, and 10^7.

7-6 A Frequency Counter

A very important and widespread use of counters is in building frequency counters. It is simple to use a counter to measure the frequency of an unknown signal. Take a system like that of Fig. 7-11. The unknown frequency is first clipped to get a square wave. This square wave is one of the inputs to the AND gate. The other input to the AND gate is an accurate 1-sec gating pulse. During the 1-sec gating time the counter counts the number of cycles that occur. At point B in time the AND gate is disabled, and the

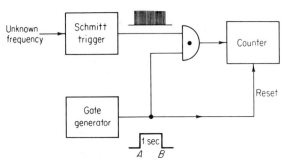

Fig. 7-11 A frequency counter.

counter stops counting. The number stored in the counter is the number of cycles in 1 sec, or simply the frequency of the unknown signal.

The period of the input signal can be measured by the system of Fig. 7-12. In this case, the unknown frequency produces the gating pulse. One cycle of the input signal causes the Schmitt trigger to go through its trip points; this in turn makes the flip-flop toggle twice per input cycle. Therefore, the width of the gating pulse equals the period of the unknown signal. Note that the upper input to the AND gate is an accurate 1-MHz clock signal.

During the gating time the counter counts the number of 1-MHz clock pulses that have occurred. At the end of the gating time the count equals the period in microseconds. For instance, suppose that the period of the unknown signal is exactly 0.1 sec. During this gating time 100,000 clock pulses are counted. Thus, at the end of the count we read 100,000 μsec, which is equivalent to 0.1 sec. As another example, suppose the period of the unknown signal is exactly 0.02 sec. The gating time will be 0.02 sec, so that 20,000 clock pulses will be counted. Thus, we read 20,000 μsec, which is equivalent to 0.02 sec.

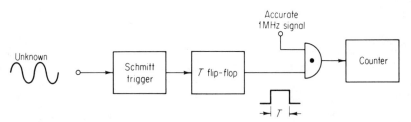

Fig. 7-12 Measuring the period of a signal.

Basic Electronic Counters

EXAMPLE 7-7

Suppose that the unknown frequency in Fig. 7-11 is 5 MHz, and that the gate generator can produce gating pulses of 0.01, 0.1, 1, and 10 sec. What would the counter read for each of these gating times?

SOLUTION

Clearly, for a 1-sec gating time the count reaches 5,000,000. We can find the other counts by using proportion.

When the gating time is 0.01 sec, the count becomes

$$0.01 \times 5,000,000 = 50,000$$

When the gating time is 0.1, the count is 500,000. And when the gating time is 10 sec, the count reaches 50,000,000.

EXAMPLE 7-8

In Fig. 7-12 suppose the unknown signal has a frequency of 5 Hz. What is the width of the gating pulse? What does the counter read?

SOLUTION

The period of the signal is

$$T = \frac{1}{5} = 0.2 \text{ sec}$$

Thus, the gating time equals 0.2 sec.

During this gating time of 0.2 sec, the counter will reach a count of

$$0.2 \times 1,000,000 = 200,000$$

Since the system of Fig. 7-12 is calibrated in microseconds, the final reading is 200,000 μsec.

7-7 A Digital Voltmeter

Another application of the counter is in digital voltmeters. Ordinary voltmeters indicate the voltage on a meter face; the accuracy of such readings is limited to a few percent. A digital voltmeter, however, shows the voltage by means of a digital display like that of a counter. With a digital voltmeter it is possible to read voltages with an accuracy in the range of 0.01 to 0.1 percent.

We will discuss one of the ways of building digital voltmeters merely to show how such voltmeters are possible. Consider the system of Fig. 7-13. A d-c voltage V is to be measured. This input voltage controls the gate

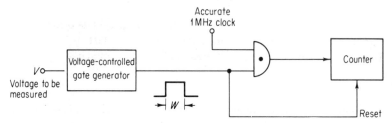

Fig. 7-13 A digital voltmeter.

generator, producing a gating time that is proportional to V. For instance, we can design a gate generator so that the width of its gating pulse is

$$W = \frac{V}{10}$$

Thus, if the input voltage V is 1 volt, the gating time is

$$W = \frac{1}{10} = 0.1 \text{ sec}$$

Or, if the input voltage V is 3.3 volts, the gating time becomes

$$W = \frac{3.3}{10} = 0.33 \text{ sec}$$

In other words, the gating time is directly proportional to the size of the input voltage V.

The gating pulse is one input to the AND gate; the other input is an accurate 1-MHz clock signal. Thus, the 1-MHz signal is gated by the pulse, and as a result, the counter reads the number of cycles that occur during the gating time. Because of this, the counter reading can be calibrated in volts. For instance, if the input voltage V is 3.3 volts, the gating time is

$$W = \frac{3.3}{10} = 0.33 \text{ sec}$$

Since the clock runs at 1 MHz, the counter will read 330,000 pulses. In the digital display we can show the decimal point after the first 3. In other words, we would read 3.30000.

The problem of building a digital voltmeter falls in the category of analog-to-digital (A/D) conversion. We will discuss A/D conversion more thoroughly in Chap. 11.

SUMMARY

Cascading flip-flops gives us a binary counter. The number of distinct states in a binary counter is 2^n.

Basic Electronic Counters

A ripple counter is the kind of counter we get when we use the output of one flip-flop to trigger the next flip-flop.

A decade counter is made by using feedback around a four-flip-flop binary counter. The idea is to make the binary counter skip through six of its states, so that the modified counter has 10 distinct states.

Logic circuits can decode the BCD number stored in a decade-counter unit, so that we can directly read out the decimal count.

We can get any counting capacity by cascading DCUs.

Gating a counter means that we turn on the counter for a specified period of time during which it counts the pulses arriving at its input.

Measuring frequency and voltage are two important applications of electronic counters.

GLOSSARY

binary counter   A counter that progresses through a straight binary sequence.

clock   Generally, this refers to an oscillator that synchronizes different operations in a digital system.

decade-counter unit (DCU)   A circuit that counts through 10 distinct states.

gating a counter   Turning the counter on for a specified period of time during which it counts the pulses arriving at its input.

ripple counter   The counter we get when we connect flip-flops in cascade, that is, with the output of one flip-flop driving the input of the next flip-flop.

REVIEW QUESTIONS

1. How many flip-flops does a binary ripple counter need if it counts to 15 before resetting?
2. Describe one way to modify a four-flip-flop counter so that it skips through six of its natural states.
3. How many distinct states does a ripple counter with six flip-flops have?
4. What does DCU stand for? What is the radix of a DCU?
5. What is a nixie tube?
6. How many DCUs are needed to count up to 99,999?
7. Describe the basic idea behind measuring frequency with a counter.
8. What is the basic idea behind the digital voltmeter discussed in this chapter?

PROBLEMS

7-1 Show the truth table for a binary ripple counter that uses five flip-flops.
7-2 How many distinct states does a nine-flip-flop binary ripple counter have?
7-3 Show the truth table for the counter of Fig. 7-14. Start with 0010.

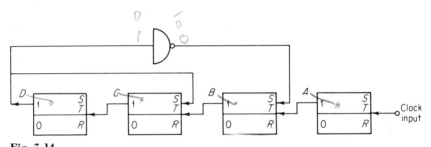

Fig. 7-14

7-4 In Fig. 7-10, suppose a 5-MHz standard is used. What frequency division do we use for an output pulse width of 0.01 sec? For 10 sec?
7-5 In the frequency counter of Fig. 7-11, suppose the unknown frequency is 248 kHz. How many pulses are counted if the gating time is 0.01 sec?
7-6 A decade-counter unit uses the 5421 BCD code. Show the decoding gates (similar to Fig. 7-6).
7-7 Show the decoding gates for a counter that uses the 6311 BCD code.
7-8 Suppose the counter in Fig. 7-12 reads 13,300 pulses during the gating time. What is the period of the unknown signal? The frequency?
7-9 In Fig. 7-13, the width of the gating pulse equals

$$W = \frac{V}{5}$$

and the 1-MHz clock signal is changed to 5 MHz. How many clock pulses are counted if V equals 7.5 volts? If V equals 2.23 volts?

CHAPTER 8

COUNTER TECHNIQUES

A counter is probably one of the most useful and versatile subsystems in a digital system. As implied by its title, this unit has the ability to count. In the previous chapter, we have seen that a counter driven by a clock can be used to count the number of clock cycles. Since the clock pulses occur at known intervals, the counter can be used as an instrument for measuring time (and therefore period or frequency).

The ripple counter previously discussed represents the simplest and most straightforward way of constructing a counter. This unit is very simple in operation and construction and usually requires a minimum of hardware. It does, however, have a speed limitation since each flip-flop is triggered by the previous flip-flop and thus the counter has a cumulative settling time. An increase in speed of operation can be achieved by the use of a parallel counter. Unfortunately, the increase in speed offered by the parallel configuration is usually accompanied by an increase in hardware. Series and parallel counters in combination can be used when a compromise between speed of operation and hardware count is indicated.

In this chapter we will discuss in detail the operation of series, parallel, and series-parallel combination counters. We will investigate the implementation of counters having any desired number of binary counts as well as one form of a BCD counter.

8-1 Binary Ripple Counter

As shown previously, a binary ripple counter can be constructed using RST flip-flops. The total number of counts or discrete states through which the counter can progress is given by 2^n, where n is the total number of flip-flops used. It is said to have a *natural count* of 2^n. For example, a basic binary counter consisting of three flip-flops is capable of counting through eight discrete states, and it is said to have a natural count of eight. A binary counter consisting of four flip-flops is capable of counting through 16 discrete states, and so on. Thus it is possible to construct counters which will count through 2, 4, 8, 16, 32, etc., states by simply using the proper number of flip-flops. A three-flip-flop counter is quite often referred to as a "modulus-8" (or mod-8) counter since it has eight states. Similarly, a four-flip-flop counter is a mod-16 counter. Thus the *modulus* of a counter defines the total number of states through which the counter can progress.

It is quite often desirable to construct counters which have moduli other than two, four, eight, etc. For example, one might like to construct a counter having a modulus of three or five or seven. A smaller-modulus counter can always be constructed from a larger-modulus counter by skipping states. Such counters are said to have a "modified count." It is first necessary to determine the total number of flip-flops required. The correct number of flip-flops can be determined by choosing the lowest natural count which is greater than the desired modified count. For example, a mod-7 counter would require three flip-flops since eight is the lowest natural count greater than the desired modified count of seven.

EXAMPLE 8-1

How many flip-flops are required to construct each of the following counters?
(a) mod-3
(b) mod-6
(c) mod-9

SOLUTION

(a) The lowest natural count greater than three is four. Two flip-flops will provide a natural count of four. Therefore, it would require a minimum of two flip-flops to construct a mod-3 counter.

(b) To construct a mod-6 counter would require at least three flip-flops since eight is the lowest natural count greater than six.

Counter Techniques

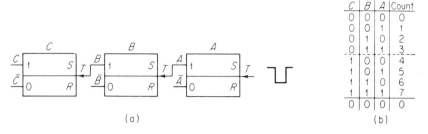

Fig. 8-1 Three-stage binary ripple counter. (*a*) Logic diagram. (*b*) Truth table showing natural count.

(*c*) A mod-9 counter would require at least four flip-flops since 16 is the lowest natural count greater than nine.

The three-flip-flop ripple counter shown in Fig. 8-1*a* has the capability of counting through the eight states shown in Fig. 8-1*b*. In order to construct a mod-7 counter it is necessary to omit, or skip, one of these states. There is a choice to be made since any one of the eight states may be omitted. If count 0 is omitted, the counter sequence would be 1-2-3-4-5-6-7. On the other hand, count 7 could be eliminated and the resulting count sequence would be 0-1-2-3-4-5-6. To construct a mod-6 counter, it would be necessary to skip two of the counts in Fig. 8-1*b*. Again there is a choice since any two states may be eliminated. For example, skipping 0 and 1, the count sequence would be 2-3-4-5-6-7.

EXAMPLE 8-2

If three flip-flops are used to build a mod-7 counter which counts in sequence, how many possible count sequences are there?

SOLUTION

To construct a mod-7 counter using three flip-flops, one count out of eight must be skipped. Since the skipped count may be any one of the eight, eight possible count sequences are available. They are:

Skipping count 0, 1-2-3-4-5-6-7
Skipping count 1, 0-2-3-4-5-6-7
Skipping count 2, 0-1-3-4-5-6-7
Skipping count 3, 0-1-2-4-5-6-7
Skipping count 4, 0-1-2-3-5-6-7
Skipping count 5, 0-1-2-3-4-6-7
Skipping count 6, 0-1-2-3-4-5-7
Skipping count 7, 0-1-2-3-4-5-6

EXAMPLE 8-3

If three flip-flops are used to build a mod-6 counter which counts in sequence, how many possible count sequences are there?

SOLUTION

A mod-6 counter using three flip-flops requires two counts out of eight to be skipped. The possible count sequences can be most easily determined by making use of the solution to Example 8-2. If count 0 is skipped, any one of the remaining seven counts may be skipped in order to form a mod-6 counter. This will then provide seven possible count sequences. If count 1 is skipped, any one of the remaining six states (2-3-4-5-6-7) may be omitted. This then provides six more possible sequences. 0 is omitted since it has already been accounted for. Skipping count 2 and any one of the remaining five counts (3-4-5-6-7) will provide five more possibilities. Skipping count 3 and any one of the four remaining states (4-5-6-7) will provide four more sequences. Skipping count 4 and any one of the remaining three counts (5-6-7) will provide three sequences. Skipping count 5 and either 6 or 7 will account for two more possibilities. Finally, skipping 6 and 7 will account for the last possible sequence. Thus a total of $7 + 6 + 5 + 4 + 3 + 2 + 1 = 28$ count sequences are possible for a mod-6 counter using three flip-flops.

The count sequence for any one particular counter is usually determined by the logical method of implementing the count (i.e., the method used to cause the counter to skip counts) since it is usually desirable to minimize the amount of hardware required.

8-2 Modified Counter Using Feedback

One method used to cause a counter to skip counts is to feed back a signal from some flip-flop to some previous flip-flop. Consider the counter shown in Fig. 8-1 and suppose it is desired to implement a mod-7 counter. Since it is desired to skip only one count, it would be convenient to discover some event which occurs only once during the natural count. Examination of the truth table in Fig. 8-1 reveals that flip-flop C changes from a 0 to a 1 only once during the natural eight counts. This represents a single unique occurrence. If C is fed back to the set input of flip-flop A, as shown in Fig. 8-2, a mod-7 counter will have been implemented. This is true since each time C goes positive (this will occur when the counter advances to state 4), flip-flop A will be set to the 1 condition and thus the counter will advance itself to state 5 immediately. Since the counter remains in state 4 only momentarily, it has essentially skipped this state and is therefore a mod-7

Counter Techniques

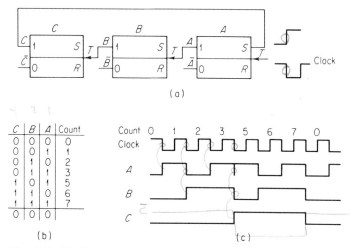

Fig. 8-2 Mod-7 counter. (a) Logic diagram. (b) Truth table. (c) Waveforms.

counter. The truth table and waveforms for this counter are shown in Fig. 8-2.

This type of counter is sometimes called a "permuting counter" since its natural count has been permuted (changed). Notice that the counter could just as easily have been implemented by feeding back a signal from $\bar{C}$. It is possible to use this type of feedback since flip-flops A and B must have completed their transitions before the feedback signal is generated. It should be noted that it is generally not desirable to feed back a signal which would cause flip-flop A to be reset since this causes a negative pulse at the input of flip-flop B which will cause it to change state, and this may cause flip-flop C to change state, etc.

EXAMPLE 8-4

Design a mod-5 and a mod-10 ripple counter using the feedback method just developed.

SOLUTION

To design a mod-5 counter will require three flip-flops. In deciding which count sequence to use, it might be convenient to use the same feedback scheme used for the mod-7 counter. In this instance it will be necessary to skip three counts. If the feedback were directed to the set input of both flip-flops A and B, a total of three counts would be skipped. This is true since the positive feedback signal developed at the beginning of count 4 would set both flip-flops A and B and the counter would immediately advance to count 7. Thus counts 4, 5, and 6 would be elimi-

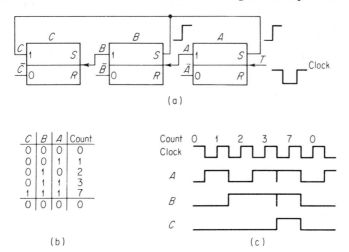

Fig. 8-3 Mod-5 counter. (a) Logic diagram. (b) Truth table. (c) Waveforms.

nated and a mod-5 counter has been formed. The complete counter is shown in Fig. 8-3. A mod-10 counter is easily implemented by adding an extra flip-flop (D) which is triggered by C as shown in Fig. 8-4. Notice that the output of flip-flop D is symmetrical. This is sometimes very convenient, as, for example, when the counter is used as a frequency divider.

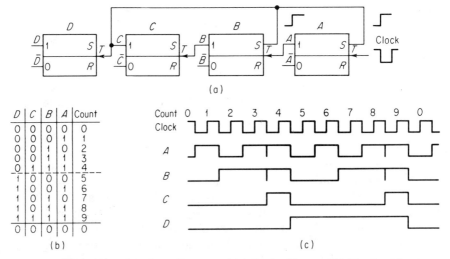

Fig. 8-4 Mod-10 counter; Decade counter. (a) Logic diagram. (b) Truth table. (c) Waveforms.

Counter Techniques

8-3 Parallel Counter

The ripple counter is the simplest to build, but it has definite limitations on its highest operating frequency. As previously discussed, each flip-flop has a delay time. In a ripple counter these delay times are additive and the total *settling time* for the counter is approximately the delay time times the total number of flip-flops. This speed limitation can be overcome by the use of a *synchronous* or parallel counter. The difference here is that every flip-flop is triggered by the clock and thus they all make their transitions simultaneously. The construction of a parallel binary counter is shown in Fig. 8-5 along with the truth table and the waveforms for the natural count sequence. The clocked JK flip-flop used will respond to a negative clock pulse at the T input and will toggle when both the J and K inputs are true. Having no connection to the J or K input is equivalent to having a positive voltage, or true level, at these inputs. Thus flip-flop A in Fig. 8-5 will change state each time a negative pulse appears at the T input. The output of NAND gate X will go low each time a clock pulse occurs and A is high. Thus flip-flop B will change state with every other clock pulse. The output of NAND gate Y will go low each time a clock pulse occurs and both A and B are high. Thus flip-flop C will change state with every fourth clock pulse. This then represents a mod-8 parallel binary counter.

The parallel counter shown in Fig. 8-5 can be used as a basis for building counters of other moduli. Suppose it is desired to construct a mod-7 parallel binary counter. It will be necessary to find some means of elimi-

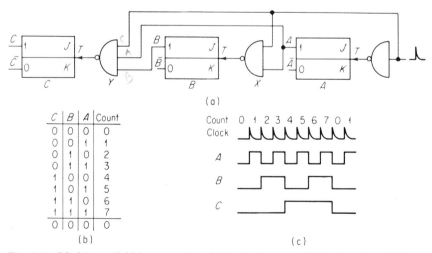

Fig. 8-5 Mod-8 parallel binary counter. (*a*) Logic diagram. (*b*) Truth table. (*c*) Waveforms.

nating one state from the natural count sequence. Examination of the truth table in Fig. 8-5b shows that all flip-flops are high during count 7. In changing from count 7 to count 0, all flip-flops change to the 0 state. This is similar to resetting all flip-flops. Now, if some means were found to prevent flip-flop A from being reset during the transition from count 7 to count 0, without affecting the operation of flip-flops B and C, the counter would progress from count 7 to count 1. Thus count 0 would have been skipped and a mod-7 counter would have been formed.

If A is high, the next clock pulse will cause flip-flop A to change state (i.e., the 1 side will go low) and this is equivalent to being reset. This will occur since the J and K inputs have no connections and are therefore considered true. On the other hand, if the K input were held low (false), the flip-flop would be set by the next clock pulse. However, the flip-flop is already set (the 1 output is high), and therefore no change will occur.

Thus to construct a mod-7 counter from the basic counter shown in Fig. 8-5, it is only necessary to ensure that the K input to flip-flop A is held low (false) during count 7. This can be easily accomplished by taking the 1 outputs of all flip-flops and using them as the inputs to a NAND gate as shown in Fig. 8-6. The output of NAND gate Z will be low only when all three flip-flop outputs are high. Thus the K input to flip-flop A will be held low only during count 7 and will be high at all other times. The truth table

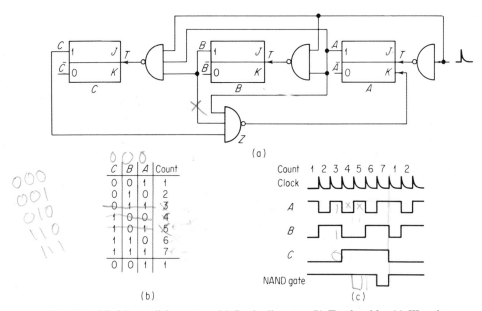

Fig. 8-6 Mod-7 parallel counter. (a) Logic diagram. (b) Truth table. (c) Waveforms.

Counter Techniques

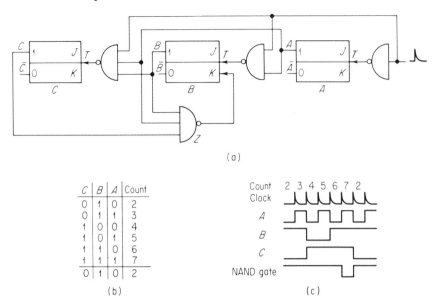

Fig. 8-7 Mod-6 parallel counter. (a) Logic diagram. (b) Truth table. (c) Waveforms.

and waveforms for the parallel mod-7 counter are shown in Fig. 8-6a and b (see Prob. 8-3).

The configuration in Fig. 8-6 can be easily modified to form a mod-6 counter. If the output of NAND gate Z is removed from flip-flop A and connected to the K input of flip-flop B, a mod-6 counter will have been formed. The proper connection for this mod-6 counter along with the appropriate truth table and waveforms are shown in Fig. 8-7. Flip-flops A and C are unaffected and will function as before. However, since the output of NAND gate Z is low during count 7, flip-flop B will be prevented from being reset as it would normally be during the transition from count 7 to count 0. Thus the counter will progress from count 7 to count 2 and states 0 and 1 will have been omitted. (In fact, the counter will work exactly as described even if the A input is left off of NAND gate Z).

Another method for forming a mod-6 counter from the configuration shown in Fig. 8-6 is simply to remove the input to NAND gate Z which comes from B. This will cause the output of NAND gate Z to be low during counts 5 and 7 and the counter will thus skip states 6 and 0.

EXAMPLE 8-5

Design a mod-5 parallel binary counter using the techniques described above.

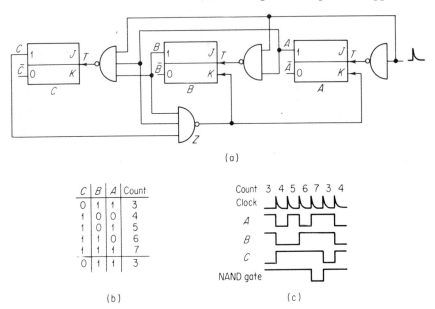

Fig. 8-8 Mod-5 parallel counter. (a) Logic diagram. (b) Truth table. (c) Waveforms.

SOLUTION

The mod-5 counter can be very easily implemented by combining the methods used to form the mod-7 and mod-6 counters. If the output of NAND gate Z is connected to the K inputs of both flip-flops A and B, a mod-5 counter will have been formed. This is true since the NAND gate output is low during count 7, which means that neither flip-flop A nor flip-flop B will be allowed to reset during the natural transition from count 7 to count 0. Flip-flop C will be allowed to reset, however, and thus the counter will advance from state 7 to state 3. In this way, counts 0, 1, and 2 will have been eliminated and a mod-5 counter is the result. The counter along with its truth table and waveforms are shown in Fig. 8-8. The A input to NAND gate Z could again be omitted.

There is a common problem which occurs in all the parallel counters so far discussed. This is the so-called "race problem" and this problem along with its solution is the topic of the next section.

8-4 The Race Problem

Generally speaking, a *race problem* may occur whenever two inputs to a gate are undergoing transitions at the same time. Consider the two-input

Counter Techniques

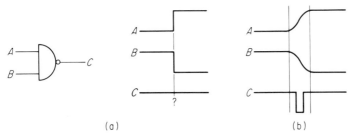

Fig. 8-9 (a) The racing problem at the input of a NAND gate. (b) Expanded waveforms.

NAND gate shown in Fig. 8-9a. When input A is low and input B is high, the output C is obviously high. Similarly when A is high and B is low, C is high. But, what happens during the transition time when A and B are both changing? The answer is, during the transition time the output C is indeterminate. The input levels to A and B must have some rise and fall times, respectively. This is shown in Fig. 8-9b. If the signals begin and end their transitions at the same time and have the same rise and fall times, a pulse may appear at C. But, if B is slightly delayed from A, there is a definite overlap of positive levels and a full-amplitude pulse will appear at the output of the gate. On the other hand, if A is slightly delayed with respect to B, no overlap of the input signals occurs and no output pulse at C appears. There are, of course, any number of possibilities in between these two extremes resulting in output pulses of various widths. Since the results of this situation are quite unpredictable, it should clearly be avoided in digital systems. For example, if the output of this gate were connected to the T input of a flip-flop, the triggering behavior of the flip-flop would be unpredictable.

Unfortunately, this exact situation is apparent at both the NAND gates (X and Y) in the parallel counters discussed in the previous section. To

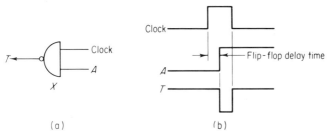

Fig. 8-10 (a) NAND gate X in Figure 8-5. (b) Waveforms going from count 0 to count 1.

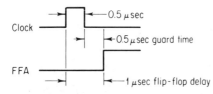

Fig. 8-11 Using flip-flop delay to avoid racing.

demonstrate this, examine the levels into and out of the X NAND gate in Fig. 8-5 as the counter is progressing from count 0 to count 1. The waveforms are shown in Fig. 8-10. Some finite delay time after the clock pulse has gone positive, A will also go positive. This overlap of pulses will cause a pulse to appear at the output of NAND gate X. This pulse goes directly to the T input of flip-flop B and it will therefore be triggered. This is most undesirable since the counter is supposed to progress to count 1 and instead it is progressing to count 3. Thus the race problem at this NAND gate has caused the counter to malfunction. As might be expected, a similar situation exists at NAND gate Y; and if all this has led you to believe that the counter will not work at all, you are quite right. It will not work as is! There are, however, a number of solutions to the problem.

First, notice that there is a definite delay between the time when the clock goes positive and when the output of the flip-flop goes positive. If it were possible to ensure that the width of the clock pulse were always less than the delay time of the flip-flop, there would be no overlap of waveforms and the undesirable trigger pulse due to the racing would be eliminated. Using clock pulses of 1 or 0.1 μsec would help the problem but not cure it since modern flip-flops have delay times of only a few tenths of a nanosecond. On the other hand, if the flip-flop had a predictable delay time of say 1 μsec (built into the flip-flop itself, or due to external circuitry), the race problem would be cured by using clock pulses having a duration of 0.5 μsec.

This situation is shown in Fig. 8-11, and it can be seen that there is a guard time of 0.5 μsec between the time the clock pulse falls and the time when the flip-flop output rises. Thus a deliberate delay to limit the response of the flip-flop will cure the race problem, and all the counters discussed in the previous section will indeed work if this solution is incorporated. This solution to the problem does have some drawbacks since deliberate delays have been introduced, and presumably the parallel counter configuration was used in the first place because greater speed was desired.

A second solution to the race problem can be achieved by the use of additional circuitry. This method, commonly referred to as "trailing-edge logic," is shown in Fig. 8-12. Examination of the waveforms in this figure shows that AND gate X has output pulses every time the clock goes positive — racing! The T input to the next flip-flop should have a trigger pulse for

Counter Techniques

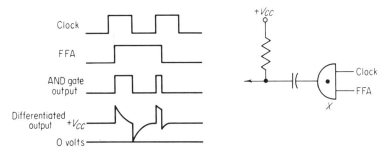

Fig. 8-12 Trailing-edge logic.

every other clock pulse, and so it is desired to eliminate one of the two existing pulses.

The differentiated output of the AND gate which appears at the T input of the next flip-flop is shown in the figure, and this waveform is achieved by choosing the proper values of resistor and capacitor. Since it requires a negative pulse to trigger the flip-flops, the positive pulses at the T input will have no effect. The first negative pulse will, however, trigger the flip-flop since it crosses the input threshold. The second negative pulse is too small to cross the threshold of the flip-flop and therefore has no effect. Thus the second pulse out of the AND gate has been eliminated and all the counters will work properly with this logic. The only difference in counter operation is the fact that the flip-flops will change state on the falling edge of the

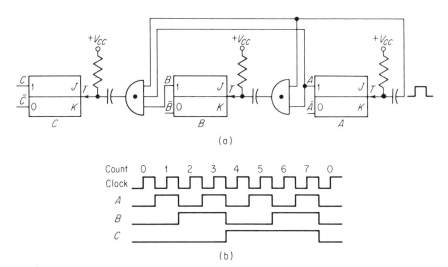

Fig. 8-13 Mod-8 parallel binary counter. (a) Logic diagram using trailing-edge logic. (b) Waveforms.

clock pulse instead of the rising edge. Hence the origin of the term trailing-edge logic. Operation of the counter may be thought of as follows: When the clock pulse goes positive, the capacitor is armed (charged). If the positive pulse is too short, as with the racing pulse or with noise pulses, the capacitor is not charged. Then, when the clock falls the capacitor is fired (discharged) and a trigger pulse is produced at the T input of the next flip-flop. The parallel binary counter of Fig. 8-5 has been redrawn in Fig. 8-13, making use of trailing-edge logic. The only disadvantage of trailing-edge logic is the additional circuitry required.

A third method for overcoming the race problem is to make use of the *master/slave* flip-flop shown in Fig. 8-14. Since the clock is connected directly to the R and S inputs of the master flip-flop, it is clear that the master is disabled and cannot be changed while the clock is low. The clock level is inverted and applied to the two AND gates at the R and S inputs to the slave flip-flop. These two AND gates are then enabled, and the slave flip-flop will be set or reset while the clock is low according to the state of the master. That is, if the 1 output of the master is high, then the S input to the slave must be high and thus the slave will be set. On the other hand, if the 0 side of the master is high, the R input to the slave will be high and the slave will be reset. When the clock goes high, the inverted clock which appears at the two slave input AND gates goes low. The two AND gates are

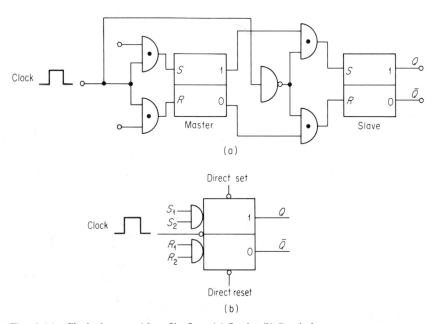

Fig. 8-14 Clocked *master/slave* flip-flop. (a) Logic. (b) Symbol.

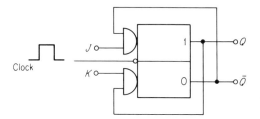

Fig. 8-15 *Master/slave* flip-flop cross-coupled to form a clocked *JK* flip-flop.

then disabled and the slave is thus disabled and cannot change state. The high-level clock, however, enables the R and S inputs to the master and thus the master will change state according to the levels at the R and S inputs. Thus the master flip-flop is either set or reset when the clock goes high. When the clock goes low again, the master is disabled and cannot change state (since its R and S inputs are disabled), but the inverted clock which appears at the two slave AND gates is now high. Thus the AND gates are enabled and the contents of the master is now shifted into the slave.

One might now ask how all this helps solve the race problem. Well, recall that the race problem occurred when the flip-flop output became high while the clock was still high. The output of the *master/slave* flip-flop can change state only after the clock has gone low. Thus the overlap of positive pulses have been avoided and the race problem for any of the counters discussed here has been solved.

The symbol for the *master/slave* flip-flop is shown in Fig. 8-14b. This is a clocked *RS* flip-flop which has two AND gated *set* inputs and two AND gated *reset* inputs. This provides for greater flexibility in design when using this configuration. There is also a *direct set* and a *direct reset* which go directly to the slave. Thus the flip-flop can be preset to either state without regard for the clock. A *JK* flip-flop can be formed from this unit by simply connecting the 1 output to either the R_1 or R_2 input and connecting the 0 output to either the S_1 or S_2 input. This connection is shown in Fig. 8-15. The flip-flop shown in this figure can be used directly to implement any of the counters previously discussed in this chapter. For example, the counter and waveforms in Fig. 8-5 will appear exactly as shown when the flip-flops used are of the master/slave variety.

This type of *master/slave* flip-flop has become so popular in industry and represents such a great savings in components and cost that it will be used almost exclusively in the remainder of this book. As a step toward simplification, the symbols shown in Fig. 8-16 will be used to distinguish the clocked *master/slave* flip-flop.

Finally, a warning should be made that the race problem is not an exclusive characteristic of parallel counters and will appear in many other places throughout a digital system. Unfortunately the problem occurs when

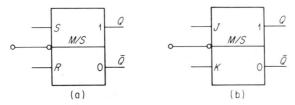

Fig. 8-16 (a) Symbol designating a clocked *master/slave* flip-flop used in the *RS* mode. (b) Symbol designating a clocked *master/slave* flip-flop which has been cross-coupled to operate in the *JK* mode.

least expected and can cause a great deal of inconvenience. Therefore, one should at all times be on guard against such occurrences.

8-5 Series-Parallel Combination Counters

Frequently a specific counter can be more easily implemented and a savings in components can be achieved by constructing it as a series-parallel combination. Consider the mod-5 counter shown in Fig. 8-17. This counter will progress through a natural binary count as shown in the truth table in

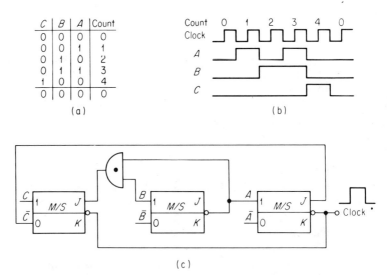

Fig. 8-17 A mod-5 binary counter. (a) Truth table. (b) Waveforms. (c) Logic diagram.

Counter Techniques

Fig. 8-17a. The waveforms shown in the same figure correspond to the desired count sequence and provide the basis for determining the proper connections to implement the counter.

Examination of the waveforms shows that flip-flop A changes state each time the clock goes negative except during the transition from count 4 to count 0. Thus flip-flop A should be triggered by the clock and must have an inhibit during count 4. That is to say, some signal must be provided during count 4 which will prevent flip-flop A from being triggered during the transition from count 4 to count 0. Notice that $\bar{C}$ is high (true) during all counts except count 4. Thus if $\bar{C}$ is connected to the J input of flip-flop A, the desired inhibit signal will be realized. This is true since the J and K inputs to flip-flop A are both true for all counts except count 4, and thus the flip-flop will trigger each time the clock goes negative. However, during count 4, the J side is low (false) and the next time the clock goes negative the flip-flop will be prevented from being set. The proper connections which will cause flip-flop A to progress through the desired sequence are shown in Fig. 8-17c.

Now, examination of the desired waveforms in Fig. 8-17b shows that flip-flop B must change state each time A goes negative. Thus the trigger input of flip-flop B will be driven by A as shown in Fig. 8-17c. If flip-flop C is triggered by the clock while the J input is held low (false) and the K input is high (no connection), every clock pulse will reset it. Now, if the J input is high only during count 3, C will be high during count 4 and low during all other counts. The necessary levels for the J input can be obtained by ANDing the 1 outputs of flip-flops A and B. Since A and B are both high only during count 3, the J input to flip-flop C will be high only during count 3. Thus when the clock goes negative during the transition from count 3 to count 4, flip-flop C will be set. At all other times, the J input to flip-flop C is low and it will be held in the reset state. The complete mod-5 counter is shown in Fig. 8-17c.

In constructing a counter of this type, it always is necessary to examine the omitted states to make sure that the counter will not malfunction. The counter in Fig. 8-17 omits states 5, 6, and 7 during its normal operating sequence. There is, however, a very real possibility that the counter may set up in one of these omitted (illegal) states when power is first applied to the system. It is necessary to check the operation of the counter when starting from each of the three illegal states to ensure that it will progress into the normal count sequence and will not become inoperative.

Begin by assuming that the counter is in state 5 ($A = 1$, $B = 0$, and $C = 1$). When the next clock pulse goes low, the following events will occur:

1. Since $\bar{C}$ is low, flip-flop A will be reset. Thus A will change from a 1 to a 0.

2. When A changes from 1 to 0, flip-flop B will trigger and B will change from 0 to 1.

3. Since the J input to flip-flop C is low, flip-flop C will be reset and C will change from 1 to 0.

Thus the counter will progress from state 5 to state 2 ($A = 0$, $B = 1$, and $C = 0$) after one clock pulse.

Now, assume that the counter starts in state 6 ($A = 0$, $B = 1$, and $C = 1$). When the next clock pulse goes low, the following events will occur:

1. Since $\overline{C}$ is low, flip-flop A will be reset. But flip-flop A is already in the reset condition and therefore A will remain at 0.

2. Since A does not change, flip-flop B will not change and B will remain at 1.

3. Since the J input to flip-flop C is low, flip-flop C will be reset and C will change from 1 to 0.

Thus the counter will progress from state 6 to state 2 after one clock pulse.

Finally, assume that the counter begins in state 7 ($A = 1$, $B = 1$, and $C = 1$). When the next clock pulse goes low, the following events will occur:

1. Since $\overline{C}$ is low, flip-flop A will be reset and A will change from a 1 to a 0.

2. Since A changes from a 1 to a 0, flip-flop B will trigger and B will change from a 1 to a 0.

3. The J input to flip-flop C is true and therefore flip-flop C will trigger.

Thus C will change from a 1 to a 0. Thus the counter will progress from state 7 to state 0 ($A = 0$, $B = 0$, and $C = 0$) after one clock pulse. Therefore, this counter configuration will automatically work itself out of any of the three illegal states after only one clock pulse.

A straight binary mod-7 counter can be easily constructed using the same techniques. The desired count sequence and waveforms are shown in Fig. 8-18 along with the complete logic diagram. The proper connections for the flip-flops can be determined by examining the waveforms. Notice that flip-flop C must change state each time B goes negative. Therefore, B can be used as the trigger input to flip-flop C.

Flip-flop B must change state each time the clock goes negative and A is high. It must also change state during the transition from state 6 to state 0. State 6 can be recognized as the time when both B and C are high. Now, if flip-flop B is triggered by the clock and the J and K inputs are tied together, the action of B will be determined by the level on the J and K input lines when the clock goes negative (when J and K are high, flip-flop B will change state; when the inputs are low, B will not change state).

Counter Techniques

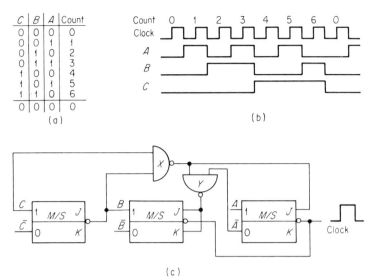

Fig. 8-18 A mod-7 binary counter. (a) Truth table. (b) Waveforms. (c) Logic diagram.

The proper action for B can be achieved by using the logic circuit shown in Fig. 8-19. Flip-flop B must change state any time A is high or any time both B and C are high. The proper logic equation is $K = J = A + BC$. Examination of Fig. 8-19 will reveal that the output of NAND gate X will be low only when both B and C are high. Any time the output of NAND gate X is low, the output of NAND gate Y (J or K) must be high and thus B will change state. Furthermore, any time $\bar{A}$ is low (A is high) the output of NAND gate Y (J or K) must be high and thus B will change state. Therefore, any time the condition $A + BC$ is true, flip-flop B will change state.

Finally, notice that flip-flop A must change state every time the clock goes negative except during the transition from count 6 to count 0. Thus the clock should be used as the trigger input to flip-flop A. A can be prevented from going positive during the transition from count 6 to count 0 by holding the J input to flip-flop A low during count 6. The output of NAND gate X is low only during count 6, and if it is connected to the J input of flip-flop A, A will progress through the desired sequence of levels.

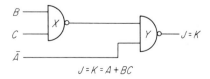

Fig. 8-19 Logic to implement the mod-7 counter in Fig. 8-18.

The only illegal state to be considered with this counter is count 7 ($A = 1$, $B = 1$, and $C = 1$). If the counter is in this state, the following events will occur the first time the clock goes low:

1. Since B and C are both high, the output of NAND gate X is low and thus flip-flop A will be reset. Therefore, A will change from a 1 to a 0.
2. Since the output of NAND gate X is low, the output of NAND gate Y will be high and flip-flop B will change state. Therefore, B will go from a 1 to a 0.
3. Since B goes low, flip-flop C will change state and C will go from a 1 to a 0.

Therefore, the counter will progress automatically from the illegal state 7 to the legal state 0 ($A = 0$, $B = 0$, and $C = 0$) the first time the clock makes a positive-to-negative transition.

EXAMPLE 8-5

Construct a mod-10 binary counter using the clocked master/slave flip-flops.

SOLUTION

It is quite simple to construct a mod-10 counter if use is made of the mod-5 counter shown in Fig. 8-17. It is only necessary to add an additional flip-flop (D) which is triggered by C. The complete counter with truth table and waveforms are shown in Fig. 8-20.

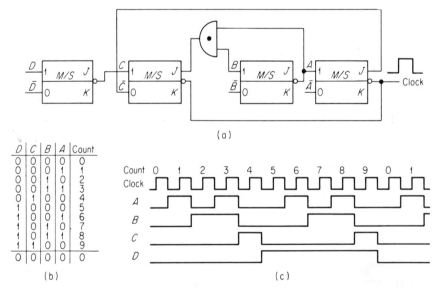

Fig. 8-20 A decade counter. (a) Logic diagram. (b) Truth table. (c) Waveforms.

8-6 Binary Decade Counter with Decoding Gates

The binary decade counter is a very useful form of counter since it provides a means of changing a count in the somewhat less familiar binary mode into an equivalent count in the more familiar decimal mode. If the 10 discrete states of the decade counter are decoded, the series of waveforms shown in Fig. 8-21 are obtained. These waveforms can be used for a multitude of purposes.

One of the most useful methods of utilizing these 10 output signals is to cause them to control 10 lamps which represent the 10 decimal numbers. Alternatively, they could be used to control the grid voltages on the popular nixie counter tubes. (A nixie counter tube is a neon-filled tube which has 10 grids in it; each grid is in the shape of a decimal number. When the proper voltage is applied to one of the grids, the neon fires and illuminates that grid. Thus any of the decimal numbers from 0 through 9 can be displayed by simply controlling the grid voltages.)

In order to produce the 10 waveforms shown in Fig. 8-21, it is necessary to decode the four flip-flops in the counter. Each of the 10 states represents a unique condition of the four flip-flops in Fig. 8-20, and requires 10 four-legged AND gates to produce these 10 output waveforms. Count 0 can be recognized as the unique time when A and B and C and D are all low. Alternatively it is the time when $\bar{A}$ and $\bar{B}$ and $\bar{C}$ and $\bar{D}$ are all high. Thus count 0 can be decoded by means of the AND gate shown in Fig. 8-22. Count 1 can be decoded by recognizing the state when A and $\bar{B}$ and $\bar{C}$ and $\bar{D}$ are all high and the four-input AND gate shown in Fig. 8-22 will provide the 1 output waveform. The remaining eight counts can be similarly decoded, and the complete system of decoding gates is shown in Fig. 8-22. The 10

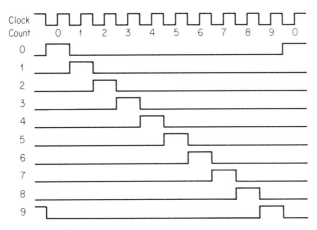

Fig. 8-21 Decade counter outputs.

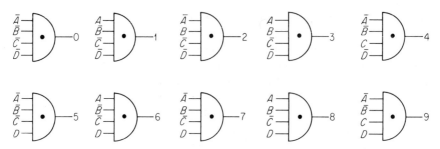

Fig. 8-22 Decoding gates for Fig. 8-20.

outputs of these gates will produce the 10 decoded waveforms shown in Fig. 8-21.

A decade counter could have been formed just as easily by using the mod-5 counter in Fig. 8-17 in conjunction with another flip-flop but connecting them as shown in Fig. 8-23. The truth table for this configuration along with the resulting waveforms are shown in the same figure. This is still a mod-10 (decade) counter since it still has 10 discrete states. If the flip-flops are decoded as in the previous counter, the 10 output waveforms will still appear exactly as shown in Fig. 8-21. However, since the counter

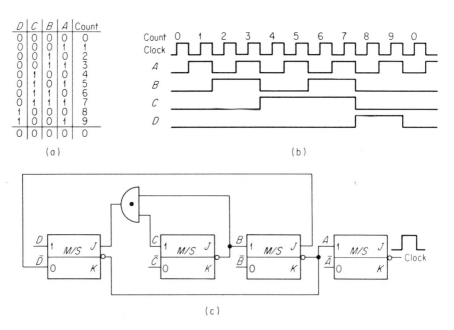

Fig. 8-23 A decade counter. (a) Truth table. (b) Waveforms. (c) Logic diagram.

Counter Techniques

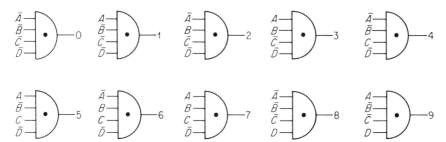

Fig. 8-24 Decoding gates for Fig. 8-23.

waveforms are different, the decoding gates will necessarily have to be different. One will still proceed in the same manner, however, and state 0 can be recognized as the time when $\bar{A}$ and $\bar{B}$ and $\bar{C}$ and $\bar{D}$ are all high. Thus the gate to decode state 0 along with the other nine decoding gates are all shown in Fig. 8-24. It is interesting to note that the counter in Fig. 8-23 counts in a true binary sequence while the counter in Fig. 8-20 does not.

EXAMPLE 8-6

Determine the gates which will be necessary to decode the mod-5 counter shown in Fig. 8-17. Draw the waveforms showing the decoded outputs.

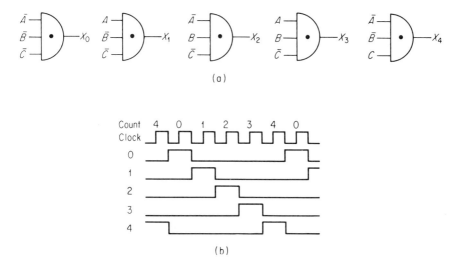

Fig. 8-25 Decoding for the mod-5 counter in Fig. 8-17. (a) Gates. (b) Decoded output waveforms.

SOLUTION

Since the mod-5 counter has five discrete states determined by three flip-flops, it will be necessary to use five three-input AND gates to decode the five states. State 0 is uniquely determined when $\bar{A}$ and $\bar{B}$ and $\bar{C}$ are all high. The logic equation is $X_0 = \bar{A}\bar{B}\bar{C}$ and the AND gate is shown in Fig. 8-25. The logic equations for the remaining four states are $X_1 = A\bar{B}\bar{C}$, $X_2 = \bar{A}B\bar{C}$, $X_3 = AB\bar{C}$, and $X_4 = \bar{A}\bar{B}C$. The correct gates along with the decoded waveforms are shown in Fig. 8-25.

8-7 Higher-modulus Counters

In the previous section, a mod-10 counter was constructed by combining a mod-5 counter and a mod-2 counter. Moreover, it was shown that the order in which the two counters were connected affected only the individual waveforms. That is to say, a mod-10 counter was still the result of combining the two counters regardless of the manner in which they were connected. The logical extension of this basic idea will make it possible to construct many different counters by simply using combinations of the counters thus far developed.

As an example, consider the implementation of a mod-6 counter by means of a mod-3 counter and a mod-2 counter (single flip-flop) interconnected. The logic diagram, truth table, and waveforms for the mod-3 counter are shown in Fig. 8-26. The mod-3 counter operates as follows: (1) A will change state every time the clock goes negative except during the transition from count 3 to count 1. It is only necessary to ensure that flip-

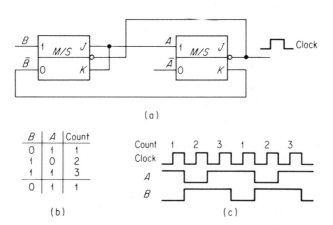

Fig. 8-26 (a) A mod-3 counter. (b) Truth table. (c) Waveforms.

Counter Techniques

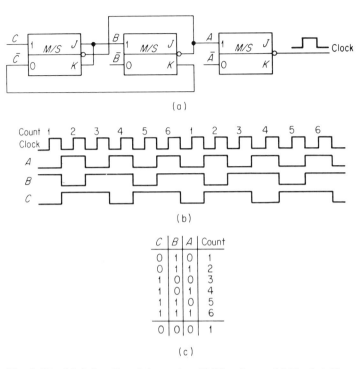

Fig. 8-27 (a) A 2 × 3 mod-6 counter. (b) Waveforms. (c) Truth table.

flop A cannot be reset during this transition. $\bar{B}$ is low during both counts 2 and 3, and if it is connected to the K input of flip-flop A, A will be prevented from going low during the transition from count 3 to count 1. (2) B changes state every time A is high and the clock goes low. If a flip-flop is now connected in series with this mod-3 counter, a mod-6 counter will result. If the flip-flop precedes the mod-3 counter, this is said to be a 2 × 3 mod-6 counter. This counter is shown in Fig. 8-27 along with the corresponding truth table and waveforms. If the mod-3 counter is used to drive the added flip-flop, a 3 × 2 mod-6 counter is formed, and this counter along with the appropriate truth table and waveforms are shown in Fig. 8-28.

This method of implementing higher-modulus counters seems quite reasonable when considering a basic flip-flop as a mod-2 counter. Then, a mod-4 counter (two flip-flops in series) is simply two mod-2 counters in series, which forms a 2 × 2 mod-4 counter. Similarly, a mod-8 counter is simply a 2 × 2 × 2 connection, and so on. Thus a great number of higher-modulus counters can be formed by using the product of any number of lower-modulus counters.

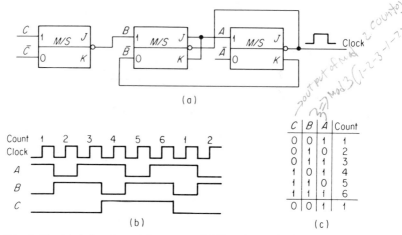

Fig. 8-28 (a) A 3 × 2 mod-6 counter. (b) Waveforms. (c) Truth table.

EXAMPLE 8-7

Outline the methods which could be used to form a modulo-12 counter using the product method just described.

SOLUTION

A mod-12 counter is divisible by 2 and 2 and 3; that is, 12 = 2 × 2 × 3. Therefore, a mod-12 counter could be implemented by using the following combinations of lower-modulus counters: (1) 2 × 2 × 3, (2) 4 × 3, (3) 2 × 6.

8-8 A BCD Counter

It is sometimes desirable to construct counters which will count in other than a straight binary code. One other very useful form would be a decade counter which counts in a modified 2421 binary-coded decimal. This basic counter was discussed in Chap. 7; because of its importance, we will consider a second means of construction using JK flip-flops. Such a counter is shown in Fig. 8-29 along with the appropriate truth table and waveforms. The operation of the counter can be described as follows:

1. A must change state each time the clock falls, and therefore the trigger input to flip-flop A should be the clock.

2. B must change state each time A goes low except during the transition from count 7 to count 8. The output of NAND gate X is low whenever C and $\bar{D}$ are both high. This will prevent B from going low during the

Counter Techniques

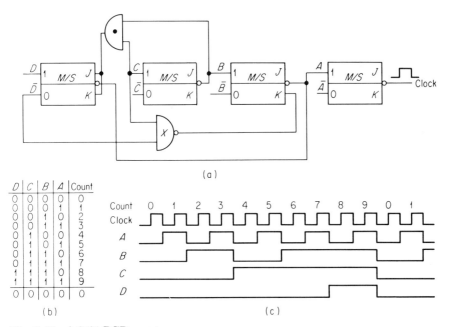

Fig. 8-29 A 2421 BCD counter.

transition from count 7 to count 8, since the output of NAND gate X holds the K input to flip-flop B low during counts 4, 5, 6, and 7. Notice that the J input to flip-flop B is always high (true) and therefore B is able to go high during the transition from count 5 to count 6.

3. The trigger input to flip-flop C is driven by B since C must change state each time B goes low.

4. D must change state each time both B and C are high and A goes low. If the J and K inputs to flip-flop D are conditioned by the output of an AND gate whose inputs are B and C, it will be possible for D to change state whenever both B and C are high. If the trigger input to flip-flop D is now driven by A, D will change states during the transition from 7 to 8 and from 9 to 0.

Since this is a decade counter and therefore has only 10 states, and since four flip-flops are used, there are six illegal (omitted) states which must be examined. The six states omitted are 1000, 1001, 1010, 1011, 1100, and 1101.

Let us examine what will happen if the counter comes up in state 1000 when the power is first applied. The first clock pulse will cause A to change state, but B, C, and D will remain unchanged. Thus the counter will advance one count to 1001. This is the second illegal state.

The second clock pulse will again cause A to change state, and since A is going low, B will change state. C and D will again remain unchanged. Thus the counter will again advance one count to 1010, which is the third illegal state.

The third clock pulse will cause A to go high, but B, C, and D will remain unchanged. Thus the counter will advance one count to state 1011, which is the fourth illegal state.

The fourth clock pulse will cause A to go low, which will cause B to go low, which will cause C to change state. D will remain unchanged and the counter will advance to 1100, which is the fifth illegal state.

The fifth clock pulse will cause A to change state while B, C, and D remain unchanged and the counter will advance to the sixth illegal state, 1101.

The sixth clock pulse will cause A to go low, which will cause B to change state while C and D remain unchanged. Thus the counter will advance to state 1110, which is count 8 in the desired count sequence.

Therefore, this counter will not lock up in any of the six illegal states and will in the worst case require six clock pulses to count itself out of an illegal state into the proper count sequence.

The BCD counter shown in Fig. 8-29 represents one method of implementing this counter, and the reader should realize that there are a number of other ways of accomplishing this task.

SUMMARY

A counter has a natural count of 2^n, where n is the number of flip-flops in the counter. Counters of any modulus can be constructed by incorporating logic which will cause certain states to be skipped over or omitted. In serial counters, the technique for skipping counts is to derive a feedback signal from certain flip-flops to reset previous flip-flops. This method takes advantage of the inherent delay in each flip-flop. In parallel counters, the technique is to precondition the logic inputs to each flip-flop in order to omit certain states. This is sometimes called look-ahead logic. It is in parallel counters that the race problem is encountered. Possible solutions to the race problem are:

1. Delaying the outputs of the flip-flops
2. Trailing-edge logic
3. The master/slave flip-flop

Higher-modulus counters can be easily constructed using combinations of lower-modulus counters. The series-parallel counter configuration is an example of this. These configurations also represent a compromise between speed and hardware count.

Counter Techniques   **211**

Feedback around a basic binary counter can also be used to construct various BCD counters.

It should be mentioned that counters can be constructed using any binary element. This includes all types of flip-flops as well as such other elements as magnetic cores. Moreover, they can be constructed to count in any desired sequence. This, however, is a topic which is covered in texts on logical design, and the reader is referred to the Bibliography for more advanced methods.

GLOSSARY

master/slave flip-flop   A clocked RS (or JK) flip-flop which is composed of two individual flip-flops wired in such a way as to avoid racing.
modulus   Defines the number of states through which a counter can progress.
natural count   The maximum number of states through which a counter can progress. Given by 2^n, where n is the number of flip-flops in the counter.
parallel counter   A counter in which all flip-flops change state simultaneously since all trigger inputs are driven by the clock.
race problem   The inability to determine the output of a gate when the inputs to the gate are making simultaneous transitions.
ripple counter   A counter constructed by triggering each flip-flop with the output of the previous flip-flop.

REVIEW QUESTIONS

1. How many possible states are there in a flip-flop composed of the following number of counters:
   (a) 3.
   (b) 5.
   (c) 6.
   (d) 7.
   (e) 10.
2. What is meant by a count sequence?
3. Explain why a parallel counter is capable of faster operation than a ripple counter.
4. Define the race problem.
5. How does a delay in the output of a flip-flop cure the race problem?
6. Of what use is trailing-edge logic?
7. Describe how the master/slave flip-flop is used to cure the race problem.

8. How many different ways can a mod-9 counter be implemented using lower-modulus counters?
9. What are the ways in which a mod-24 counter can be implemented using lower-modulus counters?

PROBLEMS

8-1 How many flip-flops would be required to build the following counters?
(a) mod-6.
(b) mod-11.
(c) mod-15.
(d) mod-19.
(e) mod-31.

8-2 Draw the logic diagram, truth table, and waveforms for the mod-7 counter shown in Fig. 8-2 if the feedback signal is taken from $\bar{C}$ instead of C.

8-3 What modulus counter is formed if the A input is removed from NAND gate Z of Fig. 8-6?

8-4 Draw the truth table and waveforms for the mod-6 counter formed by removing the B input from NAND gate Z in the counter shown in Fig. 8-6.

8-5 Construct a mod-3 parallel binary counter beginning with flip-flops A and B as shown in Fig. 8-5.

8-6 Assume the counter in Fig. 8-5 is constructed using flip-flops whose outputs are delayed by 0.5 μsec. If the clock for this counter is a 1-MHz square wave, will there be a race problem?

8-7 In Prob. 8-6 above, if the clock is modified to a series of 0.2-μsec positive pulses occurring at a 1-MHz rate, is the race problem cured? Draw the waveforms for this situation.

8-8 Redraw the mod-5 counter shown in Fig. 8-8 using trailing-edge logic and draw the resulting waveforms.

8-9 Draw the gates necessary to decode the seven outputs of the mod-7 counter shown in Fig. 8-18.

8-10 Draw the gates necessary to decode the three-stage ripple counter in Fig. 8-1.

8-11 Assume the clock for the ripple counter in Fig. 8-1 is a 1-MHz square wave and assume that each flip-flop has a 0.25-μsec delay. Carefully draw waveforms for the clock, each flip-flop, and the output decoded signals. Do you foresee any sources of difficulty?

8-12 Draw the logic diagram, truth table, and waveforms for a mod-9 counter using two mod-3 counters connected in series.

8-13 In how many different ways can a mod-18 counter be implemented using lower-modulus counters? Draw the logic diagram, truth table, and waveforms for one of these configurations.

CHAPTER 9

SPECIAL COUNTERS AND REGISTERS

A register is a very important logical block in most digital systems. Registers are quite often used to store (momentarily) binary information which appears at the output of an encoding matrix. Similarly, they are used to store (momentarily) binary data which are being decoded. Thus registers form a very important link between the main digital system and the input-output channels.

A binary register also forms the basis for some very important arithmetic operations. For example, the operations of complementation, multiplication, and division are frequently implemented by means of a register.

A shift register can be quite easily modified to form a number of different types of counters. These counters offer some very distinct advantages, and we will therefore investigate them in detail.

9-1 A Serial Shift Register

A register is nothing more than a group of flip-flops which can be used to store a binary number.* There must be one flip-flop for each bit in the

*Registers can be constructed using other binary elements; for example, the magnetic core, as we shall see in Chap. 12.

binary number. Naturally the flip-flops must be wired in some way such that the binary number can be entered (shifted) into the register and possibly shifted out. A group of flip-flops which are wired to provide either or both of these functions is called a *shift register*.

There are two general methods for shifting the binary information into a register. The first of these two methods involves shifting the information into the register one bit at a time in a series fashion and leads to the development of a *serial shift register*. The second method involves shifting all the bits into the register at the same time and leads to the development of a *parallel shift register*. The serial shift register is the main subject of interest here, and the parallel register will be discussed in a later section.

The basic idea for implementing a serial shift register can be demonstrated by considering the flip-flops shown in Fig. 9-1. Assume A and B are both low. Now, with the J and K inputs to flip-flop A held low, allow the clock to go through one cycle. During this period, A will not change state since the J and K inputs to flip-flop A are both low. Since A holds the J input to flip-flop B low and $\bar{A}$ holds the K input high, B will not change state during this period either. Therefore, both A and B will remain as before.

Now, maintain the K input to flip-flop A low but allow the J input to go high. If the clock is now allowed to go through one cycle, A will be set high but B will still remain unchanged. Thus during the second cycle of the clock, a 1 has been placed in flip-flop A (since J was high and K was low), while B remained unchanged (since its J input was low).

Now, hold the J input to flip-flop A low, hold the K input high, and allow the clock to advance through one more cycle. A will be reset low since its J input is low and its K input is high. At the same time, B will change state (since the J input to B is high). Thus during the third cycle of the clock, the 1 in flip-flop A has been effectively shifted into flip-flop B.

If the clock is now allowed to progress through one more cycle, A will remain in the low state and B will go low (since A and therefore the J input to B is low). Thus during the fourth cycle of the clock, the 1 in flip-flop B has been effectively shifted out and both flip-flops are again in their reset states.

The important points to be noted are as follows:

1. To set a 1 in flip-flop A, hold the J input to A high and the K input low and allow the clock to progress through one cycle.

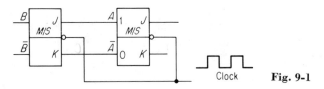

Fig. 9-1

Special Counters and Registers 215

2. To set a 0 in flip-flop A, hold the J input low and the K input high and allow the clock to progress through one cycle.

3. Any time a 1 exists in flip-flop A (A is high), that 1 will be shifted into flip-flop B during the next cycle of the clock (that is, B will go high).

4. Any time a 0 exists in flip-flop A (A is low), B will be reset low during the next clock cycle.

Consider the application of these basic ideas to form the 6-bit serial shift register shown in Fig. 9-2. It is called a 6-bit register since it has six flip-flops and is therefore capable of storing 6 bits.

Suppose it is desired to shift the binary number 101110 into this register in a serial fashion. First, notice that it will take one clock time to shift the first bit into flip-flop A. It will take a second cycle of the clock to shift the second bit into A, and at the same time the first bit will be shifted from flip-flop A into flip-flop B. A third clock pulse will shift the third bit into A, and at the same time the first bit will be shifted from B to C and the second bit will be shifted from A to B. This process must be repeated three more times, and during the sixth clock pulse the sixth bit will be shifted into A, the fifth into B, the fourth into C, the third into D, the second into E, and the first will be in F. Thus after six clock pulses, the 6 bits will have been shifted serially through the six flip-flops and the binary number is now in

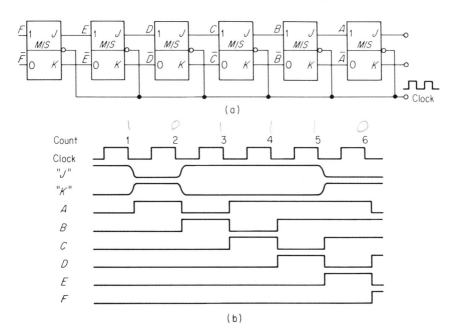

Fig. 9-2 A 6-bit serial shift register. (*a*) Logic diagram. (*b*) The six cycles required to shift the number 101110 into the register, MSB first.

the register. At this time, the clock must be stopped or a portion of the information will be lost.

It is of course necessary to control the levels of the J and K inputs to flip-flop A during the shifting operation in order to ensure that the proper information is entered into the register. It is also necessary to define the order in which the number is to be shifted into the register. That is, will the least significant bit (LSB) be entered first, or last? In this case, assume that the most significant bit (MSB) is entered first, and thus after the shifting operation the most significant bit will be in flip-flop F and the least significant bit will be in flip-flop A. The clock along with the proper J and K input levels to flip-flop A are shown in the waveforms in Fig. 9-2b. The clock is allowed to start just prior to count 1 and it is stopped at the end of count 6. During count 1, J is high and K is low and a 1 is set in flip-flop A. During count 2, the 1 in flip-flop A is shifted into flip-flop B and a 0 is set in flip-flop A (since J is low and K is high). The remaining 4 bits are shifted into the register in a similar manner, and the detailed operation can be seen by carefully examining the waveforms shown. Examination of the waveforms also shows that after the shift operation is complete (just after count 6), the desired binary number is indeed stored in the six flip-flops with the MSB in F.

Two things of importance should be stressed. First, the size of a register is determined by the size of the number to be stored. That is, there must be one flip-flop for each bit. Second, in a serial shift register it requires n clock pulses to shift an n-bit number into the register.

9-2 A Ring Counter

The serial shift register discussed in the previous section must have additional circuitry to control the J and K inputs to flip-flop A in order to ensure proper operation during the shift operation. The logic circuitry necessary to provide these control waveforms would ordinarily be derived from the control section of the system. The control section is the source of the clock and whatever other control signals are necessary, and these topics will be discussed in Chap. 14. For now, let us see if any of the feedback techniques discussed in the previous chapter can be applied to this basic shift register with any usable results.

One of the most logical applications of feedback might be to connect F to the J input of flip-flop A and at the same time connect $\overline{F}$ to the K input of flip-flop A. This configuration is shown in Fig. 9-3. Now, suppose that all flip-flops are in the reset state and the clock is allowed to run. What will happen? The answer is, nothing will happen since all J inputs are low and all K inputs are high. Therefore, every time the clock goes low, every

Special Counters and Registers

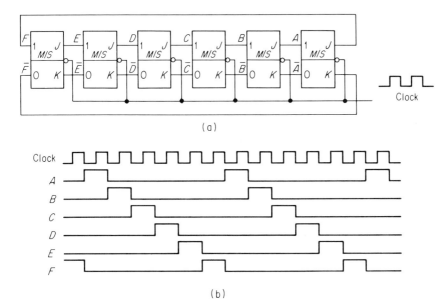

Fig. 9-3 A six-stage ring counter. (a) Logic diagram. (b) Waveforms with one 1 in the counter.

flip-flop will be reset. But all flip-flops are already reset and thus nothing will happen.

In an effort to get some action, suppose A is high and all other flip-flops are low, and then allow the clock to run. The very first time the clock goes low, the 1 in A will be shifted into B and A will be reset since F is low and $\bar{F}$ is high. All other flip-flops will remain low. The second clock pulse will shift the 1 into C and B will be reset while all other flip-flops remain low. The third clock pulse will shift the 1 into D — and so on. Thus the 1 will shift down the register, traveling from one flip-flop to the next each time the clock goes low.

The crucial point occurs when the 1 is in flip-flop F. The very next clock pulse will simply shift it into A because of the feedback connection. From that point, the process will be repeated and the 1 will simply circulate around the register as long as the clock is running. For this reason, this configuration is sometimes called a "circulating register" or "ring counter." The waveforms present in this ring counter are shown in Fig. 9-3b.

You will notice that there is a remarkable similarity between these waveforms and the decoded output waveforms of the counters in the last chapter. In the event that waveforms of this type are needed, the ring counter might in some cases be a more logical choice than a binary counter

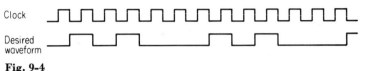

Fig. 9-4

with decoding gates. Ring counters such as this find a great many uses in developing the control waveforms previously mentioned.

There is, however, a problem with ring counters such as this. In order to produce the waveforms shown in Fig. 9-3b, it is necessary that the counter have one and only one 1 in it. The chances of this occurring naturally when power is first applied are very remote indeed. If the flip-flops should all happen to be in the reset state when power is first applied, it will not work at all, as previously discovered. On the other hand, if some of the flip-flops come up in the set state while the remainder of them come up in the reset state, a series of complex waveforms of some kind will be the result. Therefore, it is necessary to preset the counter to the desired state before it can be used.

One method of presetting the counter in Fig. 9-3 would be to tie the *direct reset* inputs of flip-flops B, C, D, E, and F to the *direct set* of flip-flop A and apply a negative pulse to this line just prior to operation of the counter. This will set a 1 in A and reset the remaining flip-flops.

Since the ring counter in Fig. 9-3 will function with more than one 1 in it, it might be desirable to operate it in this fashion at some time or other. For example, suppose the waveform in Fig. 9-4 were desired. This could be achieved by constructing a binary counter and decoding it as in the previous chapter. On the other hand, it could be quite easily accomplished by simply presetting the counter in Fig. 9-3 with a 1 in A and a 1 in C with all the other flip-flops reset. You will notice that it is immaterial where the 1s are set initially. It is only necessary to ensure that they are spaced one flip-flop apart.

EXAMPLE 9-1

How would you preset the ring counter in Fig. 9-3 in order to obtain a square-wave output which is one-half the frequency of the clock?

SOLUTION

It is only necessary to preset a 1 in every other flip-flop while the remaining flip-flops are reset.

EXAMPLE 9-2

Show how to use a five-stage ring counter in conjunction with a single flip-flop to form a decade counter.

Special Counters and Registers

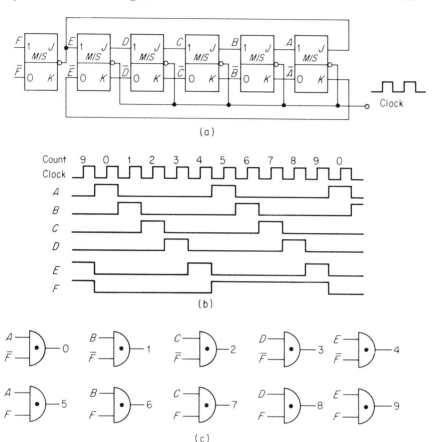

Fig. 9-5 A decade counter. (*a*) Logic diagram. (*b*) Waveforms. (*c*) Decoding gates.

SOLUTION

The circuit diagram shown in Fig. 9-5*a* will operate as a decade counter if the counter is preset with a 1 in flip-flop A and the remaining flip-flops are reset. The waveforms and the decoding gates are shown in Fig. 9-5*b* and *c*. Notice that it requires only 10 two-input AND gates to decode the 10 waveforms.

9-3 A Shift Counter

Since the feedback method applied to the shift register in the previous section was quite successful it might be worthwhile to explore one more possibility. Suppose the 1 output of the last flip-flop is returned to the K

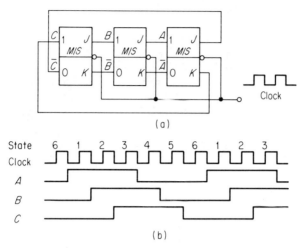

Fig. 9-6 A three-stage shift register using inverse feedback to form a shift counter. (a) Logic diagram. (b) Waveforms.

input of the first flip-flop and the 0 output is returned to the J input. This connection applied to a three-flip-flop shift register is shown in Fig. 9-6. Notice that the outputs of the shift register are first crossed and then returned to the inputs of the register. This configuration is sometimes referred to as inverse feedback.

Now assume that all flip-flops are in a reset condition and the clock is allowed to run. Since $\bar{C}$ is high and C is low, a 1 will be set in flip-flop A during the first cycle of the clock. At the same time, B and C will remain low since their J inputs are low and their K inputs are high.

During the second cycle of the clock, A will remain high since $\bar{C}$ is still high and C is still low. At the same time, B will be set high since A is now high and $\bar{A}$ is low. C will remain unchanged since B is low during this period.

During the third clock period, A and B will remain high and C will be set high since B is now high. Thus after three cycles of the clock, all three flip-flops have been changed from the low state to the high state.

During the fourth clock period, $\bar{C}$ is low and C is high and A will therefore be reset to the low state. B and C will remain high.

During the fifth cycle of the clock, A will remain low, B will be reset low (since A is now low and $\bar{A}$ is high), and C will remain high.

The sixth clock period will return the counter to the initial starting point since C will be reset low while A and B both remain low. Thus this shift register with inverse feedback has progressed through a complete cycle of counts in six clock periods.

Special Counters and Registers

Examination of the waveforms shown in Fig. 9-6b shows that the waveform of each flip-flop is a square wave which is one-sixth the frequency of the clock. Moreover, all three flip-flop outputs are identical except that they are shifted with respect to one another by one clock period. This square wave apparently shifts through the flip-flops, advancing one flip-flop each time the clock goes low. Since the operation is cyclic and the waveforms shift through the flip-flops, this configuration is commonly called a "shift counter" (it is also called a "Johnson counter").

In order to investigate this counter more carefully, let us make a truth table showing the states through which the counter progresses. This truth table, made with the aid of the counter waveforms, is shown in Fig. 9-7. For easy comparison, a straight binary truth table for three flip-flops is also shown. Notice that the three-flip-flop shift counter counts through six discrete states. The six ordered states through which the shift counter progresses correspond to the binary counts 1-3-7-6-4-0. Thus the six states of the shift counter are indeed discrete and can be decoded.

It must be noted, however, that the shift counter omits binary counts 2(010) and 5(101). Therefore, the shift counter must be examined to see whether or not it will work its way out of either of these two states, since it is possible that one of them may occur when power is first applied to the system. In fact, one of these two illegal states could occur during normal operation because of noise or some other malfunction.

First, suppose that the counter is in binary count 2(010). The next time the clock goes negative, A will go high, B will go low, and C will go high. Thus the counter will advance to binary count 5(101), which is the second illegal state.

During the second clock cycle, A will go low, B will go high, and C will go low. Thus the second clock period will advance the counter right back

C	B	A	State	Equivalent binary count
0	0	1	1	1
0	1	1	2	3
1	1	1	3	7
1	1	0	4	6
1	0	0	5	4
0	0	0	6	0
0	0	1	1	1

(a)

C	B	A	Count
0	0	0	0
0	0	1	1
0	1	0	2
0	1	1	3
1	0	0	4
1	0	1	5
1	1	0	6
1	1	1	7
0	0	0	0

(b)

Fig. 9-7 (a) Three-flip-flop shift-counter truth table. (b) Truth table for three flip-flops counting in a straight binary sequence.

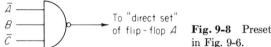

Fig. 9-8 Preset gate for the shift counter in Fig. 9-6.

into the first illegal state, binary count 2(010). Therefore, the counter will simply oscillate between the two illegal states and will not function properly.

In order to avoid this situation, some means is necessary to ensure that the counter cannot remain in one of the illegal states. One method of accomplishing this is to use the NAND gate shown in Fig. 9-8. When $\bar{A}$, B, and $\bar{C}$ are all high (corresponding to the illegal state 010), the output of the NAND gate will be low. If the NAND-gate output is applied to the *direct set* of

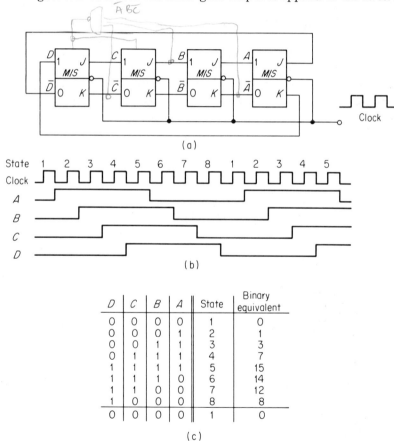

Fig. 9-9 A four-stage shift counter. (*a*) Logic diagram. (*b*) Waveforms. (*c*) Truth table.

Special Counters and Registers

flip-flop A, then A will be set high any time this condition occurs. Thus the counter will immediately advance from binary count 2(010) to binary count 3(011). This is the second state in the normal counting sequence, and the counter will therefore operate as desired (see Prob. 9-7).

EXAMPLE 9-3

Draw the diagram for a four-flip-flop shift counter. Draw the waveforms for this counter, and using these waveforms make a truth table showing the desired count sequence. List the illegal states and examine them to determine if the counter will get into an undesired mode of operation. If the counter does malfunction because of the illegal states, show a method for curing this problem.

SOLUTION

The desired counter along with the desired waveforms are shown in Fig. 9-9a and b, respectively. The truth table, derived from the desired waveforms, is shown in Fig. 9-9c. Now, since this counter is constructed of four flip-flops, there are 16 possible states. In the desired mode of operation, the counter sequences through only eight states, and there

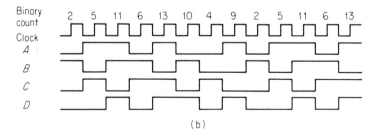

Fig. 9-10 (a) The eight illegal states for the counter shown in Fig. 9-9 and the states to which the counter will advance after one clock period. (b) Illegal-counting-sequence waveforms.

are therefore eight illegal states to be examined. The eight illegal states correspond to the binary counts of 2, 4, 5, 6, 9, 10, 11, and 13, and they are shown in the table in Fig. 9-10. The two columns to the right of the double vertical line in the table show the new state, and the new binary equivalent count, to which the counter will progress after one cycle of the clock. For example, after one clock cycle, the counter will advance from binary count 2(0010) to binary count 5(0101). Careful examination of the table will show that if the counter comes up in any one of the eight illegal states, it will count through the binary sequence 2-5-11-6-13-10-4-9-2. Thus the counter will get into an undesired mode of operation and remain there. Notice that the counter still divides by 8, which was part of the original intent. However, if it is desired to decode the counter it will be necessary to find some means of forcing the counter back into the desired count since the output waveforms in this secondary mode are quite different from the desired mode. The waveforms for the illegal count sequence are shown in Fig. 9-10b. One method of forcing the counter into the desired mode can be found by observing that the condition $\bar{A}B\bar{C}$ is true twice during the undesired count sequence (count 2 and count 10). This condition is never true during the desired count sequence, and so the NAND gate shown in Fig. 9-8 can be used here to cure the problem. It will be necessary, however, to connect the output of the NAND gate to *direct set* on flip-flops C and D. This will force the counter to state 6 whenever the output of the gate goes low (that is, when $\bar{A}B\bar{C}$ is high), and the counter will then be back into the desired count sequence.

A five-flip-flop shift counter can be constructed as shown in Fig. 9-11. This is a very useful form since it divides by 10 and can thus be used as a decade counter. The desired waveforms and corresponding truth table are shown in the same figure. Since there are five flip-flops, there are a possible 32 states in which the counter can exist. The desired count sequence uses only 10 of these possible states, and there are therefore 22 illegal states.

It can be shown that (see Prob. 9-10) if the counter comes up in any one of these illegal states it will do one of two things. First, it will continue to divide the clock by 10, but it will advance through one of the two following illegal count sequences: 2-5-11-23-14-29-26-20-8-17-2 or 4-9-19-6-13-27-22-12-25-18-4. Secondly, if the counter comes up in either count 10 or count 21, it will simply alternate back and forth between these two counts. Again, the cure to ensure that the counter operates in the proper sequence is the NAND gate shown in Fig. 9-8. The output of the gate can be connected to the *direct set* of flip-flops C, D, and E (see Prob. 9-11).

In retrospect, it can be seen that n flip-flops can be used to divide the clock by $2n$. That is, two flip-flops will divide the clock by 4, three flip-flops will divide the clock by 6, and so on. Alternatively, we can say that n flip-

Special Counters and Registers

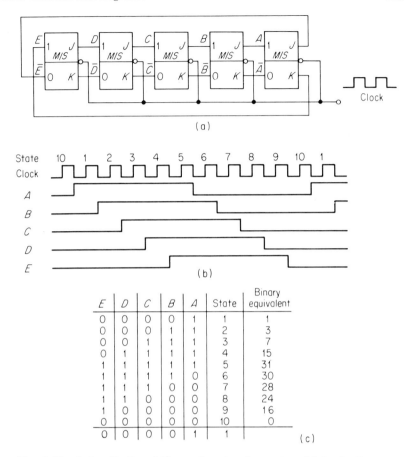

Fig. 9-11 A five-flip-flop shift counter, decade counter. (a) Logic diagram. (b) Decade-counter waveforms. (c) Truth table showing equivalent binary states.

flops connected in the shift-counter configuration will provide $2n$ discrete states through which the counter will progress. Thus it is possible to construct a counter of any even modulus by simply choosing the proper number of flip-flops and connecting them in the shift-counter configuration.

To construct odd-modulus counters from the basic shift-counter configuration is amazingly simple. Examination of the waveforms in Fig. 9-6 will show that if A is reset low one clock period earlier, B will also be reset one clock period earlier, as will C. This means that A and B and C will all be high for only two clock periods but will still be low for three periods. Therefore, the counter is now dividing by 5, instead of 6, and there are five discrete states.

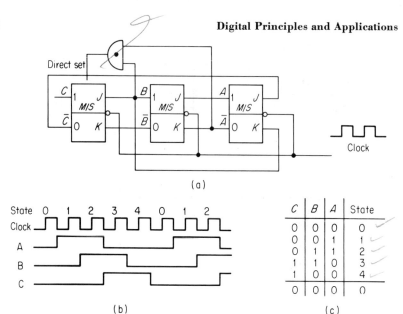

Fig. 9-12 A divide-by-5 shift counter. (*a*) Logic diagram. (*b*) Waveforms (*c*) Truth table.

The method of accomplishing this is shown in Fig. 9-12 along with the appropriate waveforms and truth table. Since it is only necessary to cause A to go low one clock period earlier than before, this means that the K input to flip-flop A must go low one period earlier. Therefore, it is only necessary to obtain the K input to flip-flop A from B instead of C. Thus any odd-modulus counter of count n can be implemented from an even counter of modulus $n+1$ by simply removing the K input of the first flip-flop from the 1 output of the last flip-flop and connecting it to the 1 output of the next to last flip-flop.

This counter still has the two illegal counts 2(010) and 5(101), and in addition the illegal count 7(111). The additional count 7 will not cause any problem since the counter will advance naturally from 7(111) to 6(110) and this is state 4 in the desired counting sequence. Furthermore it can be seen that the counter will advance from 2(010) to 5(101) and then to 6(110), which is state 3 in the desired counting sequence. Thus this counter has no permanent illegal modes (see Prob. 9-14).

9-4 A Modulo-10 Shift Counter with Decoding

The decade counter shown in Fig. 9-11 has 10 discrete states, and therefore the counter can be decoded to form the 10 waveforms as was done in the last chapter (see Fig. 8-21). In decoding the binary decade counter in the

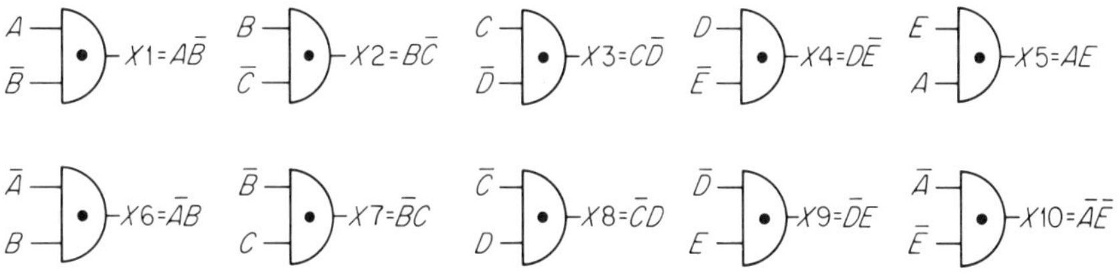

Special Counters and Registers

$X_1 = A\bar{B}$ $X_2 = B\bar{C}$ $X_3 = C\bar{D}$ $X_4 = D\bar{E}$ $X_5 = AE$

$X_6 = \bar{A}B$ $X_7 = \bar{B}C$ $X_8 = \bar{C}D$ $X_9 = \bar{D}E$ $X_{10} = \bar{A}\bar{E}$

Fig. 9-13 Decoding gates for the counter in Fig. 9-11.

last chapter, it required 10 four-input AND gates since there were four flip-flops in the counter. It would then seem reasonable to expect that it would require 10 five-input AND gates to decode the shift counter shown in Fig. 9-11 since there are five flip-flops. However, examination of the waveforms shown in this figure will show that a much simpler arrangement is possible.

Notice that state 1 can be uniquely determined as the time when A is high and B is low. Thus it requires only a two-input AND gate to decode state 1. The inputs to this AND gate are A and $\bar{B}$, and the appropriate logic equation is $X_1 = A\bar{B}$.

Similarly, state 2 is the only time when B is high and C is low. The inputs to the gate to decode 2 are B and $\bar{C}$, and the appropriate logic equation is $X_2 = B\bar{C}$. The logic equations for the remaining states are found in a similar manner. These equations along with the appropriate decoding gates are given in Fig. 9-13.

Thus another of the advantages of using a shift counter is demonstrated by the fact that it requires only two-input AND gates to decode any of the individual counter states.

EXAMPLE 9-4

The decade counter in Fig. 9-11 has been constructed and its outputs are available. It is necessary to develop the waveform shown in Fig. 9-14a to be used as a control signal. Show the logic circuitry necessary to provide this control signal.

SOLUTION

The desired control signal corresponds to a high level during states 1, 3, and 5 and is low at all other times. The most obvious solution is to use the three AND gates shown in Fig. 9-13 for decoding these states. If the outputs of these three AND gates are used as the inputs to an OR gate, as shown in Fig. 9-14b, the control signal X will be produced at the output of the OR gate.

228 Digital Principles and Applications

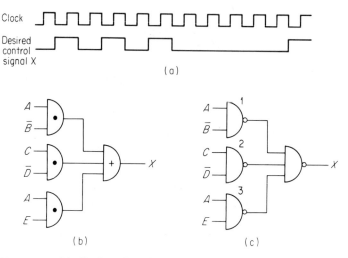

Fig. 9-14 (a) Clock and desired control signal. (b) Method for developing control signal X using AND gates and one OR gate. (c) Method for developing control signal X using NAND/NOR gates.

If the control signal X is required to drive a number of circuits, the output circuit which produces X will have to have power available to drive these circuits. A slightly different method for developing the desired waveform X is shown in Fig. 9-14c. The outputs of NAND gates 1, 3, and 5 will be low only during states 1, 3, and 5, respectively. Any time 1 or 3 or 5 is low, the output X must be high, and thus the desired control signal is developed. The output NAND gate which produces X can then be chosen such that it has the power capability to drive the desired number of circuits.

9-5 A Digital Clock

A very interesting application of counters and decoding arises in the design of a digital clock. Suppose it is desired to construct an ordinary clock which will display hours, minutes, and seconds. The power supply for this system will be the usual 120 volts a-c, 60-Hz commercial power. Since the 60-Hz frequency of most power systems is very closely controlled, it will be possible to use this signal as the basic clock frequency for our system.

In order to obtain pulses occurring at a 1-sec rate (or alternatively a 1-Hz square wave) it will be necessary to divide the 60-Hz power source by 60. If this 1-Hz square wave is again divided by 60, a 1 cycle per minute square wave will be the result. Dividing this signal by 60 will then provide a 1

Special Counters and Registers

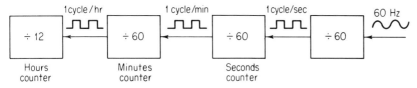

Fig. 9-15 Digital-clock block diagram.

cycle per hour square wave. This then provides the basic idea to be used in forming a digital clock.

A block diagram showing the functions to be performed is shown in Fig. 9-15. The first ÷60 counter simply divides the 60-Hz power signal down to a 1 cycle per second square wave. The second ÷60 counter changes state once each second and has 60 discrete states, and it can therefore be decoded to provide signals to display seconds. This counter is then referred to as the seconds counter. The third ÷60 counter changes state once each minute and has 60 discrete states, and it can therefore be decoded to provide the necessary signals to display minutes. This counter is then the minutes counter. The last counter will change state once each 60 min (or once per hour). Thus if it is a ÷12 counter it will have 12 states and it can be decoded to provide signals to display the hour. It is therefore called the hours counter.

As you know, there are a number of ways to implement a counter. What is desired here is to design the counters in such a way as to minimize the hardware required. The first counter must divide by 60 and it is not necessary to decode it. Therefore, it should be constructed in the easiest manner with the minimum number of flip-flops.

To obtain 60 states requires at least six flip-flops, since $2^6 = 64$. The simplest configuration is a straight binary ripple counter, and this first ÷60 counter will therefore be a simple binary ripple counter using six flip-flops. Six flip-flops will provide 64 states and we only want a total of 60 states. It will therefore be necessary to provide feedback to skip four states. The counter to provide this function is shown in Fig. 9-16.

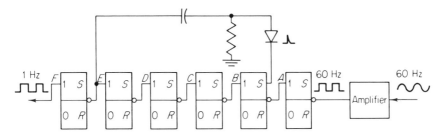

Fig. 9-16 ÷60 ripple counter to provide a 1-Hz square wave.

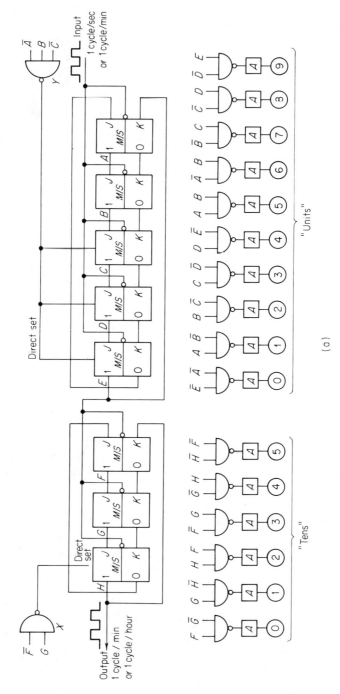

Fig. 9-17 A 10 × 6, ÷60 counter with *units* and *tens* decoding. (*a*) Logic diagram. (*b*) Waveforms (on facing page).

Special Counters and Registers

The amplifier at the input of the counter will provide a 60-Hz square wave of the proper amplitude to drive the first flip-flop. The required feedback is from flip-flop E to flip-flop B. In the normal counting sequence without feedback, the counter will progress through 64 states and E will go high twice during the sequence. Thus if E is fed back to the S input of flip-flop B, two counts will be added to the count sequence each time E goes positive. Since this happens twice during a complete cycle, a total of four counts will be added to the counter during one complete cycle. Thus the counter will essentially skip four states and will then have a total of 60 states.

The second counter in the system must also divide by 60 and could be implemented in the same way. However, the seconds counter must be decoded, and to decode this binary ripple counter would require 60 six-input AND gates. This is a total of 360 diodes! There must be a better way.

Notice that $60 = 2 \times 2 \times 3 \times 5$. We can therefore use counters of these moduli, or products of these moduli, to construct the desired $\div 60$ counter. We are interested in decoding this counter to represent each of the 60 seconds in 1 minute. This can most easily be accomplished by constructing a modulo-10 counter in series with a modulo-6 counter to form the $\div 60$. The mod-10 counter can then be decoded to represent the units digit of seconds and the mod-6 counter can be decoded to represent the tens digit of seconds. These two counters could be constructed by using feedback around ripple counters to obtain the correct counts. One such configuration requires 48 diodes for decoding. On the other hand, in the previous section

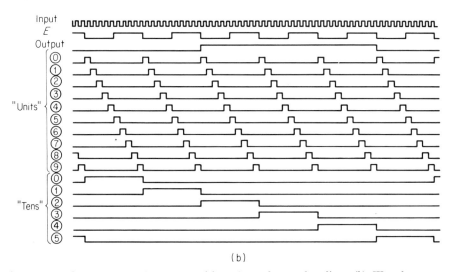

Fig. 9-17 A 10×6, $\div 60$ counter with *units* and *tens* decoding. (*b*) Waveforms.

it was shown that one of the more important advantages of a shift counter is the reduction in diodes necessary to decode the counter. Therefore, the seconds counter will be formed by connecting serially a mod-10 shift counter and a mod-6 shift counter. The entire counter is shown in Fig. 9-17.

Decoding of the counters is accomplished as discussed in the previous section and the proper decoding gates are shown below the counters. It can be seen that a total of 32 diodes are required for decoding this counter. Thus the shift-counter configuration requires one more flip-flop but provides a one-third reduction in decoding diodes. The amplifiers at the output

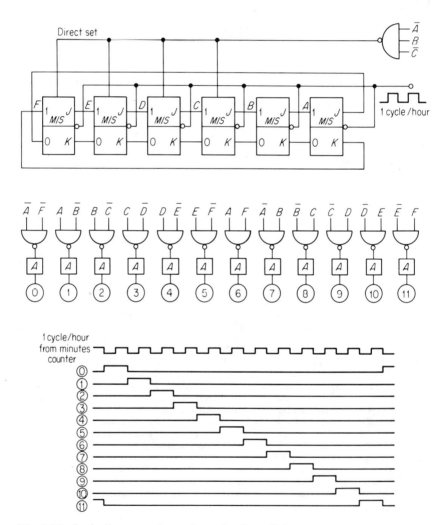

Fig. 9-18 Logic diagram and waveforms for the $\div 12$ *hours* counter.

Special Counters and Registers

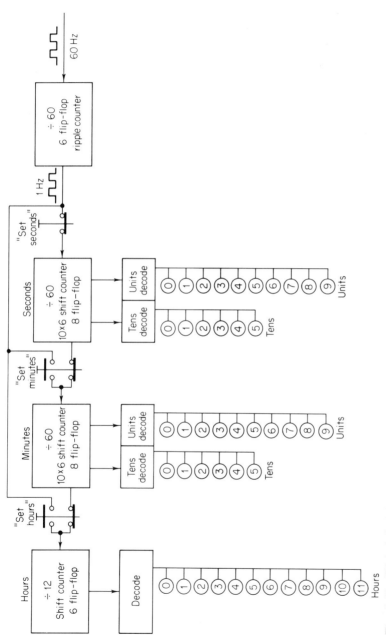

Fig. 9-19 Digital clock logic diagram.

of each of the gates provide the power necessary to drive the lamps, which are represented by the circles with numbers in them. The waveforms present in the counter are shown in the figure and the operation of the counter can be verified by examining these waveforms.

For example, if G and $\overline{H}$ are high and F is low, the lamp representing 1 in the tens column will be lit; if at the same time, B and $\overline{C}$ are high (all others are low) the lamp representing 2 in the units column will be lit. Thus in this state the counter will display 12 sec. NAND gates X and Y will ensure that the counter will always operate in the desired mode (see Prob. 9-16). The minutes counter will be exactly the same as the seconds counter except it will be driven by the 1 cycle per minute square wave from the output of the seconds counter and its output will be a 1 cycle per hour square wave. The 16 output lamps will now of course represent units and tens of minutes.

The ÷12 hours counter must be decoded into 12 states to display the hours. Again, for ease of decoding, a six-flip-flop shift counter will be used. The complete counter along with the proper decoding gates and waveforms are shown in Fig. 9-18.

The complete block diagram for the digital clock is shown in Fig. 9-19. Notice that the entire counter could have been constructed by using simple ripple counters throughout, and this would require only 22 flip-flops. This, however, would require over 700 diodes to decode the required states. Thus the six extra flip-flops used to implement the shift counters seem well justified since it now requires only 76 diodes for decoding to mention nothing of the reduction in work required to construct the clock.

Finally, some means must be found to set the clock; for when the power is turned off and then turned back on again the flip-flops will turn on in random states. Setting the clock can be quite easily accomplished by means of the *set* push buttons shown in Fig. 9-19. Depressing the *set hours* button will cause the hours counter to advance at a 1-sec rate and thus this counter can be set to the desired hour. The minutes counter can be similarly set by depressing the *set minutes* button. Depressing the *set seconds* button will remove the signal from the seconds counter, and the clock can thus be brought into synchronization.

9-6 Up-Down Counter

Up to this point, quite a large number of different counters have been considered, all of which count in an upward sequence. That is, the count sequences are 0-1-2-3-4···n-0. It is sometimes useful to have a counter which can count in a downward sequence. Any of the binary counters previously discussed can be used to implement a down counter.

Special Counters and Registers

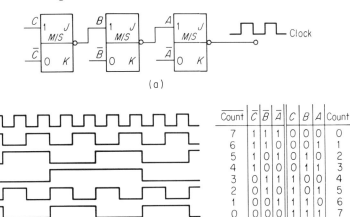

Fig. 9-20 A three-flip-flop ripple counter. (a) Logic diagram. (b) Waveforms. (c) Truth table.

The three-flip-flop ripple counter discussed in the previous chapter is redrawn in Fig. 9-20 and the waveforms for the 0 sides of the flip-flops have been added along with a new truth table. The normal count sequence, 0-1-2-3-4-5-6-7, is shown under *count* in the table and the corresponding state of the 1 side of each flip-flop is shown under A, B, and C. The 0 sides of the flip-flops are simply the negatives of the 1 sides, and these are shown under $\bar{A}$, $\bar{B}$, and $\bar{C}$. The binary states represented by $\bar{A}$, $\bar{B}$, and $\bar{C}$ are shown under $\overline{count}$.

Notice that the count sequence under $\overline{count}$ progresses in a downward fashion (7-6-5-4-3-2-1-0). Thus this counter could be decoded to provide a count-down sequence of waveforms. The proper decoding gates along with the count-down waveforms are shown in Fig. 9-21.

Notice that these decoding gates are exactly the same set of gates that would be required to produce a set of count-up waveforms. It is only necessary to change the number labels on the output of each gate (i.e., $7 \rightarrow 0$, $6 \rightarrow 1$, $5 \rightarrow 2$, $4 \rightarrow 3$, $3 \rightarrow 4$, $2 \rightarrow 5$, $1 \rightarrow 6$, and $0 \rightarrow 7$). Thus we have not truly formed a down counter yet; we have simply produced a set of decoded waveforms which correspond to a down count by rearranging the decoding gates.

A true down counter can be formed by triggering the input of each flip-flop with the 0 side of the previous flip-flop instead of the 1 side. Such a counter is shown in Fig. 9-22. Notice that A changes state each time the clock goes low as before. However, B now changes state when A goes high

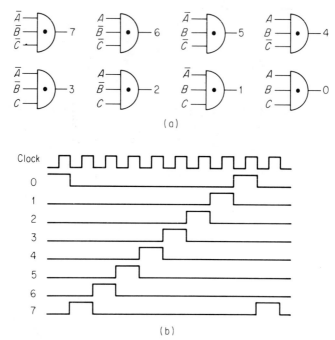

Fig. 9-21 (a) Decoding gates to provide *count down* waveforms. (b) *Count down* waveforms.

since this is the time when $\bar{A}$ goes low. Similarly, C changes state when B goes high, since this is the time when $\bar{B}$ goes low.

In order to unify this idea, a three-flip-flop *up-down counter* is shown in Fig. 9-23. In order that the counter will progress through a count up sequence,

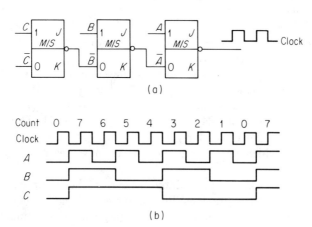

Fig. 9-22 A *down* counter. (a) Logic diagram. (b) Waveforms.

Special Counters and Registers

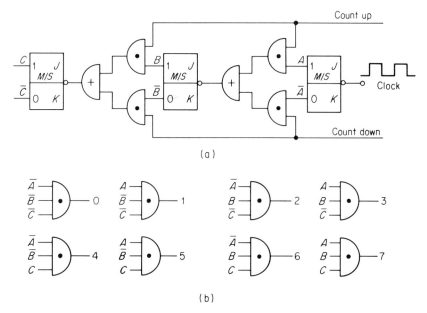

Fig. 9-23 An *up-down* counter. (*a*) Logic diagram. (*b*) Decoding gates. (Continues on page 238.)

it is necessary to trigger each flip-flop with the 1 side of the previous flip-flop. If the *count up* line is high and the *count down* line is low, this will be the case and the counter will progress through the waveforms shown in Fig. 9-23c.

In order to cause the counter to progress through a count down sequence, it is necessary to hold the *count up* line low and the *count down* line high. This will cause each flip-flop to be triggered from the 0 side of the previous flip-flop and will cause the counter to progress through the count down sequence in Fig. 9-23d.

This process can be continued to other flip-flops down the line to form an up-down counter of larger moduli. It should be noted, however, that the gates introduce additional delays and they must be taken into account when determining the maximum rate at which the counter can operate.

EXAMPLE 9-5

Show how to implement the up-down counter shown in Fig. 9-23 using NAND/NOR gates instead of AND gates and OR gates.

SOLUTION

The necessary control gate can be implemented quite simply by using the configuration shown in Fig. 9-24. When the *count down* line is low and the *count up* line is high, gate X is disabled and gate Y is enabled.

238   **Digital Principles and Applications**

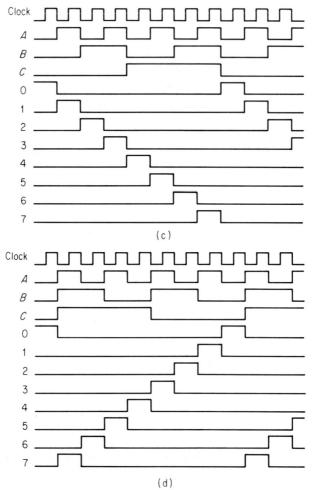

Fig. 9-23 (continued) An *up-down* counter. (*c*) *Count up* waveforms. (*d*) *Count down* waveforms.

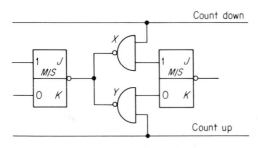

Fig. 9-24 *Up-down* counter control using NAND/NOR gates.

Special Counters and Registers

Therefore, A will not pass through X, but $\bar{A}$ will pass through Y. In passing through gate Y, $\bar{A}$ is inverted and thus appears at the input of the next flip-flop as A. With the *count down* line high and the *count up* line low, Y is disabled and X is enabled. Thus A is inverted as it passes through gate X and appears at the input of the next flip-flop as $\bar{A}$.

Care should be exercised in ORing the outputs of NAND/NOR gates (connecting the outputs together) in this fashion since some NAND/NOR gates have active pull-up transistors. This means that if one of the gates is on while the other is off, there is nearly a direct short between the supply voltage and ground.

A *parallel up-down counter* can be formed in a similar manner as shown in Fig. 9-25. This counter works in an inhibit mode since the flip-flops

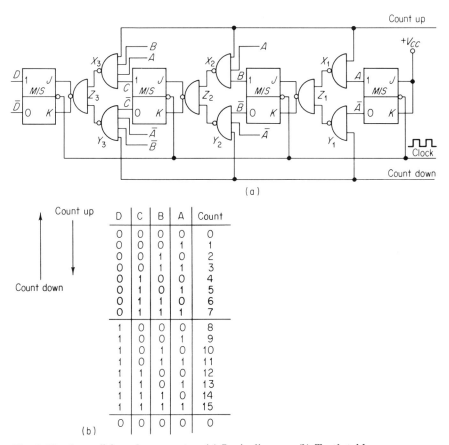

Fig. 9-25 A parallel *up-down* counter. (*a*) Logic diagram. (*b*) Truth table.

will change state when their J and K inputs are high, and will not change state when these inputs are low. You will recall that in a parallel counter the time at which any flip-flop changes state is determined by the states of all previous flip-flops in the counter. To determine the logic necessary to implement this counter, it will be convenient to refer to the truth table shown in Fig. 9-25b.

A is required to change state each time the clock goes low, and flip-flop A therefore has both its J and K inputs held in a high state. This is true in both the count up and count down modes, and therefore no other logic is necessary for this flip-flop.

In the count up mode, B is required to change state each time A goes from a 1 to a 0. Any time the *count up* line and A are both high, the output of gate X_1 will be low. Any time either of the inputs to Z_1 is low, its output will be high. Therefore, the J and K inputs to flip-flop B will be high whenever both *count up* and A are high and B will then change state each time A goes from a 1 to a 0.

In the count down mode, B must change state each time A goes from a 0 to a 1. The output of gate Y_1 will be low, and therefore the J and K inputs to flip-flop B will be high, any time $\bar{A}$ and *count down* are high. Thus in the count down mode, B will change state every time A goes from a 0 to a 1.

In the count up mode, C is required to change state every time both A and B change from a 1 to a 0 (i.e., the transitions 3 to 4, 7 to 8, 11 to 12, and 15 to 0). The output of gate X_2 will be low any time both A and B are high and the *count up* line is high. Therefore, the J and K inputs to flip-flop C will be high during these times and C will change state during the desired transitions.

In the count down mode, C is required to change state whenever both A and B change from a 0 to a 1. The output of gate Y_2 will be low any time both $\bar{A}$ and $\bar{B}$ are high and the *count down* line is high. Thus the J and K inputs to flip-flop C will be high during these times and C will then change state during the required transitions.

In the count up mode, D must change state every time A and B and C change from a 1 to a 0. The output of gate X_3 will be low, and thus the J and K inputs to flip-flop D will be high, any time A and B and C and *count up* are all high. Thus C will change states during the desired transitions.

In the count down mode, D must change state any time A and B and C go from a 0 to a 1. The output of gate Y_3 will be low, and thus the J and K inputs to flip-flop D will be high, whenever $\bar{A}$ and $\bar{B}$ and $\bar{C}$ and *count down* are high. Thus D will change states during the desired transitions.

This process can be continued for n flip-flops to make larger-modulus counters. However, notice that the X and Y gates to each flip-flop require more and more inputs. In general, flip-flop n would be preceded by X and Y gates requiring n inputs.

Special Counters and Registers

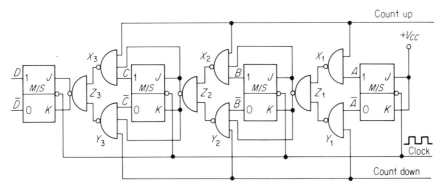

Fig. 9-26 *Up-down* counter with more efficient gating arrangement.

This problem can be minimized by noting that a part of the information required in any one gate is present in the corresponding gate in the preceding flip-flop. For example, the required inputs to gate X_3 in Fig. 9-25a are A, B, C, and *count up*. The information A, B, and *count up* is present at the output of gate X_2, but it is of the wrong polarity. However, it appears with the proper polarity at the output of gate Z_2. Therefore, the output of Z_2 could be used at the input of X_3 in place of A and B. Similarly, Z_2 can be used at the input of Y_3 in place of $\bar{A}$ and $\bar{B}$. In this fashion, all the remaining gates down the line can be limited to three inputs.

The same counter using this configuration is shown in Fig. 9-26. Notice that the *count up* line must still be used in every X gate since the Z outputs will be high at some time during both the up count and down count modes. For the same reason, the *count down* line must be used in every Y gate.

9-7 Shift-register Operations

In the first section in this chapter, a basic shift register was discussed and the method for serially shifting a binary number into the register was discussed. This method requires one cycle of the clock for each bit in the word and is therefore a relatively slow method. The transfer of information into the register could be accomplished in a much shorter period of time if it were carried out in a parallel manner. This can be accomplished using the logic shown in Fig. 9-27. Each bit is shifted into the proper flip-flop by means of the AND gates shown. The 2^0 bit is shifted into A, the 2^1 bit is shifted into B, and so on, and the 2^n bit is shifted into flip-flop $n - 1$.

Entering data into the register requires two clock cycles since the register must first be reset. Immediately after being reset, the shift pulse goes high

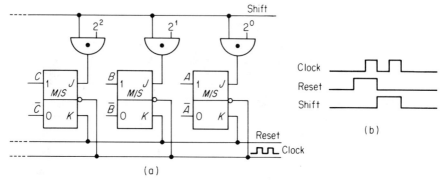

Fig. 9-27 (a) Shifting information into a shift register in parallel form. (b) Required control waveforms.

and if any of the data inputs to the AND gates (2^0, 2^1, 2^2, etc.) are high, the *set* input to the corresponding flip-flops will be high and a 1 will be set into these flip-flops. If any data input is low, that particular flip-flop will remain in the reset state. The proper control waveforms for this shifting operation are shown in Fig. 9-27b.

A similar method for entering information into a register is shown in Fig. 9-28. This method is twice as fast, since it requires no reset pulse first. It does, however, require twice the number of gates. The information is shifted into the flip-flops by means of the AND gates exactly as before. For example, if 2^0 is high, then $\overline{2^0}$ must be low. Therefore, the J input to flip-flop A must be high and the K input must be low, and A is then set high. Conversely, if 2^0 is low and $\overline{2^0}$ is high, A must be set low. The proper information will be shifted into each flip-flop regardless of the previous states of the flip-flops. Thus no reset pulse is required and it requires only one clock period to shift

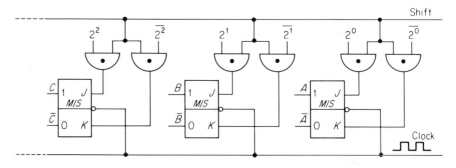

Fig. 9-28 Method for shifting data into a register in parallel in one clock period.

Special Counters and Registers

the information into the register. Notice that the clock can continue to run since when the *shift* and *reset* lines are both low, the flip-flops are prevented from changing state.

You will recall from Chap. 2 that binary arithmetic is sometimes carried out using the 1's complement of a number. It is quite easy to form the 1's complement in the register shown in Fig. 9-28. It is only necessary to hold both the J and K inputs to each flip-flop high for one cycle of the clock. Since the master/slave flip-flop will simply change state (complement) when both inputs are high, the contents of the register will be complemented. The register will then contain the 1's complement of the prior number.

Two other very useful operations which can be performed with the basic shift register are *shift right* and *shift left*. The *shift left* operation can be performed by the register shown in Fig. 9-29 by simply holding the *shift left* line high for one cycle of the clock.

In this register, shifting the data left one place is equivalent to multiplying by 2. For example, suppose the register contained the number $011_2 = 3_{10}$. If the contents of the register are shifted left one place, the register will now contain $110_2 = 6_{10}$. Thus the original number stored in the register has been effectively multiplied by 2.

As another example, suppose the number $101_2 = 5_{10}$ is stored in the register and the *shift left* operation is performed. After the shift left, the number $010_2 = 2_{10}$ is in the register. Something has gone wrong, since 2 is certainly not 2×5! Recall that a 3-bit register is capable of storing numbers from 0_{10} to 7_{10}. This defines the full capacity of the register. In the previous example, we were trying to double the number 5_{10}, which would have given 10_{10}. But this exceeds the capacity of the register and we therefore obtained an erroneous result. What actually happened was that the MSB was shifted out of the left end of the register and lost. If there were one more flip-flop to accept this bit, which represents 8_{10}, the register would then contain the number $1010_2 = 10_{10}$. Thus in the shifting operation it is always necessary to be aware of the capacity of the register.

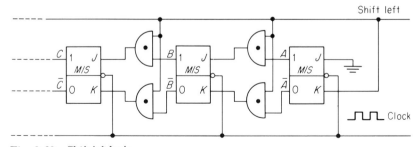

Fig. 9-29 *Shift left* logic.

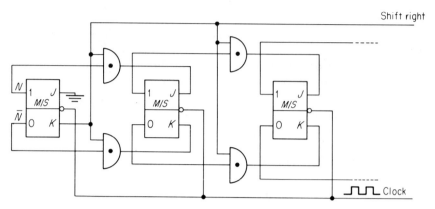

Fig. 9-30 *Shift right* logic.

In order to eliminate the necessity for stopping the clock, and to provide for a proper *shift left* control, the logic configuration shown in Fig. 9-29 can be used to shift left. The contents of the register will be shifted left one place each time the clock completes one cycle and the *shift left* line is high. Thus shifting left two places can be accomplished by holding the *shift left* line high for two clock periods, and so on.

Notice that shifting left two places is equivalent to multiplying by 4, three places by 8, etc. Notice also that A must be set low for any *shift left* operation. This is true since only 0s can be shifted into the lower significant bit positions, for if 1s were allowed the resulting number in the register would be more than twice the previous number. The J input to flip-flop A is shown grounded, but it is actually held low by other logic gates in the register (e.g., the gate used to enter information into the register serially).

The *shift right* operation is just the opposite of shifting left and results in a division by 2 of the number stored in the register. Shifting right two places will result in division by 4, and so on. The logic necessary to effect a shift right is shown in Fig. 9-30.

Notice in this case that flip-flop N, which represents the MSB, must be reset and have only 0s shifted into it in order to effect a proper divide by 2.

SUMMARY

A basic binary register is a group of flip-flops (or some other binary elements) which is used to store binary information. Information can be shifted into the register in either serial or parallel fashion. The parallel method is much faster but requires considerably more hardware.

Special Counters and Registers

Information stored in a register can be shifted left or right, and this corresponds to binary multiplication or division by 2. The register capacity must be taken into account during the *shift right* and *shift left* operations. It is also quite easy to find the complement of the information stored in the register.

Direct feedback around the basic shift register leads to the formation of a ring counter. This counter will be quite useful later on in devising control waveforms (Chap. 14). Cross-coupled feedback around the basic register leads to the formation of the shift counter. Shift counters of any modulus can be formed by taking the feedback from the proper flip-flops. The shift counter has the great advantage of simplified decoding. This type of counter does, however, have undesired states, and these must be provided for in the counter design.

Two methods for implementing up-down counters have been discussed. This type of counter will be found quite useful in such applications as digital voltmeters and analog converters (Chap. 11).

The digital clock is an interesting application and illustrates some of the methods employing counters and decoders.

GLOSSARY

register capacity Determined by the number of flip-flops in the register. There must be one flip-flop for each binary bit, and the register capacity is 2^n, where n is the number of flip-flops.

ring counter A basic shift register with direct feedback such that the contents of the register will simply circulate around the register when the clock is running.

shift counter A basic shift register with inverse feedback such that a cyclic counter is formed.

shift register A group of flip-flops connected in such a way that a binary number can be shifted into or out of the flip-flops.

up-down counter A basic counter, either serial or parallel, which is capable of counting in either an upward or a downward direction.

REVIEW QUESTIONS

1. How many flip-flops are required to store a 10-bit binary number in a register?
2. How many clock pulses are required to shift a 16-bit binary number into a 16 flip-flop serial shift register?

3. How many flip-flops would be required to construct a register capable of storing the number 2561_{10} in binary form?
4. If it is desired to shift information into the register shown in Fig. 9-2, does the register have to be cleared (reset) before the shifting operation?
5. Describe the difference between the serial and parallel methods of entering data into a register.
6. How is the complement operation performed on a register and what is the result?
7. How many flip-flops are required to construct shift counters of the following moduli?
   (a) 5.
   (b) 8.
   (c) 9.
   (d) 10.
   (e) 21.
8. How many diodes would be required to decode a seven-flip-flop shift counter?

PROBLEMS

9-1 Demonstrate that it is possible to construct the 6-bit shift register in Fig. 9-2 using master/slave RS flip-flops by drawing the logic diagram and the resulting waveforms.

9-2 How could you implement the controls for the J and K inputs to the first flip-flop in Fig. 9-2 using a single flip-flop?

9-3 Show the method for wiring the counter in Fig. 9-3 in order to preset it with a 1 in flip-flop F and 0s in all the other flip-flops.

9-4 What will happen if the counter in Fig. 9-3 has a 1 in every flip-flop just after power is applied?

9-5 How would you preset the counter in Fig. 9-3 to develop the waveform shown in Fig. 9-31?

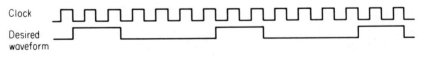

Fig. 9-31 Waveforms for Prob. 9-5.

9-6 Draw a two-flip-flop shift counter and its waveforms. Make a truth table and check for any illegal states.

Special Counters and Registers

9-7 Will the counter in Fig. 9-6 be forced into the proper mode if $\bar{C}$ is omitted from the NAND gate of Fig. 9-8?

9-8 In the worst case, how many clock periods will be required for the counter in Fig. 9-9 to get back into the proper count sequence?

9-9 Will the counter in Fig. 9-9 be forced into the proper mode if $\bar{C}$ is omitted from the NAND gate of Fig. 9-8? In the worst case how many counts will be required to get the counter into the proper sequence?

9-10 List the 22 illegal states for the decade counter in Fig. 9-11, and verify the two illegal count sequences. Also verify the results of the counter existing in either count 10 or count 21.

9-11 Verify that the NAND gate in Fig. 9-8 will force the decade counter in Fig. 9-11 into the desired mode of operation. On what illegal counts will the counter be corrected? How many clock pulses are required in the worst case to get the counter back into the proper sequence? Is the $\bar{C}$ input to the NAND gate necessary?

9-12 Draw the logic diagram, waveforms, and truth table to form a mod-9 counter out of the decade counter shown in Fig. 9-11.

9-13 Do the complete design for a mod-7 shift counter. Draw the waveforms and truth table and list the illegal states. Check the operation of the counter if it were to appear in any of the illegal states. Design a cure to place the counter back into the proper count sequence if one is needed.

9-14 Demonstrate that the mod-5 counter in Fig. 9-12 will always operate in the desired mode. Do the same for a mod-3 counter.

9-15 What states are skipped in the ÷60 counter in Fig. 9-16? Will the waveform at E be symmetrical? Would it affect the operation of the digital clock if it were not symmetrical?

9-16 Determine the method whereby NAND gates X and Y ensure proper operation of the counters in Fig. 9-17.

9-17 Draw the complete waveforms showing the results of allowing the up-down counter in Fig. 9-23 to progress through the following sequence: Starting at 0, count up five states, then down three states, then up four states, and then down six states.

9-18 Draw a three-flip-flop shift register capable of the following operations:

(a) Serial entry of data.
(b) Complement operation.
(c) Count up.
(d) Count down.

CHAPTER 10

INPUT-OUTPUT DEVICES

In any digital system it is necessary to have a link of communication between man and machine. This communication link is often called the "man-machine interface" and it presents a number of problems. Digital systems are usually capable of operating on information at speeds much greater than man, and this is one of their most important attributes. For example, a large-scale digital computer is capable of performing more than 500,000 additions per second.

The problem here then is to provide data input to the system at the highest possible rate. At the same time, there is the problem of accepting data output from the system at the highest possible rate. The problem is further magnified since most digital systems do not speak English, or any other language for that matter, and some system of symbols must therefore be used for communication (there is at present a considerable amount of research in this area, and some systems have been developed which will accept spoken commands and give oral responses on a limited basis).

Since digital systems operate in a binary mode, a number of code systems which are binary representations have been developed and are being used as the language of communication between man and machine. In this

Input-Output Devices

chapter we will discuss a number of these codes and at the same time we will consider the necessary input-output equipment.

10-1 Punched Cards

One of the most widely used media for entering data into a machine, or for obtaining output data from a machine, is a punched card. Some of the commonest examples of these cards are college registration cards, government checks, monthly oil company statements, and bank statements. It is quite simple to use this medium to represent binary information since only two conditions are required. Typically, a hole in the card represents a 1 and the absence of a hole represents a 0. Thus the card provides the means of presenting information in binary form, and it is only necessary to develop the code.

The typical punched card used in large-scale data-processing systems is $7\frac{3}{8}$ in. long, $3\frac{1}{4}$ in. wide, and 0.007 in. thick. Each card has 80 vertical columns and there are 12 horizontal rows as shown in Fig. 10-1. The columns are numbered 1 through 80 along the bottom edge of the card. Beginning at the top of the card, the rows are designated 12, 11, 0, 1, 2, 3, 4, 5, 6, 7, 8, and 9. The bottom edge of the card is the *9 edge* and the top edge of the card is the *12 edge*. Holes in the 12, 11, and 0 rows are called zone punches, and holes in the 0 through 9 rows are called digit punches. Notice that row 0 is both a zone- and a digit-punch row. Any number, any letter in the alphabet, or any of several special characters can be represented on the card by punching one or more holes in any one column. Thus the card has the capacity of 80 numbers, letters, or combinations.

Probably the most widely used system for recording information on a punched card is the *Hollerith code*. In this code the numbers 0 through 9 are represented by a single punch in a vertical column. For example, a

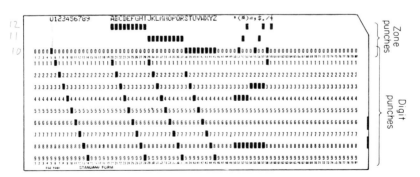

Fig. 10-1 Standard punched card using Hollerith code.

hole punched in the fifth row of column 12 represents a 5 in that column. The letters of the alphabet are represented by two punches in any one column. The letters A through I are represented by a zone punch in row 12 and a punch in rows 1 through 9. The letters J through R are represented by a zone punch in row 11 and a punch in rows 1 through 9. The letters S through Z are represented by a zone punch in row 0 and a punch in rows 2 through 9. Thus any of the 10 decimal digits and any of the 26 letters of the alphabet can be represented in a binary fashion by punching the proper holes in the card. In addition, a number of special characters can be represented by punching combinations of holes in a column which are not used for the numbers or letters of the alphabet. These characters are shown with the proper punches in Fig. 10-1.

An easy device for remembering the alphabetic characters is the phrase "JR. is 11." Notice that the letters J through R have an 11 punch, those before have a 12 punch, and those after have a 0 punch. It is also necessary to remember that S begins on a 2 and not a 1!

EXAMPLE 10-1

Decode the information punched in the card shown in Fig. 10-2.

SOLUTION

Column 1 has a zone punch in row 0 and a punch in row 3. It is therefore the letter T. Column 2 has a zone punch in row 12 and another punch in row 8. It is therefore the letter H. Continuing in this fashion, the complete message reads, "THE QUICK BROWN FOX JUMPED OVER THE LAZY DOGS BACK."

Using this card code as described, any *alphanumeric* (alphabetic and numeric) information can be used as input to a digital system. On the other hand, the system will be capable of delivering alphanumeric output information to the user. In scientific disciplines, the information might be missile

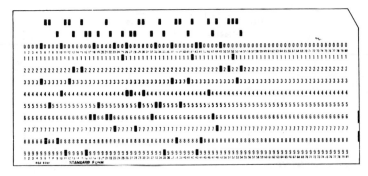

Fig. 10-2 Example 10-1.

Input-Output Devices

flight number, location, or guidance information such as pitch rate, roll rate, and yaw rate. In business disciplines, the information could be account numbers, names, addresses, monthly statements, etc. In any case, the information is punched on the card with one character per column, and the card is then capable of containing a maximum of 80 characters.

Each card is then considered one block or unit of information, and since the machine operates on one card at a time, the punched card is often referred to as a "unit record." Moreover, the digital equipment used to punch cards, read cards into a system, sort cards, etc., is referred to as "unit-record equipment."

Occasionally the information used with a digital system is entirely numeric; that is, no alphabetic or special characters are required. In this case, it is possible to input the information into the system by punching the cards in a straight binary fashion. In this system, the absence of a punch is a binary 0 and a punch is a binary 1. It is then possible to punch $80 \times 12 = 960$ bits of binary information on one card.

Many large-scale data-processing systems use binary information in blocks of 36 bits. Each block of 36 bits is called a "word." You will recall from the previous chapter that a register capable of storing a 36-bit word must contain 36 flip-flops. There is nothing magical about the 36-bit word, and there are in fact other systems which operate with other word lengths. Even so, let us consider how binary information arranged in words of 36 bits might be punched on cards.

There are in general two methods. The first method stores the information on the card horizontally by punching across the card from left to right. The first 36-bit word is punched in row 9 in columns 1 through 36. The second word is also in row 9, in columns 37 through 72. The third word is in row 8, columns 1 through 36, and so on. Thus a total of twenty-four 36-bit words can be punched in the card in straight binary form. It is then possible to store 864 bits of information on the card.

The second method involves punching the information vertically in columns rather than rows. Beginning in row 12 of column 1, the first 12 bits of the word are punched in rows 12, 11, 0, . . . , 9. The next 12 bits are punched in column 2 and the remaining 12 bits are punched in column 3. Thus a 36-bit word can be punched in every three columns. The card is then capable of containing twenty-six 36-bit words.

The most common method of entering information into punched cards initially is by means of the key-punch machine. This machine operates very much the same as a typewriter, and the speed and accuracy of the operation are entirely dependent on the operator. The information on the punched cards can then be read into the digital system by means of a card reader. The information can be entered into the system at the rate of 100 to 1000 cards per minute depending on the type of card reader used.

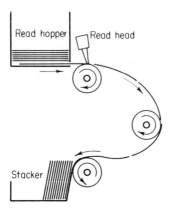

Fig. 10-3 Card-reading operation.

The basic method for changing the punched information into the necessary electrical signals is shown in Fig. 10-3. The cards are stacked in the *read* hopper and are drawn from it one at a time. Each card passes under the *read* heads, which are either brushes or photocells. There is one *read* head for each column on the card, and when a hole appears under the *read* head an electrical signal is generated. Thus each signal from the *read* heads represents a binary 1, and this information can then be used to set flip-flops which form the input storage register. The cards then pass over other rollers and are placed in the stacker. There is quite often a second *read* head which reads the data a second time to provide a validity check on the reading process.

EXAMPLE 10-2

Suppose a deck of cards has binary data punched in them. Each card has twenty-four 36-bit words. If the cards are read at a rate of 600 cards per minute, what is the rate at which data are entering the system?

SOLUTION

Since each card contains 24 words, the data rate is $24 \times 600 = 14{,}400$ words per minute. This is equivalent to $36 \times 14{,}400 = 518{,}400$ bits per minute, or $518{,}400/60 = 8640$ bits per second.

Punched cards can also be used as a medium for accepting data output from a digital system. In this case, a stack of blank cards (having no holes punched in them) are held in a hopper in a card punch which is controlled by the digital system. The blank cards are drawn from the hopper one at a time and punched with the proper information. They are then passed under *read* heads which check the validity of the punching operation and are stacked in an output hopper. Card punches are capable of operating at 100 to 250 cards per minute depending on the system used.

Input-Output Devices

Punched cards present a number of important advantages, the first of which is the fact that the cards represent a means of storing information permanently. Since the information is in machine code, and since this information can be printed on the top edge of the card, this medium represents a very convenient means of communication between man and machine, and between machine and machine. There is also a wide variety of peripheral equipment which can be used to process information stored on cards, the commonest of which are sorters, collators, calculating punches, reproducing punches, and accounting machines. Moreover, it is very easy to correct or change the information stored since it is only necessary to remove the desired card(s) and replace it(them) with the corrected one(s). Finally, these cards are usually purchased by the case and are therefore quite inexpensive.

10-2 Paper Tape

Another very widely used form of input-output medium is punched paper tape. It is used in much the same way as punched cards. Paper tape was developed initially for the purpose of transmitting telegraph messages over wires. It is now used extensively for storing information and for transmitting information from machine to machine. Paper tape differs from cards in that it is a continuous roll of paper and thus any desired amount of information can be punched into a roll. It is possible to record any alphabetic or numeric character, as well as a number of special characters, on paper tape by punching holes in the tape in the proper places.

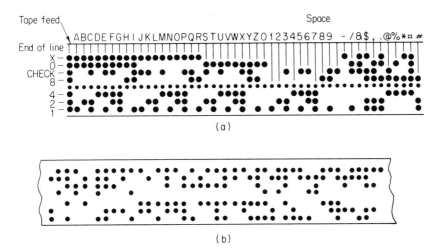

Fig. 10-4 Punched paper tape. (*a*) Eight-hole code. (*b*) Example 10-3.

There are a number of codes for punching data in paper tape, but one of the most widely used is the *eight-hole code* in Fig. 10-4a. Holes, representing data, are punched in eight parallel channels which run the length of the tape. (The channels are labeled 1, 2, 4, 8, parity, 0, X, and end of line.) Each character, numeric, alphabetic, or special, occupies one column of eight positions across the width of the tape.

Numbers are represented by punches in one or more of the channels labeled 0, 1, 2, 4, and 8, and each number is the sum of the punch positions. For example, 0 is represented by a single punch in the 0 channel; 1 is represented by a single punch in the 1 channel; 2 is a single punch in channel 2; 3 is a punch in channel 1 and a punch in channel 2, etc. Alphabetic characters are represented by a combination of punches in channels X, 0, 1, 2, 4, and 8. Channels X and 0 are used much as the zone punches in punched cards. For example, the letter A is designated by punches in channels X, 0, and 1. The special characters are represented by combinations of punches in all channels which are not used to designate either numbers or letters. A punch in the end-of-line channel is used to signify the end of a block of information, or the end of record. This is the only time a punch will appear in this channel.

As a means of checking the validity of the information punched on the tape, the parity channel is used to ensure that each character is represented by an *odd* number of holes. For example, the letter C is represented by punches in channels X, 0, 1, and 2. Since an odd number of holes is required for each character, the code for the letter C also has a punch in the parity channel, and thus a total of five punches is used for this letter.

EXAMPLE 10-3

What information is held in the perforated tape in Fig. 10-4b?

SOLUTION

The first character has punches in channels 0, 1, and 2, and this is the letter T. The second character is the letter H, since there are punches in channels X, 0, and 8. Continuing, it can be seen that the message is the same as that punched on the card in Example 10-1.

The row of smaller holes between channels 4 and 8 are guide holes which are used to guide and drive the tape under the *read* positions. The information on the tape can be sensed by brushes or photocells as shown in Fig. 10-5. The method for reading information from the paper tape and inputting it into the digital system is very similar to that used for reading punched cards. Depending on the type of reader used, information can be read into the system at a rate of 150 to 1000 characters per second. You will notice that this is only slightly faster than reading information from punched cards.

Input-Output Devices

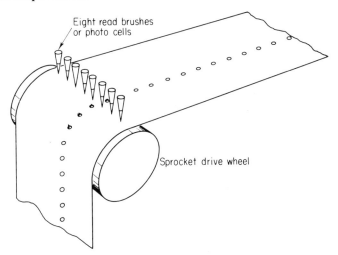

Fig. 10-5 Paper-tape drive and reading mechanism.

Paper tape can be used as a means of accepting information output from a digital system, and in this case the system will drive a tape punch which will enter the data on the tape by punching the proper holes. Typical tape punches are capable of operating at rates of 15 characters per second, and the data are punched with 10 characters to the inch. The number of characters per inch is referred to as the "data density," and in this case one would say the density is 10 characters per inch. Recording density is one of the important features of magnetic-tape recording which will be discussed in the next section.

Paper tape can also be perforated by using a manual tape punch. This unit is very similar to an electric typewriter, and indeed in some cases electric typewriters with special punching units attached are used. The accuracy and speed of this method are again a function of the machine operator. One advantage of this method is that the normal typing operation using the typewriter provides a written copy of what is punched into the tape, and this copy can be used for verification of the punched information.

10-3 Magnetic Tape

Magnetic tape has become one of the most important methods for storing large quantities of information. Magnetic tape offers a number of advantages over punched cards and punched paper tape, one of the most important of which is the fact that magnetic tape can be erased and used over and over. Reading and recording are much faster than with either cards or

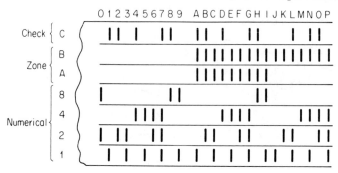

Fig. 10-6 Magnetic-tape code.

paper tape but require the use of a tape-drive unit which is much more expensive than the equipment used with cards and paper tape. On the other hand, it is possible to store up to 20 million characters on one 2400-ft reel of magnetic tape, and if a high volume of data is one of the system requirements the use of magnetic tape is well justified. Most commonly, magnetic tape is supplied on 2400-ft reels. The tape itself is a ½-in.-wide strip of plastic with a magnetic oxide coating one one side.

Data are recorded on the tape in seven parallel channels along the length of the tape. The channels are labeled 1, 2, 4, 8, A, B, and C as shown in Fig. 10-6. Since the information recorded on the tape must be digital in form,

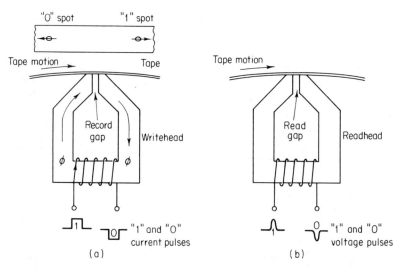

Fig. 10-7 Magnetic-tape recording and reading. (*a*) *Write* operation. (*b*) *Read* operation.

Input-Output Devices

that is, there must be two states, it is recorded by magnetizing spots on the tape in one of two directions.

A simplified presentation of the *write* and *read* operations is shown in Fig. 10-7. The magnetic spots are recorded on the tape as it passes over the *write* head as shown in Fig. 10-7a. If a positive pulse of current is applied to the *write*-head coil, as shown in the figure, a magnetic flux is set up in a clockwise direction around the *write* head. As this flux passes through the record gap, it spreads slightly and passes through the oxide coating on the magnetic tape. This will cause a small area on the tape to be magnetized with the polarity shown in the figure. If a current pulse of the opposite polarity is applied, the flux will be set up in the opposite direction and a spot magnetized in the opposite direction will be recorded on the tape. Thus it is possible to record data on the tape in a digital fashion. The spots shown in the figure are greatly exaggerated in size in order to show the direction of magnetization clearly.

In the *read* operation shown in Fig. 10-7b, a magnetized spot on the tape sets up a flux in the *read* head as the tape passes over the *read* gap. This flux will induce a small voltage in the *read*-head coil which can be amplified and used to set or reset a flip-flop. Spots of opposite polarities on the tape will induce voltages of opposite polarities in the *read* coil and thus both 1s and 0s can be sensed. There is one *read/write* head for each of the seven channels on the tape. Typically, *read/write* heads are constructed in pairs as shown in Fig. 10-8 and thus a self-checking operation can be exercised during the *write* operation. That is, data recorded on tape are immediately read as they pass over the *read* gap and can be checked for validity.

A coding system similar to that used to punch data on cards is used to record alphanumeric information on tape. Each character occupies one column of 7 bits across the width of the tape. The code is shown in Fig. 10-6. There are two independent systems for checking the validity of the information stored on the tape.

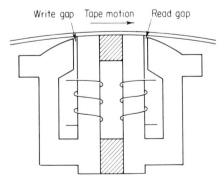

Fig. 10-8 Magnetic-tape *read/write* heads.

The first system is a vertical parity bit which is written in channel C of the tape. This is called a "character-check bit" and is written in channel C to ensure that all characters are represented by an *even* number of bits. For example, the letter A is represented by spots in channels 1, A, and B. Since this is only three spots, an additional spot is recorded in channel C to maintain even parity for this character.

The second system is the *horizontal parity-check bit*. This is sometimes referred to as the longitudinal parity bit, and it is written, when needed, at the end of a block of information or record. The total number of bits recorded in each channel is monitored, and at the end of a record, a parity bit is written if necessary to keep the total number of bits an *even* number. These two systems form an even-parity system. They could of course just as easily be implemented to form an odd-parity system. Information can also be recorded on the tape in straight binary form. In this case, a 36-bit word is written across the width of the tape in groups of 6 bits. Thus it requires six columns to record one 36-bit word.

The vertical spacing between the recorded spots on the tape is fixed by the positions of the *read/write* heads. The horizontal spacing is a function of the tape speed and the recording speed. Tape speeds vary from 50 to 200 in. per second, but 75 and 112.5 in. per second are quite common.

The maximum number of characters recorded in 1 inch of tape is called the "recording density," and it is a function of the tape speed and the rate at which data are supplied to the *write* head. Typical recording densities are 200, 556, and 800 bits per inch. Thus it can be seen that a total of $800 \times 2400 \times 12 = 23.02 \times 10^6$ characters can be stored on one 2400-ft reel of tape. This would mean that the data would have to be stored with no gaps between characters or groups of characters.

For purposes of locating information on tape, it is most common to record information in groups or blocks called "records." In between records there is a blank space of tape called the "interrecord gap." This gap is typically a 0.75-in. space of blank tape and is positioned over the *read/write* heads when the tape stops. The record gap provides the space necessary for the tape to come up to the proper speed before recording or reading of information can take place. The total number of characters recorded on a tape is then also a function of the record length (or the total number of record gaps, since they represent blank space on the tape).

The data as recorded on the tape showing records (actual data) and interrecord gaps can be represented as shown in Fig. 10-9. If there were no record gaps, the total number of characters recorded could be found by multiplying the length of the tape in inches by the recording density in characters per inch. If the record were exactly the same length as the record gap, the total storage would be cut in half. Thus it can be seen that it is desirable to keep the records as long as possible in order to use the tape most efficiently.

Input-Output Devices

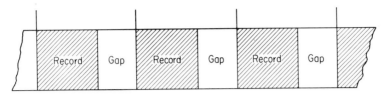

Fig. 10-9 Recording data on magnetic tape.

Given any one tape system and the recording density, it is a simple matter to determine the actual storage capacity of the tape. Consider the length of tape composed of one record and one record gap as shown in Fig. 10-9. This length of tape is repeated over and over down the length of the tape. The total number of characters that could be stored in this length of tape is the sum of the characters in the record R and the characters which could be stored in the record gap. The number of characters which could be stored in the gap is equal to the recording density D multiplied by the gap length G. Thus the total number of characters which could be stored in this length of tape is given by $R + GD$. The ratio of the characters actually recorded R to the total possible could be called a tape-utilization factor F is given by

$$F = \frac{R}{R + GD} \tag{10-1}$$

Examination of the tape-utilization factor shows that if the total number of characters in the record is equal to the number of characters which could be stored in the gap, the utilization factor reduces to 0.5. This utilization factor can be used to determine the total storage capacity of a magnetic tape if the recording density and the record length are known. Thus the total number of characters stored on a tape $CHAR$ is given by

$$CHAR = LDF \tag{10-2}$$

where L = length of tape, in.
D = recording density, characters per in.

For a standard 2400-ft reel of tape having a 0.75-in. record gap, the formula in Eq. (10-2) reduces to

$$CHAR = \frac{2400 \times 12 \times DR}{R + 0.75D} \tag{10-3}$$

EXAMPLE 10-4

What is the total storage capacity of a 2400-ft reel of magnetic tape if data are recorded at a density of 556 characters per inch and the record length is 100 characters?

Solution

The total number of characters can be found using Eq. (10-3).

$$CHAR = \frac{2400 \times 12 \times 556 \times 100}{100 + 0.75 \times 556} = 3.84 \times 10^6$$

This result can be checked by calculating the tape-utilization factor.

$$F = \frac{100}{100 + 0.75 \times 556} = \frac{1}{4.17} \cong 0.24$$

The maximum number of characters that can be stored on the tape is $2400 \times 12 \times 556 = 16.0128 \times 10^6$. Multiplying this by the utilization factor gives

$$CHAR = 16.0128 \times 10^6 \times \frac{1}{4.17} = 3.84 \times 10^6$$

10-4 Digital Recording Methods

There are a number of methods for recording data on a magnetic surface. The methods fall into two general categories called "return-to-zero" and "non-return-to-zero," and they apply to magnetic-tape recording as well as recording on magnetic disk and drum surfaces (magnetic-disk and magnetic-drum storage will be discussed in a later chapter).

In the previous section, it was stated that digital information could be recorded on magnetic tape by magnetizing spots on the tape with opposite polarities. This type of recording is known as return-to-zero, or RZ for short, recording. The technique for recording data on tape using this method is to apply a series of current pulses to the *write*-head winding as

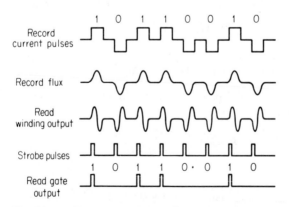

Fig. 10-10 Return-to-zero recording and reading.

shown in Fig. 10-10. The current pulses will set up corresponding fluxes in the *write* head as shown in the figure. The spots magnetized on the tape will have polarities corresponding to the direction of the flux waveform, and it is only necessary to change the direction of the input current in order to write 1s or 0s. Notice that the input current and the flux waveform return to a zero reference level between individual bits. Thus the term "return-to-zero."

When it is desired to read the recorded information from the tape, the tape is passed over the *read* heads and the magnetized spots induce voltages in the *read*-coil winding as shown in the figure. Notice that there is somewhat of a problem here since all the pulses have both positive and negative portions. One method of detecting these levels properly is to strobe the output waveform. That is, the output-voltage waveform is applied to one input of an AND gate (after being amplified) and a clock or strobe pulse is applied to the other input to the gate. The strobe pulse must be very carefully timed to ensure that it samples the output waveform at the proper time. This is one of the major difficulties of this type of recording, and it is therefore seldom used except on magnetic drums. On a magnetic drum, the strobe waveform can be recorded on one track of the drum and thus the proper timing is achieved.

A second difficulty with this type of recording is the fact that in between bits there is no record current and thus in between the spots on the tape the magnetic surface is randomly oriented. This means that if a new recording is to be made over old data, the new data will have to be recorded precisely on top of the old data. If they are not, the old data will not be erased and the tape will contain a conglomeration of information. The tape could be erased by installing another set of *erase* heads, but this is costly and unnecessary.

A method for curing these problems is to bias the record head with a current which will saturate the tape in either one direction or the other. In this system, a current pulse of positive polarity is applied only when it is desired to write a 1 on the tape as shown in Fig. 10-11. At all other times the flux in the *write* heads is sufficient to magnetize the entire track in the 0

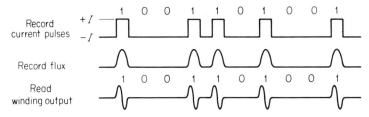

Fig. 10-11 Biased return-to-zero recording and reading.

direction. Thus recording data over old data is not a problem since the tape is effectively erased as it passes over the record heads. Moreover, the timing is not so critical since it is not necessary to record exactly over the previous data. When data are recorded in this fashion and then played back, a pulse will appear at the output of the *read* winding only when a 1 has been recorded on the tape. This makes reading the information from the tape much simpler.

The *non-return-to-zero*, or NRZ, recording technique is a variation of the RZ technique where the *write* current pulses do not return to some reference level between bits. The NRZ recording technique can be best explained by examining the record-current waveform shown in Fig. 10-12. Notice that the current is at $+I$ while recording 1s and at $-I$ while recording 0s. Since the current levels are always at either $+I$ or $-I$, the recording problems of the first RZ system are not existent.

Notice that the voltage at the *read*-winding output has a pulse only when the recorded data change from a 1 to a 0 or vice versa. Therefore, some means of sensing the recorded data is necessary for the *read* operation. If the *read*-winding voltage is amplified and used to set or reset a flip-flop as shown in the figure, the 1 side of the flip-flop will be high during each time that a 1 is being read. It will be low during any time when the data being read is a 0. Thus if the 1 output of the flip-flop is used as a control signal at one input of an AND gate, while the other input is a clock, the output of the AND gate will be an exact replica of the digital data being read. Notice that the clock must be carefully synchronized with the data train from the *read*-head winding. Notice also that the maximum rate of flux changes occurs when recording (or reading) alternate 1s and 0s.

In comparing with the RZ recording methods, it can be seen that the NRZ method offers the distinct advantage that the maximum rate of flux changes occurs at only one-half that for RZ recording. Thus the *read/write*

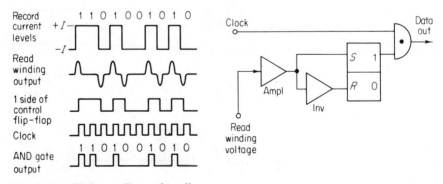

Fig. 10-12 NRZ recording and reading.

Input-Output Devices 263

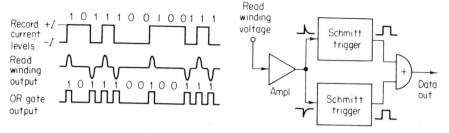

Fig. 10-13 NRZI recording and reading.

heads and associated electronics can have reduced requirements for operation at the same rates, or will be capable of operating at twice the rate for the same specifications.

A variation on this basic form of NRZ recording is shown in Fig. 10-13. This technique is quite often called "non-return-to-zero-inverted," NRZI, since both 1s and 0s are recorded at both the high and low saturation-current levels. The key to this method of recording is that a 1 is sensed whenever there is a flux change, whether it be positive or negative. If the *read*-winding output voltage is amplified and presented to the OR gate as shown in the figure, the output of the gate will be the desired data train. The upper Schmitt trigger is sensitive only to positive pulses while the lower one is sensitive only to negative pulses. Both outputs of the Schmitt triggers are low until a pulse arrives. At this time the output goes positive for a fixed duration and generates the desired output pulse.

10-5 Other Peripheral Equipment

A wide variety of peripheral equipment has been developed for use with digital systems. Only a cursory description of some of the various equipment will be given here, and the reader is encouraged to study equipment of particular interest by consulting the data manuals from the various manufacturers.

One of the simplest means of inputting information into a digital system is by the use of switches. These switches could be push-button, toggle, etc., but the important thing is the fact that they are capable of representing binary information. A row of 10 switches could, for example, be switched to represent the 10 binary bits in a 10-bit word.

Similarly, one of the simplest means of reading data out of a digital system is to put lights on the outputs of the flip-flops in a storage register. Admittedly, this is a rather slow means of communication since the operator must convert the displayed binary data into something more meaningful.

Nevertheless, this represents an inexpensive and practical means of communication between man and machine.

A much more sophisticated method for reading data out of a digital system is by means of a cathode-ray tube. One type of cathode-ray tube used is very similar to the tube used in oscilloscopes, and the operation of the tube is nearly the same. The unit is generally used to display curves representing information which has been processed by the system, and a camera can be attached to some units to photograph the display for a permanent record. The information displayed might be such things as the transient response of an electrical network or a guided-missile trajectory.

A second type of cathode-ray tube used for display which is called a "charactron" has the ability to display alphanumeric characters on the face of the screen. This tube operates by shooting an electron beam through a matrix (mask) which has each of the characters cut in it. As the beam passes through the matrix it is shaped in the form of the character through which it passes, and this shaped beam is then focused on the face of the screen. Since the operation of the electron beam is very fast, it is possible to write information on the face of the tube, and the operator can then read the display.

Some tubes of this type which are used in large radar systems have matrices with the proper characters to display map coordinates, friendly aircraft, unfriendly aircraft, etc., and the operator thus sees a display of the surrounding area complete with all aircraft, properly designated, in the vicinity. These systems usually have an additional accessory called a "light pen" which enables the operator to input information into the digital system by placing the light pen on the surface of the tube and activating it. The operator can do such things as expand an area of interest, request information on an unidentified flying object and designate certain aircraft as targets.

A somewhat more common piece of equipment, but nevertheless useful when large quantities of data are being handled, is the printer. Printers are available which will print the output data in straight binary form, octal form, or all the alphanumeric characters. The typical printer has the ability to print information on a 120-space line at rates from a few hundred lines up to over 1200 lines per minute. The simplest printers are simply converted, or specially made, electric typewriters known as "character-at-a-time printers." They are relatively slow and operate at speeds of 10 to 30 characters per second.

A more sophisticated printer is known as the "line-at-a-time printer" since an entire line of 120 characters is printed in one operation. These types of printers are capable of operation at rates of around 250 lines per minute.

Somewhat faster operation is possible with machines which use a print wheel. The print-wheel printer is composed of 120 wheels, one for each

Input-Output Devices 265

position on the line to be printed. These wheels rotate continuously, and when the proper character is under the print position a hammer strikes an inked ribbon against the paper, which contacts the raised character on the print wheel. Wheel printers are capable of operation at the rate of 1250 lines per minute, and have a maximum capacity of 160 characters per line

One other very important piece of peripheral equipment is the digital plotter. These units are being used more and more in a wide variety of ways including automatic drafting, numerical control, production artwork masters (used to manufacture integrated circuits), charts and graphs for management information, maps and contours, biomedical information, and traffic analysis, as well as a host of other applications. A somewhat hybrid form of digital plotting is formed when the digital output of a system is converted to analog form (digital-to-analog conversion is the subject of the next chapter) to drive servomotors which position a cursor or pen. A piece of graph paper is positioned on a flat plotting surface, and the pen is caused to move across the paper in response to the numbers received from the digital system.

Another digital plotting system, which is more truly a digital plotter, makes use of bidirectional stepping motors to position the pen and thus plot the information on the graph paper. In this system, which is known as a "digital incremental plotter," the necessity for digital-to-analog conversion is eliminated, and these systems are usually less expensive and smaller in size. Digital incremental plotters are capable of plotting increments as small as 0.0025 in. and offer much greater accuracies than the hybrid model previously discussed. Furthermore, these plotters are capable of plotting at the rate of $4\frac{1}{2}$ in. per second and in addition provide a complete system of annotation and labeling.

10-6 Encoding and Decoding Matrices

Encoding and decoding matrices are often used to alter the form of the data being entered into a system or being taken out of a system. A decoding matrix is used to decode the binary information in a digital system by changing it into some other number system. For example, in a previous chapter the binary output of a register was decoded into decimal form by means of AND gates and the decoded output was used to drive nixie tubes. Encoding information is just the reverse process and could, for example, involve changing decimal signals into equivalent binary signals for entry into a digital system.

The most straightforward way of decoding information is simply to construct the necessary AND gates as was done for the nixie tubes. Decoding in this fashion is quite simple and is most easily accomplished by using the truth table or waveforms for the signals involved. The decoding of a four-flip-flop counter would, for example, require 16 four-input AND gates, since there are 16 possible states determined by the four flip-flops. This type of decoding then requires $n \times 2^n$ diodes, where n is the number of flip-flops, for the complete decoding network.

EXAMPLE 10-5

Draw the 16 gates necessary to decode a four-flip-flop counter.

SOLUTION

The necessary gates can be best implemented by using a truth table to determine the necessary gate connections. The gates are shown in Fig. 10-14.

There is a second method of decoding which can be used to realize a savings in diodes. This method is referred to as "tree decoding," and it

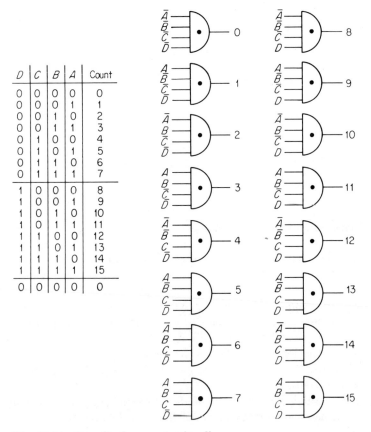

Fig. 10-14 Four-flip-flop counter decoding.

Input-Output Devices

results in a reduction of the number of required diodes by grouping the states to be decoded. Decoding of the four-flip-flop counter discussed in the previous example can be accomplished by separating the counts into four groups. These groups are 0,1,2,3; 4,5,6,7; 8,9,10,11; and 12,13,14,15. Notice that the first group can be distinguished by an AND gate whose output is $\bar{D}\bar{C}$, the second group by $\bar{D}C$, the third group by $D\bar{C}$, and the last group by DC. Each of these four groups can then be divided in half by using B or $\bar{B}$. These eight subgroups can then be further divided into the 16 counts by using A and $\bar{A}$. The complete decoding network is shown in Fig. 10-15.

A savings of 8 diodes has been achieved since the previous decoding scheme required 64 diodes and this method only requires 56. The savings in diodes here is not very spectacular, but the construction of a matrix in this manner to decode five flip-flops would result in a savings of 40 diodes. As the number of flip-flops to be decoded increases, the savings in diodes increases very rapidly.

This type of decoding matrix does have the disadvantage that the decoded signals must pass through more than one level of gates (in the

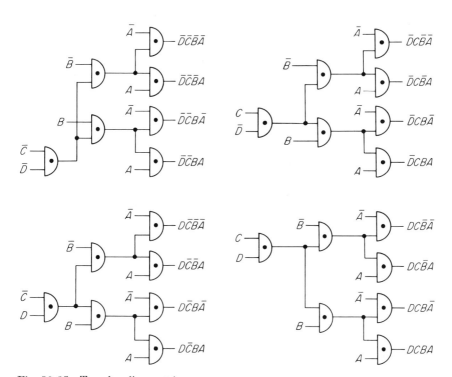

Fig. 10-15 Tree decoding matrix

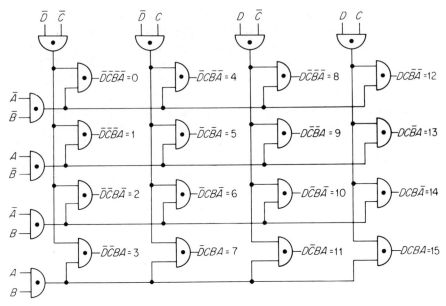

Fig. 10-16 Balanced multiplicative decoder.

previous method the signal passes through only one gate) and the output signal level may therefore suffer considerable reduction in amplitude. Furthermore, there may be a speed limitation due to the number of gates through which the decoded signals must pass.

A third type of decoding network is known as a "balanced multiplicative decoder." This will always result in the minimum number of diodes required for the decoding process. The idea is much the same as a tree decoder since the counts to be decoded are divided into groups. However, in this system the flip-flops to be decoded are divided into groups of two and the results are then combined to give the desired output signals. To decode the four flip-flops discussed previously, four groups are formed by combining flip-flops C and D just as before. In addition, flip-flops B and A are combined in a similar arrangement. The outputs of these eight gates are then combined in 16 AND gates to form the 16 output signals. The results are shown in Fig. 10-16. It can be seen that a total of 48 diodes are required and a savings of 16 diodes is then realized over the first method, while a savings of 8 diodes is realized over the tree method. This scheme again has the same disadvantages of signal-level degradation and speed limitation as the tree decoder.

Encoding a number is just the reverse process of decoding. One of the simplest examples of encoding would be the use of a thumb-wheel switch (a 10-position switch) which is used to enter data into a digital system. The

Input-Output Devices

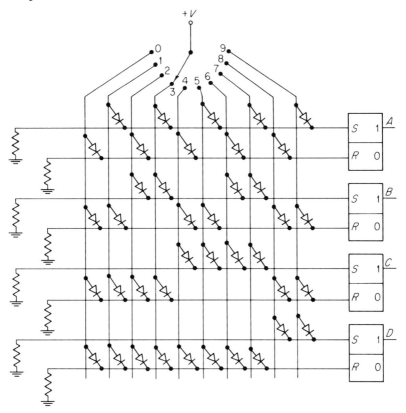

Fig. 10-17 Decimal encoding matrix.

operator can set the switch to any one of 10 positions which represent decimal numbers. The output of the switch is then transformed by a proper encoding matrix which changes the decimal number into an equivalent binary number.

An encoding matrix which will change a decimal number into an equivalent binary number and store it in a register is shown in Fig. 10-17. Setting the switch to a position places a positive voltage on the line connected to that position. Notice that the R and S input to each flip-flop is essentially the output of an OR gate.

For example, if the switch is set to position 1, the diodes connected to that line have a positive voltage on their plates (they are therefore forward-biased). Thus the *set* input to flip-flop A goes high while the *reset* inputs to flip-flops B, C, and D go high. This then sets the binary number 0001 in the flip-flops, where A is the least significant bit. Notice that this encoding matrix requires 40 diodes. As might be expected, it is possible to reduce the

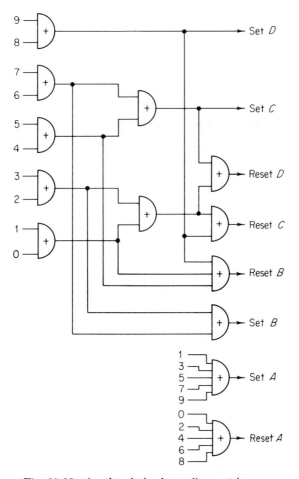

Fig. 10-18 Another decimal encoding matrix.

number of diodes required by combining the input functions as was done with decoding matrices. One method of doing this is shown in Fig. 10-18, and it can be seen that a savings of 7 diodes is effected since this scheme requires only 33 diodes.

SUMMARY

Punched cards provide one of the most useful and widely used media for storing binary information. Each card is considered as a block or unit of information and is therefore referred to as a unit record. Furthermore, punched-card equipment (punches, sorters, readers, etc.) is commonly called "unit-record equipment."

Alphanumeric information, as well as special characters, can be punched into cards by means of a code. The most common code in use is the Hollerith code.

A similar medium for information storage is punched paper tape. Alphanumeric and special characters are recorded by perforating the tape according to a code. There are a number of codes, but the one most commonly used is the eight-hole code. A perforated role of paper tape is a continuous record and is thus distinct from the unit record (punched card).

For handling large quantities of information, magnetic tape is a most convenient recording medium. Magnetic tape offers the advantages of much higher processing rate and much greater recording densities. Moreover, magnetic tape can be erased and used over and over.

The three most common methods for recording on magnetic tape are the return-to-zero (RZ), the non-return-to-zero (NRZ), and the non-return-to-zero-inverted (NRZI). The NRZ and NRZI methods effectively erase or clean the tape automatically during the record operation and thus eliminate one of the problems of RZ recording. These two methods also lend themselves to higher recording rates.

Encoding and decoding matrices form an important part of input-output equipment. These matrices are generally used to change information from one form to another, for example, binary to octal, or binary to decimal, or decimal to binary.

There is a wide variety of digital peripheral equipment including unit-record equipment, printers, cathode-ray-tube displays, plotters, etc. The choice of peripheral equipment to be used with any one system is a major engineering decision. The decision involves establishing the system requirements, studying the available equipment, meeting with the equipment manufacturers, and then making the decision based on operational characteristics, delivery time, and cost.

GLOSSARY

alphanumeric information   Information composed of the letters of the alphabet, the numbers, and special characters.

bit   One binary digit.

character   A number, letter, or symbol represented by a combination of bits.

decoding matrix   A matrix used to alter the format of information taken from the output of a system.

encoding matrix A matrix used to alter the format of information being entered into a system.
Hollerith code The system for representing information by punching holes in a prescribed manner in a punched card.
NRZ Non-return-to-zero recording.
NRZI Non-return-to-zero inverted recording.
parity The method of using an additional punched hole (or magnetic spot for magnetic recording) to ensure that the total number of holes (or spots) for each character is even or odd.
record gap A blank piece of tape between recorded information.
recording density The number of characters recorded per inch of tape.
tape-utilization factor The ratio of the characters actually recorded to the maximum number of characters that could be recorded.
unit record A punched card represents a unit record since each card contains a unit or block of information.

REVIEW QUESTIONS

1. Describe some of the problems of the man-machine interface.
2. Give a description of a typical punched card (size, number of columns, number of rows).
3. Which rows are the zone punches on a punched card?
4. Which rows are the digit punches on a punched card?
5. What is the Hollerith code? What does "JR. is 11" signify?
6. How is binary information represented on a card, i.e., what does a hole represent and what does the absence of a hole represent?
7. What is the meaning of unit record?
8. Name three pieces of unit-record peripheral equipment and give a brief description of how they are used.
9. Describe the eight-hole code used to punch information into paper tape.
10. Describe how 1s and 0s are recorded on magnetic tape by means of a magnetic record head.
11. How is alphanumeric information recorded on magnetic tape?
12. How is binary information recorded on magnetic tape?
13. Explain the dual-parity system used in magnetic-tape recording.
14. What is the purpose of an interrecord gap on magnetic tape?
15. How can the tape-utilization factor be used to determine the total number of characters stored on a magnetic tape?
16. Describe the operation of the RZ recording method. What are some of the difficulties with this system?
17. Describe the operation of the NRZ recording method. What advantages does this method offer over RZ recording?

Input-Output Devices   273

18. Describe the NRZI recording technique.
19. Why is a digital incremental plotter a true digital plotting system?
20. What is the difference between an encoding and a decoding matrix?

PROBLEMS

10-1 Make a sketch of a punched card and code your name, address, and social security number using the Hollerith code. Use a dark spot to represent a hole.

10-2 Change your social security number to the equivalent binary number. Make a sketch of a punched card and record this number on the card in the horizontal binary fashion.

10-3 Repeat Prob. 10-2, but record the number on the card in the vertical fashion.

10-4 Assume that alphanumeric information is being punched into cards at the rate of 250 cards per minute. If the cards have an average of 65 characters each, at what rate in characters per second is the information being processed?

10-5 Make a sketch of a length of paper tape. Using the eight-hole code, record your name, address, and social security number on the tape. Use a dark spot to represent a hole.

10-6 What length of paper tape is required for the storage of 60,000 characters of alphanumeric information using the eight-hole code? Assume no record gaps.

10-7 What length of magnetic tape would be required to store the information in Prob. 10-6 if the recording density is 500 bits per inch? Assume no record gaps.

10-8 Assume that data are recorded on magnetic tape at a density of 200 bits per inch. If the record length is 200 characters, and the interrecord gap is 0.75 in., what is the tape-utilization factor? Using this scheme, how many characters can be stored in 1000 ft of tape?

10-9 Verify the solution to Prob. 10-8 above by using Eq. (10-3). Notice that the 2400 in the equation must be replaced by 1000 since this is the tape length.

10-10 Repeat Probs. 10-8 and 10-9 for a density of 800 bits per inch.

10-11 What length of magnetic tape is required to store 10^5 characters recorded at a density of 800 bits per inch with a record length of 500 characters?

10-12 Can you explain why it is desired to keep the record length as long as possible in magnetic-tape recording?

10-13 Draw the eight three-input AND gates necessary to decode a three-flip-flop counter. Now draw the gates necessary to decode this counter using the tree decoding scheme. How many diodes are saved?

10-14 Draw the gates necessary to change a 3-bit binary number to an octal number.

10-15 How would you change a 4-bit binary number to an octal number? Draw the necessary gates.

10-16 Draw the gates necessary to form an octal-to-binary encoder.

10-17 See if you can simplify the gates in Prob. 10-16 by grouping the eight inputs.

10-18 What are the advantages and disadvantages of a tree and a balanced multiplicative decoder?

CHAPTER 11

D/A AND A/D CONVERSION

Digital-to-analog (D/A) and analog-to-digital (A/D) conversion form two very important aspects of digital data processing. D/A conversion involves translating digital information into equivalent analog information. As an example, the output of a digital system might be changed to analog form for the purpose of driving a pen recorder. Similarly an analog signal might be required for the servomotors which drive the cursor arms of a plotter, as discussed in the preceding chapter. In this respect, a D/A converter is sometimes considered a decoding device since it operates on the output of a digital system.

Quite often the opposite conversion is needed. The process of changing an analog signal to an equivalent digital signal is accomplished by the use of an A/D converter. For example, an A/D converter might be used to change the analog output signals from transducers (measuring temperature, pressure, vibration, etc.) into equivalent digital signals. These signals would then be in a form suitable for entry into a digital system for processing. An A/D converter is often referred to as an encoding device since it is commonly used to encode signals for entry into a digital system.

D/A conversion is a straightforward process and is considerably easier than A/D conversion. In fact, a D/A converter is usually an integral part of any A/D converter. For this reason, we will consider the digital-to-analog conversion process first.

11-1 Variable-resistor Network

The basic problem in converting a digital signal into an equivalent analog signal is to change the n digital voltage levels into one equivalent analog voltage. This can be most easily accomplished by designing a resistive network which will change each of the digital levels into an *equivalent binary weighted voltage* (or current).

As an example of what is meant by equivalent binary weight, consider the truth table for the 3-bit binary signal shown in Fig. 11-1. Assume that it is desired to change the eight possible digital signals in this figure into equivalent analog voltages. The smallest number represented is 000, and let us make this equal to 0 volts. The largest number is 111, and let us make this equal to $+7$ volts. This then establishes the range of the analog signal which will be developed. (There is nothing special about the voltage levels chosen; they were simply selected for convenience.)

Now, notice that between 000 and 111 there are seven discrete levels to be defined. Therefore, it will be convenient to divide the analog signal into seven levels. The smallest incremental change in the digital signal is represented by the least significant bit LSB (2^0). Thus we would like to have this bit cause a change in the analog output which is equal to $\frac{1}{7}$ of the full-scale analog output voltage. The resistive divider will then be designed such that a 1 in the 2^0 position will cause $+7 \times \frac{1}{7} = +1$ volts at the output.

Since $2^1 = 2$ and $2^0 = 1$, it can be clearly seen that the 2^1 bit represents a number which is twice the size of the 2^0 bit. Therefore, a 1 in the 2^1 bit position must cause a change in the analog output voltage which is twice the size of the LSB. The resistive divider must then be constructed such that a 1 in the 2^1 bit position will cause a change of $+7 \times \frac{2}{7} = +2$ volts in the analog output voltage.

2^2	2^1	2^0
0	0	0
0	0	1
0	1	0
0	1	1
1	0	0
1	0	1
1	1	0
1	1	1

Fig. 11-1

D/A and A/D Conversion

Bit	Weight
2^0	1/7
2^1	2/7
2^2	4/7
Sum	7/7

(a)

Bit	Weight
2^0	1/15
2^1	2/15
2^2	4/15
2^3	8/15
Sum	15/15

(b)

Fig. 11-2 Binary equivalent weights.

Similarly, $2^2 = 4 = 2 \times 2^1 = 4 \times 2^0$, and thus the 2^2 bit must cause a change in the output voltage which is 4 times that of the LSB. The 2^2 bit must then cause an output-voltage change of $+7 \times 4/7 = +4$ volts.

The process can be continued, and it will be seen that each successive bit must have a value which is twice the preceding bit. Thus the LSB is given a binary equivalent weight of $1/7$ or 1 part in 7. The next least significant bit is given a weight of $2/7$, which is twice the LSB, or 2 parts in 7. The MSB (in the case of this 3-bit system) is given a weight of $4/7$, which is 4 times the LSB or 4 parts in 7. Notice that the sum of the weights must equal 1. Thus $1/7 + 2/7 + 4/7 = 7/7 = 1$. In general, the binary equivalent weight assigned to the LSB is $1/(2^n - 1)$, where n is the number of bits. The remaining weights are found by multiplying by 2, 4, 8, ..., etc.

EXAMPLE 11-1

Find the binary equivalent weight of each bit in a 4-bit system.

SOLUTION

The LSB has a weight of $1/(2^4 - 1) = 1/(16 - 1) = 1/15$, or 1 part in 15. The second LSB has a weight of $2 \times 1/15 = 2/15$. The third LSB has a weight of $4 \times 1/15 = 4/15$, and the MSB has a weight of $8 \times 1/15 = 8/15$. As a check, the sum of the weights must equal 1. Thus, $1/15 + 2/15 + 4/15 + 8/15 = 15/15 = 1$. The binary equivalent weights for a 3-bit and a 4-bit system are summarized in Fig. 11-2.

What is now desired is a resistive divider having three digital inputs and one analog output as shown in Fig. 11-3a. Assume that the digital input levels are $0 = 0$ volts and $1 = +7$ volts. Now for an input of 001, the output will be $+1$ volt. Similarly, an input of 010 will provide an output of $+2$ volts and an input of 100 will provide an output of $+4$ volts. The digital input 011 is seen to be a combination of the signals 001 and 010. If the $+1$ volt from the 2^0 bit is added to the $+2$ volts from the 2^1 bit, the desired $+3$ volt output for the 011 input is achieved. The other desired

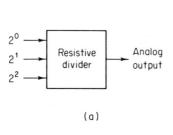
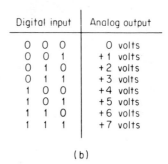

Fig. 11-3

voltage levels are shown in Fig. 11-3b, and it can be seen that they too are additive combinations of voltages.

Thus the resistive divider must do two things in order to change the digital input into an equivalent analog output voltage:

1. The 2^0 bit must be changed to $+1$ volt; the 2^1 bit must be changed to $+2$ volts; the 2^2 bit must be changed to $+4$ volts.
2. These three voltages representing the digital bits must be summed together to form the analog output voltage.

A resistive divider which will perform the above functions is shown in Fig. 11-4. The resistors R_0, R_1, and R_2 form the divider network. R_L represents the load to which the divider is connected and is considered to be large enough that it does not load the divider network.

Assume that the digital input signal 001 is applied to this network. Recalling that $0 = 0$ volts and $1 = +7$ volts, the equivalent circuit shown in Fig. 11-5 can be drawn. R_L is considered large and is neglected. The analog output voltage V_A can be most easily found by use of Millman's theorem. Millman's theorem states that the voltage appearing at any node

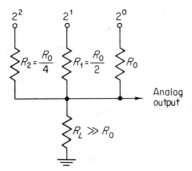

Fig. 11-4 Resistive ladder.

D/A and A/D Conversion

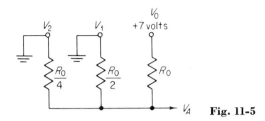

Fig. 11-5

in a resistive network is equal to the summation of the currents entering the node (found by assuming the node voltage is zero) divided by the summation of the conductances connected to the node. In equation form, Millman's theorem is

$$V = \frac{E_1/R_1 + E_2/R_2 + E_3/R_3 + \cdots}{1/R_1 + 1/R_2 + 1/R_3 + \cdots}$$

Applying Millman's theorem to Fig. 11-5 yields

$$V_A = \frac{V_0/R_0 + V_1/(R_0/2) + V_2/(R_0/4)}{1/R_0 + 1/(R_0/2) + 1/(R_0/4)}$$

$$= \frac{7/R_0}{1/R_0 + 2/R_0 + 4/R_0} = \frac{7}{7} = +1 \text{ volt}$$

Drawing the equivalent circuits for the other seven input combinations and applying Millman's theorem will lead to the table of voltages shown in Fig. 11-3 (see Prob. 11-3).

EXAMPLE 11-2

For a four-input resistive divider (0 = 0 volts, 1 = +10 volts) find the following:
1. The full-scale output voltage
2. The output-voltage change due to the least significant bit
3. The analog output voltage for a digital input of 1011

SOLUTION

1. The maximum output voltage occurs when all the inputs are at +10 volts. If all four inputs are at +10 volts, the output must also be at +10 volts (ignoring the effects of R_L).

2. For a 4-bit digital number, there are 16 possible states. There are 15 steps between these 16 states, and the LSB must be equal to one-fifteenth of the full-scale output voltage. Therefore, the change in output voltage due to the LSB is $+10 \times \frac{1}{15} = +\frac{2}{3}$ volt.

3. Using Millman's theorem, the output voltage for a digital input of 1011 is

$$V_A = \frac{10/R_0 + 10/(R_0/2) + 0/(R_0/4) + 10/(R_0/8)}{1/R_0 + 1/(R_0/2) + 1/(R_0/4) + 1/(R_0/8)}$$

$$= \frac{110}{15} = 22/3 = +7\frac{1}{3} \text{ volts}$$

To summarize, a resistive divider can be built to change a digital voltage into an equivalent analog voltage. The following criteria can be applied to this divider:

1. There must be one input resistor for each digital bit.
2. Beginning with the LSB, each following resistor value is one-half the size of the previous resistor.
3. The full-scale output voltage is equal to the + voltage of the digital input signal (the divider would work equally well with input voltages of 0 and $-V$ volts).
4. The LSB has a weight of $1/(2^n - 1)$, where n is the number of input bits.
5. The change in output voltage due to a change in the LSB is equal to $V/(2^n - 1)$, where V is the digital input-voltage level.
6. The output voltage V_A can be found for any digital input signal using the following modified form of Millman's theorem:

$$V_A = \frac{V_0 2^0 + V_1 2^1 + V_2 2^2 + V_3 2^3 + \cdots + V_{n-1} 2^{n-1}}{2^n - 1} \quad (11\text{-}1)$$

where $V_0, V_1, V_2, V_3, \ldots, V_n$ are the digital input voltage levels (0 or V volts) and n is the number of input bits.

EXAMPLE 11-3

For a 5-bit resistive divider, determine the following:
1. The weight assigned to the LSB.
2. The weight assigned to the second and third LSB.
3. The change in output voltage due to a change in the LSB, the second LSB, and the third LSB.
4. The output voltage for a digital input of 10101. Assume $0 = 0$ volts and $1 = +10$ volts.

SOLUTION

1. The LSB weight is $1/(2^5 - 1) = \frac{1}{31}$.
2. The second LSB weight is $\frac{2}{31}$ and the third LSB weight is $\frac{4}{31}$.
3. The LSB causes a change in the output voltage of $\frac{10}{31}$ volts. The second LSB causes an output-voltage change of $\frac{20}{31}$ volts and the third LSB causes an output-voltage change of $\frac{40}{31}$ volts.

D/A and A/D Conversion

4. The output voltage for a digital input of 10101 is

$$V_A = \frac{10 \times 2^0 + 0 \times 2^1 + 10 \times 2^2 + 0 \times 2^3 + 10 \times 2^4}{2^5 - 1}$$

$$= \frac{10(1 + 4 + 16)}{32 - 1} = \frac{210}{31} = +6.77 \text{ volts}$$

This resistive divider has two serious drawbacks. The first is the fact that each resistor in the network is a different value. Since these dividers are usually constructed using precision resistors, the added expense in obtaining and handling many different sizes of resistors becomes unattractive. Moreover, the resistor used for the MSB will be required to handle a much greater current than the LSB resistor. For example, in a 10-bit system, the current through the MSB resistor will be approximately 500 times as large as the current through the LSB resistor (see Prob. 11-5). For these reasons, a second type of resistive network called a "ladder" has been developed.

11-2 Binary Ladder

The binary ladder is a resistive network whose output voltage is a properly weighted sum of the digital inputs. Such a ladder, designed for 4 bits, is shown in Fig. 11-6. It is constructed of resistors having only two values and thus overcomes one of the objections to the resistive divider previously discussed. The left end of the ladder is terminated in a resistance of $2R$, and we will assume for the moment that the right end of the ladder (the output) is open-circuited.

Let us now examine the resistive properties of the network, assuming that all the digital inputs are at ground. Beginning at node A, the total resistance looking into the terminating resistor is $2R$. The total resistance looking out toward the 2^0 input is also $2R$. These two resistors can be combined to form an equivalent resistor of value R as shown in Fig. 11-7a.

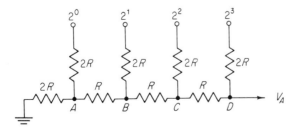

Fig. 11-6 Binary ladder.

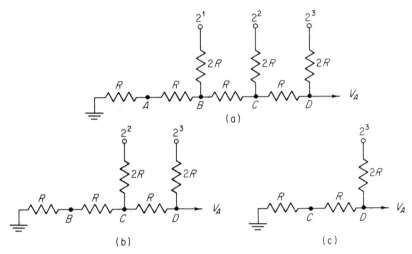

Fig. 11-7

Now, moving to node B, the total resistance looking into the branch toward node A is seen to be $2R$, as is the total resistance looking out toward the 2^1 input. These three resistors can be combined to simplify the network as shown in Fig. 11-7b.

From this figure, it can be seen that the total resistance looking from node C down the branch toward node B or out the branch toward the 2^2 input is still $2R$. The circuit shown in Fig. 11-7b can then be reduced to the equivalent shown in Fig. 11-7c.

From this equivalent circuit, it is clear that the resistance looking back toward node C is $2R$ as is the resistance looking out toward the 2^3 input.

From the above, we can conclude that the total resistance looking from any node back toward the terminating resistor or out toward the digital input is $2R$. Notice that this is true regardless of whether the digital

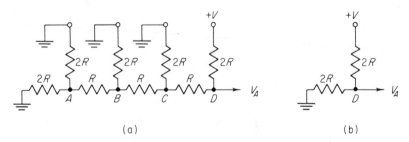

Fig. 11-8 (a) Binary ladder for a digital input of 1000. (b) Equivalent circuit for digital input of 1000.

D/A and A/D Conversion

inputs are at ground or $+V$ volts. The justification for this statement is the fact that the internal impedance of an ideal voltage source is zero ohms, and we are assuming that the digital inputs are ideal voltage sources.

We can use the resistance characteristics of the ladder to determine the output voltages for the various digital inputs. First, assume that the digital input signal is 1000. With this input signal, the binary ladder can be drawn as shown in Fig. 11-8a. Since there are no voltage sources to the left of node D, the entire network to the left of this node can be replaced by a resistance of $2R$ to form the equivalent circuit shown in Fig. 11-8b. From this equivalent circuit, it can be easily seen that the output voltage is

$$V_A = +V\frac{2R}{2R + 2R} = \frac{+V}{2}$$

Thus a 1 in the MSB position will provide an output voltage of $+V/2$.

To determine the output voltage due to the second most significant bit, assume a digital input signal of 0100. This can be represented by the circuit shown in Fig. 11-9a. Since there are no voltage sources to the left of node C, the entire network to the left of this node can be replaced by a resistance of $2R$, as shown in Fig. 11-9b. Let us now replace the network to the left of node C with its Thévenin equivalent by cutting the circuit on the jagged line shown in Fig. 11-9b. The Thévenin equivalent is clearly a resistance R in series with a voltage source $+V/2$. The final equivalent

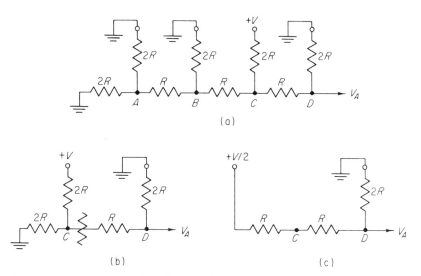

Fig. 11-9 (a) Binary ladder for digital input of 0100. (b) Partially reduced equivalent circuit. (c) Final equivalent circuit using Thévenin's theorem.

circuit with the Thévenin included is shown in Fig. 11-9c. From this circuit, the output voltage is clearly

$$V_A = \frac{+V}{2} \frac{2R}{R + R + 2R} = \frac{+V}{4}$$

Thus the second MSB provides an output voltage of $+V/4$.

This process can be continued, and it can be shown that the third MSB will provide an output voltage of $+V/8$, the fourth MSB will provide an output voltage of $+V/16$, and so on. The output voltages for the binary ladder are summarized in Fig. 11-10, and it can be seen that each digital input is transformed into a properly weighted binary output voltage.

EXAMPLE 11-4

What are the output voltages caused by each bit in a 5-bit ladder if the input levels are $0 = 0$ volts and $1 = +10$ volts?

SOLUTION

The output voltages can be easily calculated using Fig. 11-10. They are

$$\text{MSB} \quad V_A = \frac{V}{2} = \frac{+10}{2} = +5 \text{ volts}$$

$$\text{Second MSB} \quad V_A = \frac{V}{4} = \frac{+10}{4} = +2.5 \text{ volts}$$

$$\text{Third MSB} \quad V_A = \frac{V}{8} = \frac{+10}{8} = +1.25 \text{ volts}$$

$$\text{Fourth MSB} \quad V_A = \frac{V}{16} = \frac{+10}{16} = +0.625 \text{ volt}$$

$$\text{LSB} = \text{Fifth MSB} \quad V_A = \frac{V}{32} = \frac{+10}{32} = +0.3125 \text{ volt}$$

Bit position	Binary weight	Output voltage
MSB	1/2	V/2
2nd MSB	1/4	V/4
3rd MSB	1/8	V/8
4th MSB	1/16	V/16
5th MSB	1/32	V/32
6th MSB	1/64	V/64
7th MSB	1/128	V/128
⋮	⋮	⋮
Nth MSB	$1/2^N$	$V/2^N$

Fig. 11-10 Binary ladder output voltages.

D/A and A/D Conversion

Since this ladder is composed of linear resistors, it is a linear network and the principle of superposition can be used. This means that the total output voltage due to a combination of input digital levels can be found by simply taking the sum of the output levels caused by each digital input individually.

In equation form, the output voltage is given by

$$V_A = \frac{V}{2} + \frac{V}{4} + \frac{V}{8} + \frac{V}{16} + \cdots + \frac{V}{2^n} \qquad (11\text{-}2)$$

where n is the total number of bits at the input.

This equation can be simplified somewhat by factoring and collecting terms. The output voltage can then be given in the form

$$V_A = \frac{V_0 2^0 + V_1 2^1 + V_2 2^2 + V_3 2^3 + \cdots + V_{n-1} 2^{n-1}}{2^n} \qquad (11\text{-}3)$$

where $V_0, V_1, V_2, V_3, \ldots, V_{n-1}$ are the digital input-voltage levels. Equation (11-3) can be used to find the output voltage from the ladder for any digital input signal.

Example 11-5

Find the output voltage from a 5-bit ladder which has a digital input of 11010. Assume $0 = 0$ volts and $1 = +10$ volts.

Solution

Using Eq. (11-3),

$$V_A = \frac{0 \times 2^0 + 10 \times 2^1 + 0 \times 2^2 + 10 \times 2^3 + 10 \times 2^4}{2^5}$$

$$= \frac{10(2 + 8 + 16)}{32} = \frac{10 \times 26}{32} = +8.125 \text{ volts}$$

This solution can be checked by adding the individual bit contributions calculated in Example 11-4 above.

Notice that Eq. (11-3) is very similar to Eq. (11-1), which was developed for the resistive divider. They are in fact identical with the exception of the denominators. This is a subtle but very important difference. Recall that the full-scale voltage for the resistive divider is equal to the voltage level of the digital input 1. On the other hand, examination of Eq. (11-2) shows that the full-scale voltage for the ladder is given by

$$V_A = V(\tfrac{1}{2} + \tfrac{1}{4} + \tfrac{1}{8} + \tfrac{1}{16} + \cdots + \tfrac{1}{2^n})$$

The terms inside the brackets form a geometric series whose sum approaches 1, given enough terms. However, it never quite reaches 1. Therefore, the full-scale output voltage of the ladder approaches V in the limit, but never quite reaches it.

EXAMPLE 11-6

What is the full-scale output voltage of the 5-bit ladder in Example 11-4 above?

SOLUTION

The full-scale output voltage is simply the sum of the individual bit voltages. Thus

$$V_A = 5 + 2.5 + 1.25 + 0.625 + 0.3125 = +9.6875 \text{ volts}$$

To keep the ladder in perfect balance and to maintain symmetry, the output of the ladder should be terminated in a resistance of $2R$. This will result in a lowering of the output voltage, but if the $2R$ load is maintained constant, the output voltages will still be a properly weighted sum of the binary input bits. If the load is varied, the output voltage will not be a properly weighted sum and care must be exercised to ensure that the load resistance is constant.

Terminating the output of the ladder with a load of $2R$ will also ensure that the input resistance to the ladder seen by each of the digital voltage sources is constant. With the ladder balanced in this manner, the resistance looking into any branch from any node will have a value of $2R$. Thus the input resistance seen by any input digital source will be $3R$. This is a definite advantage over the resistive divider since the digital voltage sources can now all be designed for the same load.

EXAMPLE 11-7

Suppose that the value of R for the 5-bit divider described in Example 11-4 above is 1000 ohms. Determine the current that each input digital voltage source must be capable of supplying. Also determine the full-scale output voltage if the ladder is terminated with a load resistance of 2000 ohms.

SOLUTION

The input resistance into the ladder seen by each of the digital sources is $3R = 3000$ ohms. Thus, for a voltage level of $+10$ volts, each source must be capable of supplying $I = 10/(3 \times 10^3) = 3\frac{1}{3}$ ma (without the $2R$ load resistor, the resistance looking into the MSB terminal is actually $4R$). The no-load output voltage of the ladder has already been determined in Example 11-6. This open-circuit output voltage along with the open-circuit output resistance can be used to form a Thévenin equivalent circuit for the output of the ladder. The resistance looking back

D/A and A/D Conversion

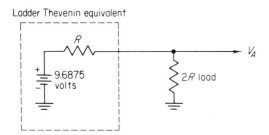

Fig. 11-11 Example 11-7.

into the ladder is clearly $2R = 2000$ ohms. Thus the Thévenin equivalent is as shown in Fig. 11-11. From this figure, the output voltage is seen to be

$$V_A = +9.6875 \frac{2R}{2R + R} = +6.4583 \text{ volts}$$

11-3 D/A Converter

Either the resistive divider or the ladder can be used as the basis for a digital-to-analog converter. It is in the resistive network that the actual translation from a digital signal to an analog voltage takes place. There is, however, the need for additional circuitry to complete the design of the D/A converter.

As an integral part of the D/A converter there must be a register which can be used to store the digital information. This register could be any one of the many types discussed in the previous chapter. The simplest register is formed using RS flip-flops, with one flip-flop per bit. There must also be level amplifiers between the register and the resistive network to ensure that the digital signals presented to the network are all of the same level and are constant. Finally, there must be some form of gating on the input of the register such that the flip-flops can be set with the proper information from the digital system. A complete D/A converter in block-diagram form is shown in Fig. 11-12.

Let us expand on the block diagram shown in this figure by drawing the complete schematic for a 4-bit D/A converter as shown in Fig. 11-13. You will recognize that the resistor network used is of the ladder type.

The level amplifiers each have two inputs: one input is the $+10$ volts from the precision voltage source, and the other is from a flip-flop. The amplifiers work in such a way that when the input from a flip-flop is high, the output of the amplifier is at $+10$ volts. When the input from the flip-flop is low, the output is 0 volts.

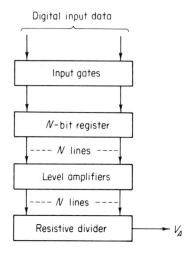

Fig. 11-12   D/A converter.

The four flip-flops form the register necessary for storing the digital information. The flip-flop on the right represents the MSB and the flip-flop on the left represents the LSB. Each flip-flop is of the simple RS type and requires a positive level at the R or S inputs to reset or set it. The gating

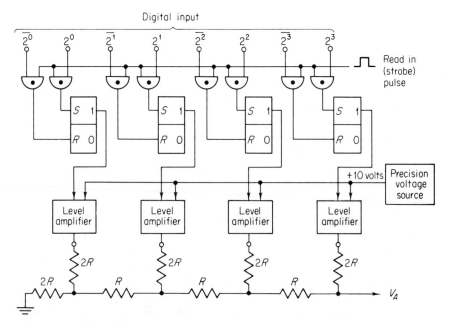

Fig. 11-13   4-bit D/A converter.

D/A and A/D Conversion

scheme for entering information into the register was discussed in the preceding chapter and should be familiar. You will recall that with this gating scheme the flip-flops need not be reset (or set) each time new information is entered. When the *read in* line goes high, one of the two gates connected to each flip-flop is true and the flip-flop is set or reset accordingly. Thus data are entered into the register each time the *read in* (strobe) pulse occurs.

Quite often there is the necessity for decoding more than one signal, for example, the X and Y coordinates for a plotting board. In this event, there are two ways in which to decode the signals.

The first and most obvious method is simply to use one D/A converter for each signal. This method, shown in Fig. 11-14a, has the advantage that each signal to be decoded is held in its register and the analog output voltage is then held fixed. The digital input lines are connected in parallel to each converter. The proper converter is then selected for decoding by the select lines.

The second method involves using only one D/A converter and switching its output. This is called "multiplexing," and such a system is shown in

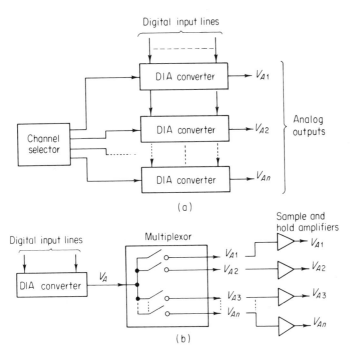

Fig. 11-14 Decoding a number of signals. (*a*) Channel-selection method. (*b*) Multiplex method.

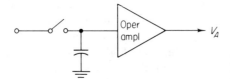

Fig. 11-15 Sample-and-hold amplifier.

Fig. 11-14b. The disadvantage here is that the analog output signal must be held between sampling periods, and the outputs must therefore be equipped with sample-and-hold amplifiers.

A sample-and-hold amplifier can be approximated by a capacitor and high-gain amplifier (operational amplifier) as shown in Fig. 11-15. When the switch is closed, the capacitor charges to the D/A converter output voltage. When the switch is opened, the capacitor holds the voltage level until the next sampling time. The operational amplifier provides a large input impedance so as not to discharge the capacitor appreciably and at the same time offers gain to drive external circuits.

When the D/A converter is used in conjunction with a multiplexor, the maximum rate at which the converter can operate must be considered. Each time data are shifted into the register, transients will appear at the output of the converter. This is due mainly to the fact that each flip-flop has different rise and fall times. Thus a settling time must be allowed between the time data are shifted into the register and the time when the analog voltage is read out. This settling time is the main factor in determining the maximum rate of multiplexing the output. The worst case is when all bits change (e.g., from 1000 to 0111).

Naturally, the capacitors on the sample-and-hold amplifiers are not capable of holding a voltage indefinitely, and the sampling rate must therefore be great enough to ensure that these voltages do not decay appreciably between samples. The sampling rate is a function of the capacitors as well as the frequency of the analog signal which is expected at the output of the converter.

Two simple but important tests which can be performed to check for proper operation of the D/A converter are the steady-state accuracy test and the monotonicity test.

The steady-state accuracy test involves setting a known digital number in the input register, measuring the analog output with an accurate meter, and comparing with the theoretical value.

Checking for monotonicity means checking that the output voltage increases regularly as the input digital signal increases. This can be easily accomplished by using a counter as the digital input signal and observing the analog output on an oscilloscope. For proper monotonicity, the output waveform should be a perfect staircase waveform as shown in Fig. 11-16.

D/A and A/D Conversion

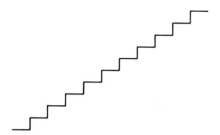

Fig. 11-16 Correct output voltage waveform for monotonicity test.

The steps on the staircase waveform must be equally spaced and of the exact same amplitude. Missing steps, steps of different amplitude, or steps in a downward fashion indicate malfunctions.

The monotonicity test does not check the system for accuracy, but if the system passes the test, it is relatively certain that the converter error is less than ± 1 LSB. Converter accuracy and resolution are the subjects of the next section.

Example 11-8

Suppose that in the course of checking out the 4-bit converter shown in Fig. 11-13 the waveform shown in Fig. 11-17 is observed during the monotonicity check. What is the probable malfunction in the converter?

Solution

There is obviously some malfunction since the actual output waveform is not continuously increasing as it should be. Directly below the waveform the actual digital inputs are shown. Notice that the converter

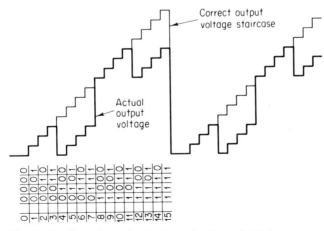

Fig. 11-17 Irregular output voltage for Example 11-8.

functions correctly up to count 3. At count 4, however, the output should be 4 units in amplitude. Instead it drops to 0. It remains 4 units below the correct level until it reaches count 8. Then from count 8 to count 11 the output level is correct. But again at count 12 the output falls 4 units below the correct level, and remains there for the next four levels. If you examine the waveform carefully, you will note that the output is 4 units below normal during the time when the 2^2 bit is supposed to be high. This then suggests that the 2^2 bit is being dropped (i.e., the 2^2 input to the ladder is not being held high). This means that the 2^2 level amplifier is malfunctioning, or the 2^2 flip-flop is not being set, or the 2^2 AND gate is not operating properly. In any case, the monotonicity check has clearly shown that the second MSB is not being used and that the converter is not operating properly.

11-4 D/A Accuracy and Resolution

Two very important aspects of the D/A converter are the resolution and the accuracy of the conversion. There is a definite distinction between the two and the differences should be clearly understood.

The accuracy of the D/A converter is primarily a function of the accuracy of the precision resistors used in the ladder and the precision of the reference voltage supply used. Accuracy is a measure of how close the actual output voltage is to the theoretical output voltage.

For example, suppose that the theoretical output voltage for a particular input is supposed to be $+10$ volts. An accuracy of ± 10 percent means that the actual output voltage must be somewhere between $+9$ and $+11$ volts. Similarly, if the actual output voltage were somewhere between $+9.9$ and $+10.1$ volts, this would imply an accuracy of ± 1 percent.

Resolution, on the other hand, defines the smallest increments in voltage that can be discerned. Resolution is primarily a function of the number of bits in the digital input signal. That is to say, the smallest increment in output voltage is determined by the LSB.

In a 4-bit system using a ladder, for example, the LSB has a weight of $\frac{1}{16}$. This means that the smallest increment in output voltage is $\frac{1}{16}$ of the input voltage. To make the arithmetic easy, let us assume that this 4-bit system has input voltage levels of $+16$ volts. Since the LSB has a weight of $\frac{1}{16}$, a change in the LSB will result in a change of 1 volt in the output. Thus the output voltage will change in steps (or increments) of 1 volt. The output voltage of this converter is then the staircase shown in Fig. 11-16 and ranges from 0 to $+15$ volts in 1-volt steps. This converter can be used to represent analog voltages from 0 to $+15$ volts but it cannot resolve voltages into increments smaller than 1 volt. Thus if we desired to

D/A and A/D Conversion

produce +4.2 volts using this converter, the actual output voltage would be +4.0 volts. Similarly, if we desired a voltage of +7.8 volts the actual output voltage would be +8.0 volts. It is clear then that this converter is not capable of distinguishing voltages finer than 1 volt — which is the resolution of the converter.

If we desired to represent voltages to a finer resolution, we would have to use a converter with more input bits. As an example, the LSB of a 10-bit converter has a weight of $\frac{1}{1024}$. Thus the smallest incremental change in the output of this converter is approximately $\frac{1}{1000}$ of the full-scale voltage. If this converter has a +10-volt full-scale output, the resolution would be approximately $+10 \times \frac{1}{1000} = 10$ mv. This converter is then capable of representing voltages to within ± 10 mv.

EXAMPLE 11-9

What is the resolution of a 9-bit D/A converter which uses a ladder network? What is this resolution expressed as a percent? If the full-scale output voltage of this converter is +5 volts, what is the resolution in volts?

SOLUTION

The LSB in a 9-bit system has a weight of $\frac{1}{512}$. Thus this converter has a resolution of 1 part in 512. The resolution expressed as a percentage is $\frac{1}{512} \times 100$ percent $\cong 0.2$ percent. The voltage resolution is obtained by multiplying the weight of the LSB by the full-scale output voltage. Thus the resolution in volts is $\frac{1}{512} \times 5 \cong 10$ mv.

EXAMPLE 11-10

How many bits are required at the input of a converter if it is required to resolve voltages to 5 mv and the ladder has +10 volts full scale?

SOLUTION

The LSB of an 11-bit system has a resolution of $\frac{1}{2048}$. This would provide a resolution at the output of $\frac{1}{2048} \times (+10) \cong +5$ mv.

It is an important point to realize that resolution and accuracy in a system should be compatible. For example, in the 4-bit system previously discussed the resolution was found to be 1 volt. Clearly it would be unjustified to construct such a system to an accuracy of 0.1 percent. This would mean that the system would be accurate to ± 16 mv but would be capable of distinguishing only to the nearest volt.

Similarly, it would be wasteful to construct the 11-bit system described in Example 11-10 to an accuracy of only ± 1 percent. This would mean that the output voltage would be accurate only to ± 100 mv whereas it is capable of distinguishing to the nearest 5 mv.

11-5 A/D Converter — Simultaneous Conversion

The process of converting an analog voltage into an equivalent digital signal is known as analog-to-digital conversion. This operation is somewhat more complicated than the converse operation of D/A conversion. As such, a number of different methods have been developed, the simplest of which is probably the simultaneous method.

The simultaneous method of A/D conversion is based on the use of a number of comparator circuits. One such system using three comparator circuits is shown in Fig. 11-18. The analog signal to be digitized serves as one of the inputs to each comparator. The second input is a standard reference voltage. The reference voltages used are $+V/4, +V/2$, and $+3V/4$. The system is then capable of accepting an analog input voltage between 0 and $+V$ volts.

If the analog input signal exceeds the reference voltage to any comparator, that comparator turns on (let us assume that this means that the output of the comparator goes high). Now, if all the comparators are off, the analog input signal must be between 0 and $+V/4$ volts. If C_1 is high (comparator C_1 is on), and C_2 and C_3 are low, then the input must be between $+V/4$ and $+V/2$ volts. If C_1 and C_2 are high while C_3 is low, the input must be between $+V/2$ and $+3V/4$. Finally, if all comparator outputs are high, the input signal must be between $+3V/4$ and $+V$ volts. The comparator output levels for the various ranges of input voltages are summarized in Fig. 11-18b.

Examination of Fig. 11-18 shows that there are four voltage ranges that can be detected by this converter. Four ranges can be effectively discerned

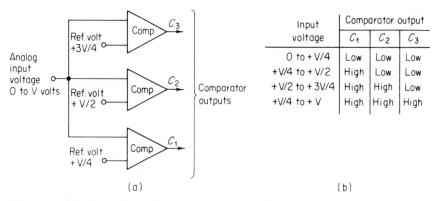

Fig. 11-18 Simultaneous A/D conversion. (a) Logic circuit. (b) Comparator outputs for input voltage ranges.

D/A and A/D Conversion 295

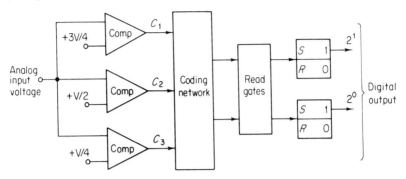

Fig. 11-19 2-bit simultaneous A/D converter.

by 2 digital bits. The three comparator outputs can then be fed into a coding network to provide 2 bits which will be equivalent to the input analog voltage. The bits out of the coding network can then be entered into a flip-flop register for storage. The complete block diagram for such an A/D converter is shown in Fig. 11-19.

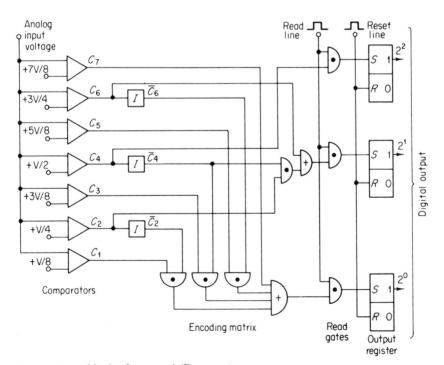

Fig. 11-20 3-bit simultaneous A/D converter.

In order to gain a clear understanding of the operation of the simultaneous A/D converter let us investigate the 3-bit converter shown in Fig. 11-20. Notice that in order to convert the input signal to a digital signal having 3 bits, it is necessary to have seven comparators (this allows a division of the input into eight ranges). Recall that for the 2-bit converter it required three comparators to define four ranges. In general it can be said that it will require $2^n - 1$ comparators to convert to a digital signal having n bits. Some of the comparators have inverters at their outputs since both C and $\overline{C}$ are needed for the encoding matrix.

The encoding matrix must accept seven inputs levels and encode them into a 3-bit binary number (having eight possible states). Operation of the encoding matrix can be most easily understood by examination of the table of outputs shown in Fig. 11-21.

The 2^2 bit is easiest to determine since it must be high (the 2^2 flip-flop must be set) whenever C_4 is high.

The 2^1 line must be high whenever C_2 is high and $\overline{C}_4$ is high, or whenever C_6 is high. In equation form, we can write $2^1 = C_2\overline{C}_4 + C_6$.

The logic equation for the 2^0 bit can be found in a similar manner, and it is $2^0 = C_1\overline{C}_2 + C_3\overline{C}_4 + C_5\overline{C}_6 + C_7$.

The transfer of data from the encoding matrix into the register must be carried out in two steps. First, a positive reset pulse must appear on the *reset* line to reset all the flip-flops low. Then, a positive *read* pulse will allow the proper *read* gates to go high and thus transfer the digital information into the flip-flops.

The construction of a simultaneous A/D converter is quite straightforward and relatively easy to understand. However, as the number of bits in the desired digital number increases, the number of comparators increases very rapidly ($2^n - 1$) and the problem soon gets out of hand. Even though this method is simple and is indeed capable of extremely fast conversion rates, there are preferable methods for digitizing numbers having more than 3 or 4 bits.

Input voltage	Comparator for level							Binary output		
	C_1	C_2	C_3	C_4	C_5	C_6	C_7	2^2	2^1	2^0
0 to V/8	Low	Low	Low	Low	Low	Low	Low	0	0	0
V/8 to V/4	High	Low	Low	Low	Low	Low	Low	0	0	1
V/4 to 3V/8	High	High	Low	Low	Low	Low	Low	0	1	0
3V/8 to V/2	High	High	High	Low	Low	Low	Low	0	1	1
V/2 to 5V/8	High	High	High	High	Low	Low	Low	1	0	0
5V/8 to 3V/4	High	High	High	High	High	Low	Low	1	0	1
3V/4 to 7V/8	High	High	High	High	High	High	Low	1	1	0
7V/8 to V	High	High	High	High	High	High	High	1	1	1

Fig. 11-21 Logic table for the simultaneous converter in Fig. 11-20.

11-6 A/D Converter — Counter Method

A higher-resolution A/D converter using only one comparator could be constructed if a variable reference voltage were available. This reference voltage could then be applied to the comparator, and when it became equal to the input analog voltage the conversion would be complete.

To construct such a converter, suppose we begin with a simple binary counter. The digital output signals will be taken from this counter, and therefore we will want it to be an n-bit counter, where n is the desired number of bits. Now let us connect the output of this counter to a standard binary ladder to form a simple D/A converter. If a clock is now applied to the input of the counter, the output of the binary ladder will be the familiar staircase waveform shown in Fig. 11-16. This waveform is just exactly the reference voltage signal we would like to have for the comparator! With a minimum of gating and control circuitry, this simple D/A converter can be changed into the desired A/D converter.

Figure 11-22 shows the block diagram for a counter-type A/D converter. The operation of the counter is as follows: First, the counter is reset to all 0s. Then when a convert signal appears on the *start* line, the gate opens and clock pulses are allowed to pass through to the input of the counter. The counter advances through its normal binary count sequence and the staircase waveform is generated at the output of the ladder. This waveform is applied to one side of the comparator and the analog input voltage is applied to the other side. When the reference voltage equals (or exceeds) the input analog voltage, the gate is closed, the counter stops, and the conversion is complete. The number stored in the counter is now the digital equivalent of the analog input voltage.

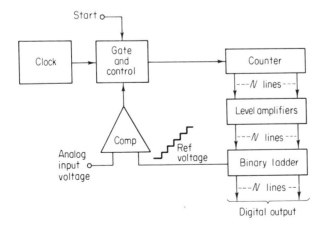

Fig. 11-22 Counter-type A/D converter.

Notice that this converter is composed of a D/A converter (the counter, level amplifiers, and binary ladder), one comparator, a clock, and the gate and control circuitry. This can really be considered as a closed-loop control system. An error signal is generated at the output of the comparator by taking the difference between the analog input signal and the feedback signal (staircase reference voltage). The error is detected by the control circuit and the clock is allowed to advance the counter. The counter advances in such a way as to reduce the error signal by increasing the feedback voltage. When the error is reduced to zero, the feedback voltage is equal to the analog input signal, the control circuitry stops the clock from advancing the counter, and the system comes to rest.

The counter-type A/D converter provides a very good method for digitizing to a high resolution. This method is much simpler than the simultaneous method for high resolution, but the conversion time required is longer. Since the counter always begins at zero and counts through its normal binary sequence, it may require as many as 2^n counts before conversion is complete. The average conversion time is, of course, $2^n/2$ or 2^{n-1} counts.

The counter advances one count for each cycle of the clock, and the clock therefore determines the conversion rate. Suppose, for example, we have a 10-bit converter. It will then require 1024 clock cycles for a full-scale count. If we are using a 1-MHz clock, the counter will advance 1 count every microsecond. Thus to count full scale will require $1024 \times 10^{-6} \cong 1.024$ msec. The converter will reach one-half full scale in half this time, or in 0.512 msec. The time required to reach one-half full scale can be considered the average conversion time for a large number of conversions.

Example 11-11

Suppose the converter shown in Fig. 11-22 is an 8-bit converter driven by a 500-kHz clock. Find:
(a) The maximum conversion time
(b) The average conversion time
(c) The maximum conversion rate

Solution

(a) An 8-bit converter has a maximum of $2^8 = 256$ counts. With a 500-kHz clock, the counter will advance at the rate of 1 count each 2 μsec. To advance 256 counts will require $256 \times 2 \times 10^{-6} = 512 \times 10^{-6} = 0.512$ msec.

(b) The average conversion time is one-half the maximum conversion time. Thus it is $\frac{1}{2} \times 0.512 \times 10^{-3} = 0.256$ msec.

(c) The maximum conversion rate is determined by the longest conversion time. Since the converter has a maximum conversion time of 0.512 msec, it will be capable of making at least $1/(0.512 \times 10^{-3}) \cong 1953$ conversions per second.

Figure 11-23 shows one method of implementing the necessary control circuitry for the converter shown in Fig. 11-22. The waveforms for one conversion are also shown. A conversion is initiated by the receipt of a *start* signal. The positive edge of the *start* pulse is used to reset all the

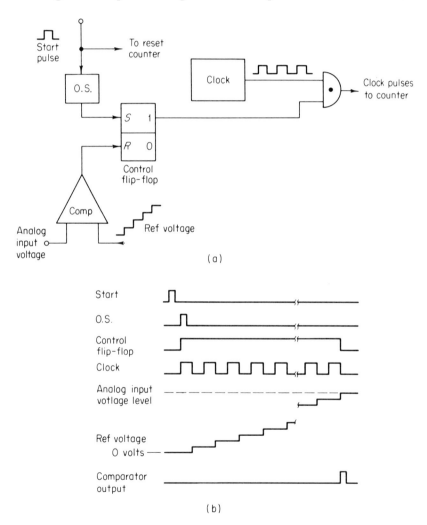

Fig. 11-23 Operation of the A/D converter in Fig. 11-22. (a) Control logic. (b) Waveforms.

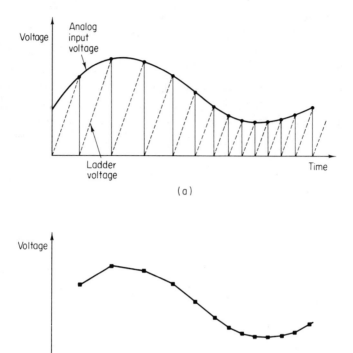

Fig. 11-24 (a) Digitizing an output voltage. (b) Reconstructed signal from digital data.

flip-flops in the counter and also to trigger the one-shot. The output of the one-shot sets the *control* flip-flop, which makes the AND gate true and allows clock pulses to advance the counter.

The delay between the *reset* pulse to the flip-flops and the beginning of the clock pulses (ensured by the one-shot) is to ensure that all flip-flops are indeed reset before counting begins. This is a definite attempt to avoid any racing problems.

With the *control* flip-flop set, the counter advances through its normal count sequence until the staircase voltage from the ladder is equal to the analog input voltage. At this time, the comparator output changes state, generating a positive pulse, which resets the *control* flip-flop. Thus the AND gate is closed and counting ceases. The counter now holds a digital number which is equivalent to the analog input voltage. The converter will remain in this state until another conversion signal is received.

D/A and A/D Conversion

If a new start signal is generated immediately after each conversion is completed, the converter will be operating at its maximum rate. The converter could then be used to digitize a signal as shown in Fig. 11-24a. Notice that the conversion times in digitizing this signal are not constant but depend on the amplitude of the input signal. The analog input signal can be reconstructed from the digital information by drawing straight lines from each digitized point to the next. Such a reconstruction is shown in Fig. 11-24b, and it is indeed a reasonable representation of the original input signal. Take careful note of the fact that in this case the conversion times are small compared with the transient time of the input waveform.

On the other hand, if the transient time of the input waveform approaches the conversion time, the reconstructed output signal will not be quite so accurate. Such a situation is shown in Fig. 11-25a and b. In this case, the input waveform changes at a rate faster than the converter is

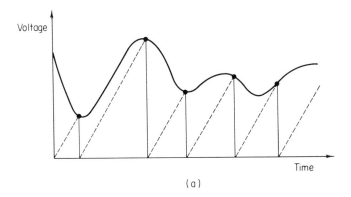

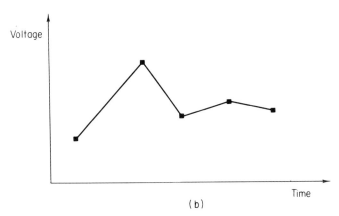

Fig. 11-25 (a) Digitizing an output voltage. (b) Reconstructed signal from digital data.

capable of recognizing. Thus the need for reducing conversion time becomes apparent.

11-7 Advanced A/D Techniques

One of the most obvious methods for speeding up the conversion of the signal as shown in Fig. 11-25 is to eliminate the need for resetting the counter each time a conversion is made. If this were done, the counter would not begin at zero each time, but instead would begin at the value of the last converted point. This means that the counter would have to be capable of counting either up or down. This is no problem since we are already familiar with the operation of up-down counters (Chap. 9).

There is, however, the need for additional logic circuitry, since we must decide whether to count up or down by examining the output of the comparator. An A/D converter which uses an up-down counter is shown in Fig. 11-26. This method is known as continuous conversion, and thus this is called a "continuous-type A/D converter."

The D/A portion of this converter is the same as those previously discussed with the exception of the counter. It is an up-down counter and has the *up* and *down count* control lines in addition to the *advance* line at its input.

The output of the ladder is fed into a comparator which has two outputs instead of one as before. When the analog voltage is more positive than the ladder output, the *up* output of the comparator is high. When the analog voltage is more negative than the ladder output, the *down* output is high.

If the *up* output of the comparator is high, the AND gate at the input of the *up* flip-flop is open, and the first time the clock goes positive, the *up* flip-flop will be set. If we assume for the moment that the *down* flip-flop is reset, the AND gate which controls the *count up* line of the counter will be true and the counter will advance one count. The counter can advance only one count since the output of the one-shot resets both the *up* and *down* flip-flops just after the clock goes low. This can then be considered as one *count up* conversion cycle.

Notice that the AND gate which controls the *count up* line has inputs of *up* and $\overline{down}$. Similarly, the *count down* line AND gate has inputs of *down* and $\overline{up}$. This could be considered an exclusive-OR arrangement, and ensures that the *count down* and *count up* lines cannot both be high at the same time.

As long as the *up* line out of the comparator is high, the converter will continue to operate one conversion cycle at a time. At the point where the ladder voltage becomes more positive than the analog input voltage, the *up* line of the comparator will go low and the *down* line will go high. The

D/A and A/D Conversion

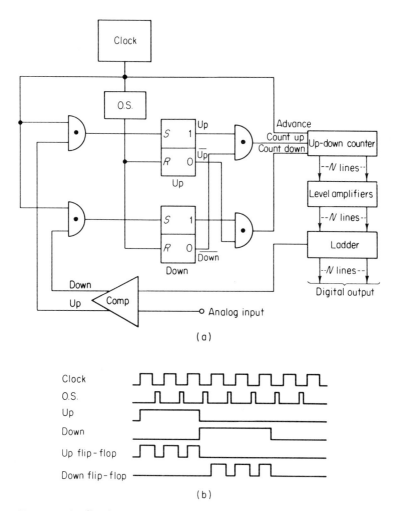

Fig. 11-26 Continuous-type A/D converter. (*a*) Logic diagram. (*b*) Typical waveforms.

converter will then go through a *count down* conversion cycle. At this point, the ladder voltage is within one LSB of the analog voltage and the converter will oscillate about this point. This is not desirable since we want the converter to cease operation and not jump around the final value. The trick here is to adjust the comparator such that its outputs do not change at the same time.

We can accomplish this by adjusting the comparator such that the *up* output will not go high unless the ladder voltage is more than ½ LSB below the analog voltage. Similarly the *down* output will not go high unless the ladder voltage is more than $\frac{1}{2}$ LSB above the analog voltage. This is called centering on the LSB and provides a digital output which is within $\pm\frac{1}{2}$ LSB.

A waveform typical of this type of converter is shown in Fig. 11-27, and it can be seen that this converter is capable of following input voltages which change at a much faster rate.

EXAMPLE 11-12

Quite often additional circuitry is added to a continuous converter to ensure that it cannot count off scale in either direction. For example, if the counter contained all 1s, it would be undesirable to allow it to progress through a *count up* cycle since the next count would advance it to all 0s. We would like to design the logic necessary to prevent this.

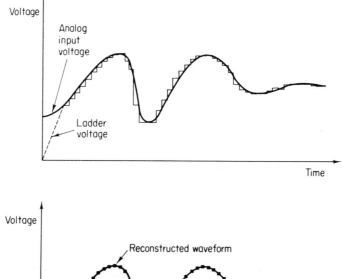

Fig. 11-27 Continuous A/D conversion.

D/A and A/D Conversion

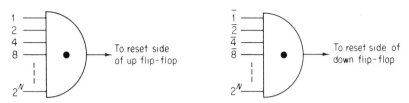

Fig. 11-28 Count-limiting gates for the converter in Fig. 11-26.

SOLUTION

The two limit points which must be detected are all 1s and all 0s in the counter. Suppose we construct an AND gate having the 1 sides of all the counter flip-flops as its inputs. The output of this gate will be true whenever the counter contains all 1s. If the gate is then connected to the *reset* side of the *up* flip-flop, the counter will be unable to count beyond all 1s.

Similarly we can construct an AND gate having the 0 sides of all the counter flip-flops as its inputs. The output of this gate can be connected to the *reset* side of the *down* flip-flop and the counter will then be unable to count beyond all 0s. The gates are shown in Fig. 11-28.

There is quite a wide variety of other methods for digitizing analog signals — too numerous to discuss in detail. Nevertheless, we will take the time to examine some of the methods and the reasons for their importance.

Probably the most important single reason for investigating other methods of conversion is to determine ways to reduce the conversion time. You will recall that the simultaneous converter has a very fast conversion time but becomes unwieldy for more than a few bits of information. The counter converter is simple logically but has a relatively long conversion time. The continuous converter has a very fast conversion time once it is locked on the signal but loses this advantage when multiplexing inputs.

If multiplexing is required, the successive-approximation converter is most useful. The block diagram for this type of converter is shown in Fig. 11-29a. The converter operates by successively dividing the voltage ranges in half. The counter is first reset to all 0s and the MSB is then set. The MSB is then left in or taken out (by resetting the MSB flip-flop) depending on the output of the comparator. Then the second MSB is set in and a comparison is made to determine whether or not to reset the second MSB flip-flop. The process is repeated down to the LSB, and at this time the desired number is in the counter. Since the conversion involves operating on one flip-flop at a time, beginning with the MSB, a ring counter may be used for flip-flop selection.

The successive-approximation method then is the process of approximating the analog voltage by trying 1 bit at a time beginning with the MSB. The operation is shown in diagram form in Fig. 11-29b. It can be clearly seen from this diagram that each conversion takes the same time and requires one conversion cycle for each bit. Thus the total conversion time is equal to the number of bits n times the time required for one conversion cycle. One conversion cycle normally requires one cycle of the clock. As an example, a 10-bit converter operating with a 1-MHz clock has a conversion time of $10 \times 10^{-6} = 10^{-5} = 10\ \mu\text{sec}$.

When dealing with conversion times of this short a duration, it is usually necessary to take into account the other delays in the system (e.g., switching time of the multiplexor, settling time of the ladder network, comparator delay, and settling time).

Another method for reducing the total conversion time of a simple counter converter is to divide the counter into sections. Such a configuration is called a "section counter." To determine how the total conversion time might be reduced by this method assume we have a standard 8-bit counter. If this counter is divided up into two equal counters of 4 bits each, we have a section converter. The converter operates by setting the section containing the four LSBs to all 1s and then advancing the other section until the ladder voltage exceeds the input voltage. At this point the four LSBs are all reset, and this section of the counter is then advanced until the ladder voltage equals the input voltage.

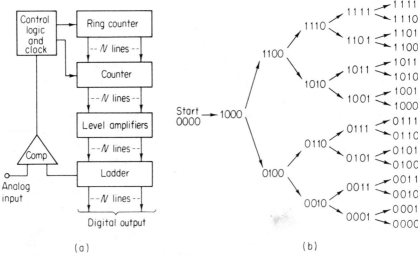

Fig. 11-29 Successive-approximation converter. (a) Logic diagram. (b) Operation diagram.

Notice that it requires a maximum of $2^4 = 16$ counts for each section to count full scale. Thus this method requires only $2 \times 2^4 = 2^5 = 32$ counts to reach full scale. This is a considerable reduction over the $2^8 = 256$ counts required for the straight 8-bit counter. There is, of course, some extra time required to set the counters initially and to switch from counter to counter during the conversion. This logical operation time is very small, however, compared with the total time saved by this method.

This type of converter is quite often used for digital voltmeters since it is very convenient to divide the counters by counts of 10. Each counter will then be used to represent one of the digits of the decimal number appearing at the output of the voltmeter.

11-8 A/D Accuracy and Resolution

Since the A/D converter is a closed-loop system involving both analog and digital systems, the overall accuracy must include errors from both the analog and digital positions. In determining the overall accuracy it is easiest to separate the two sources of error.

If we assume that all components are operating properly, the source of the digital error is simply determined by the resolution of the system. In digitizing an analog voltage, we are trying to represent a continuous analog voltage by an equivalent set of digital numbers. When the digital levels are converted back into analog form by the ladder, the output is the familiar staircase waveform. This waveform is a representation of the input voltage but is certainly not a continuous signal. It is in fact a discontinuous signal composed of a number of discrete steps. In trying to reproduce the analog input signal, the best we can do is to get on the step which most nearly equals the input voltage in amplitude.

The simple fact that the ladder voltage has steps in it leads to the digital error in the system. The smallest digital step, or *quantum*, is due to the LSB and can be made smaller only by increasing the number of bits in the counter. This inherent error is often called the "quantization error" and is commonly ± 1 bit. If the comparator is centered, as with the continuous converter, the quantization error can be made $\pm \frac{1}{2}$ LSB.

The main source of analog error in the A/D converter is probably the comparator. Other sources of error are the resistors in the ladder, the reference-voltage supply, ripple, and noise. These can, however, usually be made secondary to the sources of error in the comparator.

The sources of error in the comparator are centered around variations in the d-c switching point. The d-c switching point is the difference between the two input-voltage levels which causes the output to change state. Variations in switching are primarily due to offset, gain, and linearity of the amplifier used in the comparator. These parameters will usually vary slightly with input-voltage levels and quite often with temperature. It is these changes which give rise to the analog error in the system.

An important measure of converter performance is given by the differential linearity. *Differential linearity* is a measure of the variation in voltage step size which causes the converter to change from one state to the next. It is usually expressed as a percent of the average step size. This performance characteristic is also a function of the conversion method and is best for the converters having counters which count continuously. The counter-type and continuous-type converters usually have better differential linearity than the successive-approximation-type converters. This is true since, in the one case, the ladder voltage is always approaching the analog voltage from the same direction. In the other case, the ladder voltage is first on one side of the analog voltage and then on the other. The comparator is then being used in both directions and the net analog error from the comparator is thus greater.

The next logical question which might be asked is, what should be the relative order of magnitudes of the analog and digital errors? As mentioned previously, it would be hard to justify constructing a 15-bit converter which has an overall error of ± 1 percent. On the other hand, it would be equally difficult to justify building a 6-bit converter to an accuracy of 0.1 percent. In general, it is considered good practice to construct converters having analog and digital errors of approximately the same magnitudes. There are many arguments pro and con for this, and any final argument would have to depend on the specific situation. As an example, an 8-bit converter would have a quantization error of $\frac{1}{256} \cong 0.4$ percent. It would then seem reasonable to construct this converter to an accuracy of 0.5 percent in an effort to achieve an overall accuracy of 1.0 percent. This might mean constructing the ladder to an accuracy of 0.1 percent, the comparator to an accuracy of 0.2 percent, etc., since these errors are all accumulative.

Example 11-13

What overall accuracy could one reasonably expect from the construction of a 10-bit A/D converter?

Solution

A 10-bit converter has a quantization error of $\frac{1}{1024} \cong 0.1$ percent. If the analog portion can be constructed for an accuracy of 0.1 percent, it would seem reasonable to strive for an overall accuracy of 0.2 percent.

11-9 Electromechanical A/D Conversion

There is another area of application in which A/D conversion is very important. This involves the translation of the angular position of a shaft into digital information. A very common application of this type of conversion is found in large radar installations where the azimuth and elevation information are determined directly from shaft position. Many other examples are evident in aircraft and aerospace applications. The method is not necessarily limited to rotational information since rectilinear information can be translated into rotational information by means of a gearing arrangement.

In any case, the task involves changing position information (which can usually be considered analog information since it is continuous) into equivalent digital information. This is most generally accomplished by the use of a code wheel as shown in Fig. 11-30. This particular wheel is coded in straight binary fashion and represents 3 bits. The wheel is divided into three concentric bands, each band representing 1 bit. The innermost band is divided into two equal segments and is the MSB. The middle band is divided into four equal segments and represents the second MSB. The outer band has eight equal segments and is the LSB.

If the light areas on the wheel are transparent and the dark areas are opaque, the digital information can be obtained by placing light sources and photosensors on opposite sides of the disk as shown in the figure. The outputs of the photosensors will be high if light is sensed and low if no light is sensed. These outputs will then represent 1s and 0s and can be amplified, passed through logic circuits, and used to set flip-flops. Such a system is called an "optical encoder." The sensing could also be accomplished by

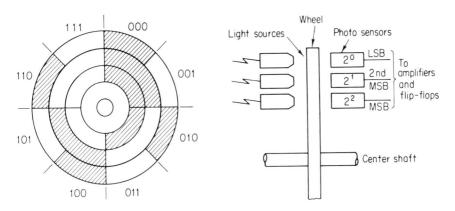

Fig. 11-30 3-bit binary-code wheel.

placing a brush on each band and making the light areas a conducting material and the dark areas an insulating material.

Now let us suppose that the wheel is positioned under the sensors such that the output reads 011. Further, let us assume that the wheel is very near the dividing line between 100 and 011. If there is an ambiguity in reading the LSB (i.e., the sensor cannot decide whether to read a 1 or a 0 since it is right on the dividing line), the output may vary between 011 and 010. This is not too bad since the output is jumping only from one adjacent position to the next. However, if the ambiguity is in the MSB, the results could be disastrous since the output is jumping from 011 to 111, which is 180° of error. Consider the results of this error if this wheel were the source of azimuth information which directs the firing of a large missile installation! Obviously something must be done to overcome this ambiguity.

Since it might be quite difficult to resolve the problem of sensing when the wheel stops on the dividing line, it might be more fruitful to investigate a different code. What we would like is a code which changes only 1 bit at a time when going from any position to the next. This would then ensure an ambiguity of only one position, and would eliminate the 180° ambiguity previously demonstrated. Such a code, as you might recall from Chap. 3, is the Gray code. A code wheel constructed using Gray code is shown in Fig. 11-31. By examination of the wheel you can see that the greatest error caused by a reading ambiguity is one segment of rotation.

The 3-bit wheel discussed has the ability to digitize shaft position into an equivalent 3-bit binary number. This implies eight positions around the wheel, and thus each position represents a 45° segment. To determine shaft position to the nearest 45° is of course not very accurate, and one would generally like a better measurement.

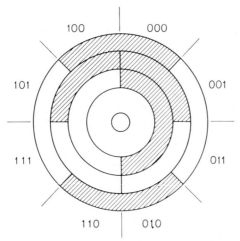

Fig. 11-31 3-bit Gray-code wheel.

D/A and A/D Conversion

To obtain a closer reading, it is only necessary to add extra bands to the code wheel and thus extra bits to the digital number. In general, the degree of resolution obtained is given by $360°/2^n$, where n is the number of bits in the binary number (or bands on the wheel).

EXAMPLE 11-14

What degree of resolution can be obtained using an 8-bit optical encoder?

SOLUTION

The resolution is $360°/2^8 = 360°/256 \cong 1.4°$.

The optical encoder as described will convert an analog position signal into an equivalent digital signal in Gray code. If this digital information is going to be entered into a digital system for processing, it might be more convenient to have it in straight binary form. One method of performing this conversion is to use the Gray code as the input to a series of gates which are wired to perform a Gray-to-binary conversion. The outputs of these gates will then be the equivalent binary number.

Gray code			Binary code		
A	B	C	2^2	2^1	2^0
0	0	0	0	0	0
0	0	1	0	0	1
0	1	1	0	1	0
0	1	0	0	1	1
1	1	0	1	0	0
1	1	1	1	0	1
1	0	1	1	1	0
1	0	0	1	1	1

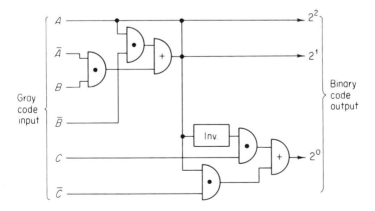

Fig. 11-32 Gray-to-binary encoder.

The logic gates necessary to perform a Gray-to-binary conversion for the 3-bit encoder are shown in Fig. 11-32. The 2^2 bit (binary) is the same as the A bit (Gray). Thus the 2^2 bit is simply wired directly to the A bit.

The 2^1 bit must be high whenever $\bar{A}$ and B are high and also when A and $\bar{B}$ are high. Thus the 2^1 bit is given by $2^1 = \bar{A}B + A\bar{B}$.

The 2^0 bit can be encoded by making use of the 2^1 bit. Notice that 2^0 must be high whenever 2^1 and $\bar{C}$ are high and also when $\overline{2^1}$ and C are high. Thus the 2^0 bit can be formed by $2^0 = 2^1\bar{C} + \overline{2^1}C$.

SUMMARY

D/A conversion, the process of converting digital input levels into an equivalent analog output voltage, is most easily accomplished by the use of resistance networks. The binary ladder was found to have definite advantages over the resistance divider. The complete D/A converter consists of a binary ladder (usually) and a flip-flop register to hold the digital input information.

The simultaneous method for A/D conversion is very fast but becomes cumbersome for more than a few bits of resolution. The counter-type A/D converter is somewhat slower but represents a much more reasonable solution for digitizing high-resolution signals. The continuous-converter method, the successive-approximation method, and the section-counter method are all variations of the basic counter-type A/D converter which lead to a much faster conversion time.

Optical encoders provide a convenient means for digitizing shaft position information (and possibly rectilinear position information). Gray code is used to code the optical encoder wheel to eliminate the large ambiguity error found in binary-coded wheels.

GLOSSARY

A/D conversion The process of converting an analog input voltage into a number of equivalent digital output levels.

D/A conversion The process of converting a number of digital input signals into one equivalent analog output voltage.

differential linearity A measure of the variation in size of the input voltage to an A/D converter which causes the converter to change from one state to the next.

equivalent binary weight The value assigned to each bit in a digital number, expressed as a fraction of the total. The values are assigned in binary

D/A and A/D Conversion

fashion according to the sequence 1,2,4,8, ..., 2^n, where n is the total number of bits.

Millman's theorem   A theorem from network analysis which states that the voltage at any node in a resistive network is equal to the sum of the currents entering the node divided by the sum of the conductances connected to the node, all determined by assuming the voltage at the node is zero.

quantization error   The error inherent in any digital system due to the size of the LSB.

REVIEW QUESTIONS

1. Why is a D/A converter usually considered a decoder?
2. Why is an A/D converter usually considered an encoder?
3. How is a binary equivalent weight determined?
4. Describe how Millman's theorem is used to find the output voltage of a resistive divider.
5. What is the principle of superposition? How can it be used to find the output voltage of a binary ladder?
6. What are some of the advantages of the binary ladder over the resistive divider?
7. What is monotonicity?
8. What is the difference between accuracy and resolution?
9. What is meant by the statement "resolution and accuracy should be compatible"?
10. How does a simultaneous A/D converter operate?
11. Describe the operation of a counter-type A/D converter.
12. Why is a counter-type A/D converter better than the simultaneous type for high-resolution conversions?
13. Explain why the counter-type A/D converter is slower than the simultaneous type.
14. Where does the staircase waveform appear in a counter-type A/D converter?
15. What does the staircase waveform have to do with monotonicity?
16. How does a continuous-type A/D converter aid conversion time?
17. Explain how it is possible to perform A/D conversion to $\pm \frac{1}{2}$ LSB.
18. Why is a section-counter-type A/D converter useful for digital voltmeters?
19. Describe how Gray code eliminates large ambiguity errors in reading optical encoders.
20. Explain why you would or would not make the innermost ring of the code wheel for an optical encoder the LSB.

PROBLEMS

11-1 What is the binary equivalent weight of each bit in a 6-bit resistive divider?

11-2 Draw the schematic for a 6-bit resistive divider.

11-3 Verify the voltage-output levels for the network of Fig. 11-5 using Millman's theorem. Draw the equivalent circuits.

11-4 Assume the divider in Prob. 11-2 has +10 volts full-scale output and find the following:
  (a) The change in output voltage due to a change in the LSB.
  (b) The output voltage for an input of 110110.

11-5 A 10-bit resistive divider is constructed such that the current through the LSB resistor is 100 μa. Determine the maximum current that will flow through the MSB resistor.

11-6 What is the full-scale output voltage of a 6-bit binary ladder if 0 = 0 volts and 1 = +10 volts? What is it for an 8-bit ladder?

11-7 Find the output voltage of a 6-bit binary ladder with the following inputs:
  (a) 101001.
  (b) 111011.
  (c) 110001.

11-8 Check the results of Prob. 11-7 by adding up the individual bit contributions.

11-9 What is the resolution of a 12-bit D/A converter which uses a binary ladder? If the full-scale output is +10 volts, what is the resolution in volts?

11-10 How many bits are required in a binary ladder to achieve a resolution of 1 mv if full scale is +5 volts?

11-11 How many comparators are required to build a 5-bit simultaneous A/D converter?

11-12 Redesign the encoding matrix and *read* gates of Fig. 11-20 using NAND gates.

11-13 Find the following for a 12-bit counter-type A/D converter using a 1-MHz clock:
  (a) Maximum conversion time.
  (b) Average conversion time.
  (c) Maximum conversion rate.

11-14 What clock frequency must be used with a 10-bit counter-type A/D converter if it must be capable of making at least 7000 conversions per second?

11-15 What is the conversion time of a 12-bit successive-approximation-type A/D converter using a 1-MHz clock?

D/A and A/D Conversion 315

11-16 What is the conversion time of a 12-bit section-counter-type A/D converter using a 1-MHz clock? The counter is divided into three equal sections.

11-17 What overall accuracy could you reasonably expect from a 12-bit A/D converter?

11-18 What degree of resolution can be obtained using a 12-bit optical encoder?

11-19 Redesign the Gray-to-binary encoder in Fig. 11-32 using NAND gates.

CHAPTER 12

MAGNETIC DEVICES AND SYSTEMS

There is a large class of devices and systems which are useful as digital elements because of their magnetic behavior. A ferromagnetic material can be magnetized in a particular direction by the application of a suitable magnetizing force (a magnetic flux resulting from a current flow). The material will then remain magnetized in that direction after the removal of the excitation. Application of a magnetizing force of the opposite polarity will then switch the material, and it will remain magnetized in the opposite direction after removal of the excitation. Thus the ability to store information in two different states is available, and a large class of binary elements has been devised using these principles. In this chapter we will investigate a number of these devices and the systems constructed using them.

12-1 Magnetic Cores

One of the most widely used magnetic elements is the magnetic core. The typical core is toroidal (doughnut-shaped), as shown in Fig. 12-1, and is

Magnetic Devices and Systems

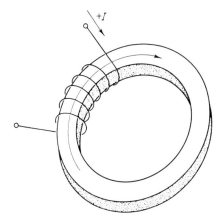

Fig. 12-1 A magnetic core.

usually constructed in one of two ways. The metal-ribbon core is constructed by winding a very thin metallic ribbon on a ceramic-core form. A popular ribbon is $\frac{1}{8}$ mil thick 4-79 molybdenum-permalloy (known as ultrathin ribbon), and a typical core might consist of 20 turns of this ribbon wound on a 0.2-in.-diameter ceramic form.

Ferrite cores are constructed from a finely powdered mixture of magnetite, various bivalent metals such as magnesium or manganese, and a binder material. The powder is pressed into the desired shape and fired. During firing, the powder is fused into a solid, homogeneous, polycrystalline form. Ferrite cores such as this are commonly constructed with 50 mil outside diameters and 30 mil inside diameters.

Ferrite cores can be constructed in smaller dimensions than metal-ribbon cores and usually have a better uniformity and a lower cost. Furthermore, ferrite cores typically have resistivities greater than 10^5 ohm-cm, which means eddy-current losses are negligible and thus core heating is reduced. For these reasons, they are widely used as the principal memory or storage element in large-scale digital computers.

Metal-ribbon cores, on the other hand, have very good magnetic characteristics and generally require a smaller driving current for switching. They are somewhat better for the construction of logic circuits and shift registers.

The binary characteristics of a core can be most easily seen by examining the *hysteresis curve* for a typical core. Hysteresis comes from the Greek work *hysterein*, which means to lag behind. A magnetic core exhibits a lag-behind characteristic in the hysteresis curve shown in Fig. 12-2a. In this figure, the *magnetic flux density* **B** is plotted as a function of the *magnetic force* **H**. However, since the flux density **B** is directly proportional to the flux ϕ, and since the magnetic field **H** is directly proportional to

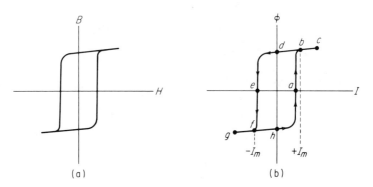

Fig. 12-2 Ferrite-core hysteresis curves. (*a*) Magnetic flux density **B** versus magnetic field **H**. (*b*) Magnetic flux ϕ versus current I.

current producing it I, a plot of ϕ versus I is a curve of the same general shape. A plot of flux in the core ϕ versus driving current I is shown in Fig. 12-2*b*, and we will base our discussion on this curve since it is generally easier to talk in terms of these quantities.

Now, suppose that a current source is attached to the windings on the core shown in Fig. 12-1 and a positive current is applied (current flows into the upper terminal of the winding). This will create a flux in the core in the clockwise direction shown in the figure (remember the *right-hand rule*). If the drive current is just slightly greater than I_m shown in Fig. 12-2, the operating point of the core will be somewhere between points *b* and *c* on the ϕI curve. The magnitude of the flux can then be read from the ϕ axis in this figure.

If the drive current is now removed, the operating point will move along the ϕI curve through point *b* to point *d*. The core is now storing energy with no input signal since there is a remaining or *remanent flux* in the core at this point. This property is known as *remanence*, and this point is known as a *remanent point*.

The repeated application of positive current pulses will simply cause the operating point to move between points *d* and *c* on the ϕI curve. Notice that the operating point will always come to rest at point *d* when all drive current is removed.

If a negative drive current somewhat greater than $-I_m$ is now applied to the winding (in a direction opposite to that shown in Fig. 12-1), the operating point will move from *d* down through *e* and will stop at a point somewhere between points *f* and *g* on the ϕI curve. At this point the flux has switched in the core and is now directed in a counterclockwise direction in Fig. 12-1. If the drive current is now removed, the operating point will come to rest at point *h* on the ϕI curve of Fig. 12-2. Notice that the flux

Magnetic Devices and Systems

has approximately the same magnitude but is the negative of what it was previously. This indicates that the core has been magnetized in the opposite direction.

Repeated application of negative drive currents will simply cause the operating point to move between points g and h on the ϕI curve, but the final resting place with no applied current will be point h. Point h then represents a second remanent point on the ϕI curve.

By way of summary, a core has two remanent states: point d after the application of one or more positive current pulses, point h after the application of one or more negative current pulses. For the core in Fig. 12-1, point d corresponds to the core magnetized with flux in a clockwise direction and point h corresponds to magnetization with flux in the counterclockwise direction.

EXAMPLE 12-1

Cores can be magnetized by utilizing the magnetic field surrounding a current-carrying wire by simply *threading* the cores on the wire. For the two possible current directions in the wire shown in Fig. 12-3, what are the corresponding directions of magnetization for the core?

SOLUTION

Using the right-hand rule, a current of $+I$ will magnetize the core with the flux in a clockwise direction around the core. A current of $-I$ will magnetize the core with flux in a counterclockwise direction around the core.

It is now quite easy to see how a magnetic core is used as a binary storage device in a digital system. The core has two states, and we can simply define one of the states as a 1 and the other state as a 0. It is perfect-

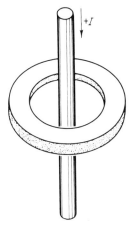

Fig. 12-3

ly arbitrary which is which, but for our discussion purposes let us define point d as a 1 and point h as a 0. This means that a positive current will record a 1 and result in clockwise flux in the core in Fig. 12-1. A negative current will record a 0 and result in a counterclockwise flux in the core.

We now have the means for recording or writing a 1 or a 0 in the core but we do not as yet have any means of detecting the information stored in the core. A very simple technique for accomplishing this is to apply a current to the core which will switch it to a known state and detect whether or not a large flux change occurs. Consider the core shown in Fig. 12-4. Application of a drive current of $-I$ will switch the core to the 0 state. If the core has a 0 stored in it, the operating point will move between points g and h on the ϕI curve (Fig. 12-2) and very little flux change will occur. This small change in flux will induce a very small voltage across the sense-winding terminals. On the other hand, if the core has a 1 stored in it, the operating point will move from point d to point h on the ϕI curve, resulting in a much larger flux change in the core. This change in flux will induce a much larger voltage in the sense winding, and we can thus detect the presence of a 1.

To summarize, we can detect the contents of a core by applying a *read* pulse which resets the core to the 0 state. The output voltage at the sense winding is much greater when the core contains a 1 than when it contains a 0, and we can therefore detect a 1 by distinguishing between the two output-voltage signals. You will notice that we could set the core by applying a *read* current of $+I$ and detect the larger output voltage at the sense winding as a 0.

The output voltage appearing at the sense winding for a typical core is also shown in Fig. 12-4. Notice that there is a difference of about 3 to 1 in output-voltage amplitude between a 1 and a 0 output. Thus a 1 can be detected by using simple amplitude discrimination in an amplifier. In large systems where many cores are used on common windings (such as the large memory systems in digital computers) the 0 output voltage may become considerably larger because of additive effects. In this case, ampli-

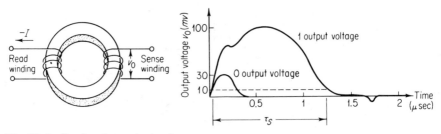

Fig. 12-4 Sensing the contents of a core.

Magnetic Devices and Systems

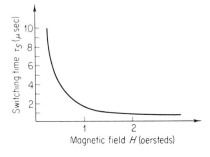

Fig. 12-5 Magnetic-core switching time characteristics.

tude discrimination is quite often used in combination with a strobing technique. Even though the amplitude of the 0 output voltage may increase because of additive effects, the width of the output will not increase appreciably. This means that the 0 output-voltage signal will have decayed and will be very small before the 1 output voltage has decayed. Thus if we were to strobe the *read* amplifiers some time after the application of the *read* pulse (for example, between 0.5 and 1.0 μsec in Fig. 12-4), this should improve our detection ability.

The *switching time* of the core is commonly defined as the time required for the output voltage to go from 10 percent up through its maximum value and back down to 10 percent again (see Fig. 12-4). The switching time for any one particular core is a function of the drive current as shown in Fig. 12-5. It is evident from this curve that an increased drive current results in an increased switching time. In general, the switching time for a core is dependent on the physical size of the core, the type of core, and the materials used in its construction, as well as the manner in which it is used. It will be sufficient for our purposes to know that cores are available with switching times from around 0.1 μsec up to milliseconds, with drive currents of 100 ma to 1 amp.

12-2 Magnetic-core Logic

Since a magnetic core is a basic binary element, it can be used in a number of ways to implement logical functions. Because of its inherent ruggedness, the core is a particularly useful logical element in applications where environmental extremes are experienced, for example, the temperature extremes and radiation exposure experienced by space vehicles.

Since the core is essentially a storage device and its content is detected by resetting the core to the 0 state, any logic system constructed using cores must necessarily be a dynamic system. The basis for using the core as

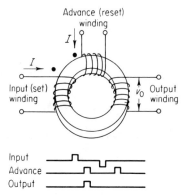

Fig. 12-6 Basic magnetic-core logic element.

a logical element is shown in Fig. 12-6. A 1 input to the core is represented by a current of $+I$ at the *input* winding, and this will set a 1 in the core (magnetize it in a clockwise direction). An *advance* pulse will occur sometime after the *input* pulse has disappeared. Logical operations will be carried out during the time when the *advance* pulse appears at the *advance* (*reset*) winding, since at this time the core is forced into the 0 state and a pulse will appear at the output winding only if the core previously stored a 1. The current in the output winding can then be used as the input for other cores or other logical elements.

There will be some energy loss in the core during switching. For this reason, the output winding will normally have more turns than either the *input* or *advance* windings in order that the output will be capable of driving one or more cores.

Notice that a 0 can be set in the core by application of a current of $-I$ at the *input* winding. Alternatively, a 0 could be stored by a current of $+I$

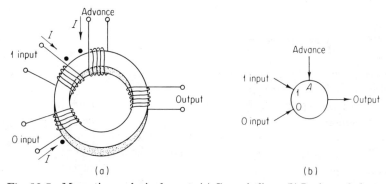

Fig. 12-7 Magnetic-core logic element. (*a*) Core windings. (*b*) Logic symbol.

into the undotted side of the *input* winding. The important thing to notice is that either a 1 or a 0 can be stored in the core by application of a current to the proper terminal of the *input* winding.

To simplify our discussion, and the logic diagrams, we will adopt the symbols for the core and its windings shown in Fig. 12-7. A pulse at the 1 input will set a 1 in the core; a pulse at the 0 input will set a 0 in the core; during the *advance* pulse, a pulse will appear at the output only if the core previously held a 1. Let us now consider some of the basic logic functions using the symbol shown in Fig. 12-7b.

A method for implementing the OR function is shown in Fig. 12-8a. A current pulse at either the X OR Y inputs will set a 1 in the core. Sometime after the input pulse(s) have been terminated, an *advance* pulse occurs. If the core has been set to the 1 state, a pulse will appear at the output winding. Notice that this is truly an OR function since a pulse at either the X OR Y input or *both* will set a 1 in the core.

The method shown in Fig. 12-8b provides the means for obtaining the complement of a variable. The *set input* winding to the core has a 1 input. This means that during the *input* pulse time this winding will always have a *set input* current. If there is no current at the X input (signifying $X = 0$) the core will be set. Then when the *advance* pulse occurs a 1 will appear at the output, signifying that $\overline{X} = 1$. On the other hand, if $X = 1$, a current will appear at the X input during the *set* time and the effects of the X input

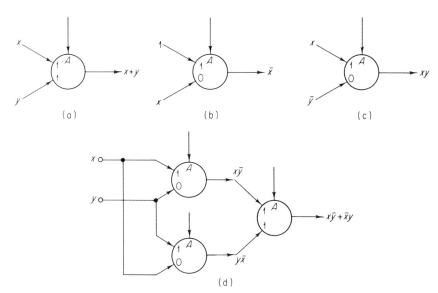

Fig. 12-8 Basic core logic functions. (a) OR. (b) Complement. (c) AND. (d) Exclusive-OR.

current and the 1 input current will cancel one another. The core will then remain in the reset state (recall that the core is reset during the *advance* pulse). In this case no pulse will appear at the output during the *advance* pulse since the core previously contained a 0. Thus the output represents $\overline{X} = 0$.

The AND function can be implemented using a core as shown in Fig. 12-8c. The two inputs to the core are X and $\overline{Y}$, and there are four possible combinations of these two inputs. Examining these input combinations in detail yields:

1. $X = 0$, $Y = 0$. Since $X = 0$, the core cannot be set. Since $Y = 0$, $\overline{Y} = 1$ and the core will then be reset. Thus this input combination will reset the core and it will store a 0.

2. $X = 0$, $Y = 1$. Since $X = 0$, the core still cannot be set. $Y = 1$ and therefore $\overline{Y} = 0$. In this input combination, there is no input current in either winding and the core cannot change state. Thus the core will remain in the 0 state because of the previous *advance* pulse.

3. $X = 1$, $Y = 0$. The current in the X *winding* will attempt to set a 1 in the core. However, $\overline{Y} = 1$ and this current will attempt to reset the core. These two currents will offset one another and the core will not change states. It will then remain in the 0 state because of the previous advance pulse.

4. $X = 1$, $Y = 1$. The current in the X *winding* will set a 1 in the core since $\overline{Y} = 0$ and there is no current in the Y *winding*. Thus this combination will store a 1 in the core.

In summary, the input X AND $\overline{Y}$ is the only combination which will result in a 1 being stored in the core. Thus this is truly an AND function.

An exclusive-OR function can be implemented as shown in Fig. 12-8d by ORing the outputs of two AND-function cores.

EXAMPLE 12-2

Make a truth table for the exclusive-OR function shown in Fig. 12-8d.

SOLUTION

X	Y	$X\overline{Y}$	$\overline{X}Y$	$X\overline{Y} + \overline{X}Y$
0	0	0	0	0
0	1	0	1	1
1	0	1	0	1
1	1	0	0	0

One of the major problems of core logic becomes apparent in the operation of the exclusive-OR shown in Fig. 12-8d. This is the problem of the time required for the information to shift down the line from one core to the next. For the exclusive-OR, the inputs X and Y appear at time t_1, and the AND cores are *set* or *reset* at this time. At time t_2 an *advance* pulse is applied to the AND cores and their outputs are used to *set* the OR core. Then at time t_3 an *advance* pulse is applied to the OR core and the final output appears. It should be obvious from this discussion that the operation time for more complicated logic functions may become excessively long.

A second difficulty with this type of logic is the fact that the *input* pulses must be of exactly the same width. This is particularly true for functions such as the COMPLEMENT and the AND since the input signals are at times required to cancel one another. It is apparent that if one of the input signals is wider than the other, the core may contain erroneous data after the *input* pulses have disappeared.

You will recall that in order to switch a core from one state to another a certain minimum current I_m is required. This is sometimes referred to as the *select current*. The core arrangement shown in Fig. 12-8a can be used to implement an AND function if the X and Y inputs are each limited to one-half the select current $\frac{1}{2}I_m$. In this way, the only time the core can be set is when both X and Y are present since this is the only time the core receives a full select current I_m. Core logic functions can be constructed using the half-select current ideas. This idea is quite important and forms the basis of one type of large-scale memory system which we will discuss later in this chapter.

12-3 Magnetic-core Shift Register

A review of the previous section will reveal that a magnetic core exhibits at least two of the major characteristics of a flip-flop: First, it is a binary device capable of storing binary information; second, it is capable of being set or reset. Thus it would seem reasonable to expect that the core could be used to construct a shift register or a ring counter. Cores are indeed frequently used for these purposes, and in this section we will consider some of the necessary precautions and techniques.

The main idea involves connecting the output of each core to the input of the next core. When a core is reset (or set), the signal appearing at the output of that core is used to set (or reset) the next core. Such a connection between two cores, called a "single-diode transfer loop," is shown in Fig. 12-9.

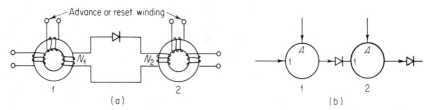

Fig. 12-9 Single-diode transfer loop. (a) Circuit. (b) Symbolic representation.

There are three major problems to overcome when using the single-diode transfer loop. The first problem is the gain through the core. This is similar to the problem discussed previously, and the solution is the same. That is, the losses in signal through the core can be overcome by constructing the output winding with more turns than the input winding. This will ensure that the output signal will have sufficient amplitude to switch the next core.

The second problem concerns the polarity of the output signal. A signal will appear at the output when the core is set or when the core is reset. These two signals will of course have opposite polarities and both will be capable of switching the next core. In general, it is desirable that only one of the two output signals be effective, and this can be achieved by the use of the diode shown in Fig. 12-9. In this figure, the current produced in the output winding will go through the diode in the forward direction (and thus set the next core) when the core is reset from the 1 state to the 0 state. On the other hand, when the core is being set to the 1 state, the diode will prevent current flow in the output and thus the next core cannot be switched. Notice that the opposite situation could be realized by simply reversing the diode.

The third problem arises from the fact that resetting core 2 induces a current in winding N_2 which will pass through the diode in the forward direction and thus tend to set a 1 in core 1. This constitutes the transfer of information in the reverse direction and is highly undesirable. Fortunately the solution to the first problem (that of gain) will result in a solution for this problem as well. That is, since N_2 has fewer windings than N_1, this reverse signal will not have sufficient amplitude to switch core 1. With this understanding of the basic single-diode transfer loop, let us investigate the operation of a simple core shift register.

A basic magnetic-core shift register in symbolic form is shown in Fig. 12-10. There are two sets of advance windings which are necessary for shifting information down the line. The advance pulses occur alternately as shown in the figure. A_1 is connected to cores 1 and 3 and would be connected to all *odd*-numbered cores for a larger register. A_2 is connected

Magnetic Devices and Systems 327

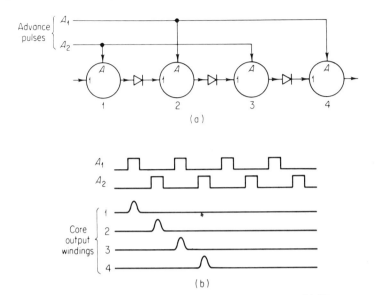

Fig. 12-10 Four-core shift register. (*a*) Symbolic circuit. (*b*) Waveforms.

to cores 2 and 4 and would be connected to all *even*-numbered cores. If we assume that all cores are reset with the exception of core 1, it is clear that the advance pulses will shift this 1 down the register from core to core until it is shifted "out the end" when core 4 is reset. The operation is as follows: The first A_1 pulse which occurs will reset core 1 and thus set core 2. This is followed by an A_2 pulse which will reset core 2 and thus set core 3. The next A_1 pulse will reset core 3 and set core 4, and the following A_2 pulse will shift the 1 "out the end" by resetting core 4. Notice that the two phases of advance pulses are required since it is not possible to set a core while an advance (or reset) pulse is present.

The output of each core winding can be used as an input to an amplifier to produce the waveforms shown in Fig. 12-10*b*. Notice that after four advance pulses the 1 has been shifted completely through the register and the output lines all remain low after this time.

The need for a two-phase clock or advance pulse system could be eliminated if some delay were introduced between the output of each core and the input of the next core. Suppose that a delay greater than the width of the advance pulses were introduced between each core. In this case, it would be possible to drive every core with the same advance pulse since the output of any one core could not arrive at the input to the next core until after the advance pulse had disappeared.

One method for introducing a delay between cores is shown in Fig. 12-11. The advance-pulse amplitude is several times the minimum required to switch the cores and will reset all cores to the 0 state. If a core previously contained a 0, no switching occurs and thus no signal appears at the output winding. On the other hand, if a core previously contained a 1, current will flow in the output winding and will charge the capacitor. Some current will flow through the set winding of the next core, but it is small because of the presence of the resistor and furthermore it is overridden by the magnitude of the advance pulse. However, at the cessation of the advance pulse, C remains charged. Thus C will discharge through the input winding and R and will set core 2 to the 1 state.

In this system, the amplitude of the advance pulses is not too critical, but the width must be matched to the RC time constant of the loop. If the advance pulses are too long, or alternatively if the RC time constant is too short, the capacitor will discharge too much during the advance pulse time and will be incapable of setting the core at the cessation of the advance pulse. The RC time constant may limit the upper frequency of operation; it should be noted, however, that resetting a core induces a current in its input winding in a direction which will tend to discharge the capacitor.

The arrangements we have discussed here are called *one-core-per-bit* registers. There are numerous other methods (too many to discuss here) for implementing registers and counters, and the reader is referred to the references listed for more advanced techniques. Some of the other methods include *two-core-per-bit* systems, *modified-advance-pulse* systems, *modified-winding-core* systems, *split-winding-core* systems, and *current-routing-transfer* systems.

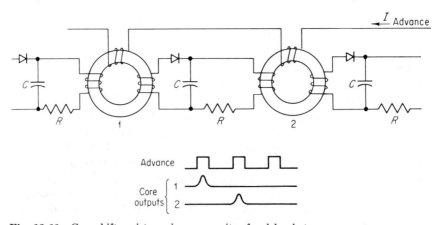

Fig. 12-11 Core shift register using a capacitor for delay between cores.

Magnetic Devices and Systems

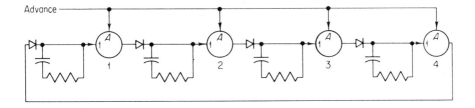

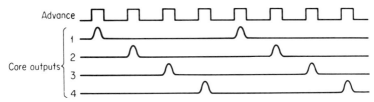

Fig. 12-12 Four-stage ring counter for Example 12-3.

EXAMPLE 12-3

Using core symbols and the capacitor-delay technique, draw the diagram for a four-stage ring counter. Show the expected waveforms.

SOLUTION

A ring counter can be formed from a simple shift register by using the output of the last core as the input for the first core. Such a system along with the expected waveforms is shown in Fig. 12-12.

12-4 Coincident-current Memory

The core shift register discussed in the previous section suggests the possibility of using an array of magnetic cores for storing words of binary information. For example, a 10-bit core shift register could be used to store a 10-bit word. The operation would be serial in form, much like the 10-bit flip-flop shift register discussed earlier. It would, however, be subject to the same speed limitations observed in the serial flip-flop register. That is, since each bit must travel down the register from core to core, it will require n clock periods to shift an n-bit word into or out of the register. This shift time may become excessively long in some cases, and a faster method must then be developed. Much faster operation can be achieved if the information is written into and read out of the cores in a parallel manner. Since all the bits are processed simultaneously, an entire word can be transferred in only one clock period. A straight parallel system would, however, require one input wire and one output wire for each core. For a large number of cores the total number of wires makes this arrangement impractical, and some other form of core selection must be developed.

The most popular method for storing binary information in parallel form using magnetic cores is the *coincident-current drive* system. Such memory systems are widely used in all types of digital systems from small-scale special-purpose machines up to large-scale digital computers. The basic idea involves arranging cores in a matrix and using two *half-select currents;* the method is shown in Fig. 12-13.

The matrix consists of two sets of drive wires: the X drive wires (vertical) and the Y drive wires (horizontal). Notice that each core in the matrix is threaded by one X wire and one Y wire. Suppose one half-select current $\frac{1}{2}I_m$ is applied to line X_1 and one half-select current $\frac{1}{2}I_m$ is applied to line Y_1. Then the core which is threaded by both lines X_1 and Y_1 will have a total of $\frac{1}{2}I_m + \frac{1}{2}I_m = I_m$ passing through it, and it will switch states. The remaining cores which are threaded by X_1 or Y_1 will each receive only $\frac{1}{2}I_m$, and they will therefore not switch states. Thus we have succeeded in switching one of the 16 cores by selecting two of the input lines (one of the X lines and one of the Y lines). We will designate the core that switched in this case as core X_1Y_1, since it was switched by selecting lines X_1 and Y_1. The designation X_1Y_1 is called the *address* of the core since it specifies its

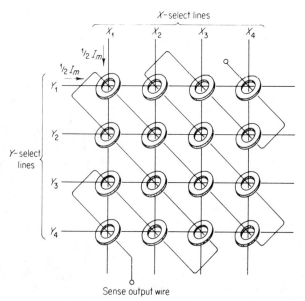

Fig. 12-13 Magnetic-core coincident-current memory.

Magnetic Devices and Systems

location. We can then switch any core X_aY_b located at address X_aX_b by applying $\frac{1}{2}I_m$ to lines X_a and Y_b. For example, the core located in the lower right-hand corner of the matrix is at the address X_4Y_4 and can be switched by applying $\frac{1}{2}I_m$ to lines X_4 and Y_4.

In order that the selected core will switch, the directions of the half-select currents through the X line and the Y line must be additive in the core. In Fig. 12-13, the X select currents must flow through the X lines from the top toward the bottom, while the Y select currents flow through the Y lines from left to right. Application of the *right-hand rule* will demonstrate that currents in this direction will switch the core such that the core flux is in a clockwise direction (looking from the top). We will define this as switching the core to the 1 state. It is obvious then that reversing the directions of both the X and Y line currents will switch the core to the 0 state. Notice that if the X and Y line currents are in a subtractive direction the selected core will receive $\frac{1}{2}I_m - \frac{1}{2}I_m = 0$ and the core will not change state.

With this system we now have the ability to switch any one of 16 cores by selecting any two of eight wires. This is a savings of 50 percent over a direct parallel selection system. This savings in input wires becomes even more impressive if we enlarge the existing matrix to 100 cores (a square matrix with 10 cores on each side). In this case, we are able to switch any one of 100 cores by selecting any two of only 20 wires. This represents a reduction of 5 to 1 over a straight parallel selection system.

At this point we need to develop a method of sensing the contents of a core. This can be very easily accomplished by threading one *sense wire* through every core in the matrix. Since only one core is selected (switched) at a time, any output on the sense wire will be due to the changing of states of the selected core, and we will know which core it is since the core address is prerequisite to selection. Notice that the sense wire passes through half of the cores in one direction and through the other half of the cores in the opposite direction. Thus the output signal may be either a positive or a negative pulse. For this reason, the output from the sense wire is usually amplified and rectified to produce an output pulse which will always appear with the same polarity.

EXAMPLE 12-4

From the standpoint of construction, the core matrix shown in Fig. 12-14 is more convenient. Explain the necessary directions of half-select currents in the X and Y lines for proper operation of the matrix.

SOLUTION

Core X_1Y_1 is exactly similar to the previously discussed matrix in Fig. 12-13. Thus a current passing down through X_1 and to the right

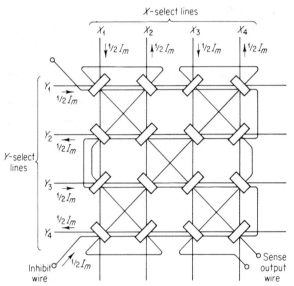

Fig. 12-14 Coincident-current memory matrix (one plane).

through Y_1 will set core X_1Y_1 to the 1 state. To set core X_1Y_2 to the 1 state, current will pass down through line X_1, but current must pass from the right to the left through line Y_2 (check with the *right-hand rule*). Proceeding in this fashion, we see that core X_1Y_3 is similar to X_1Y_1. Therefore, current must pass through line Y_3 from left to right. Similarly, core X_1Y_4 is similar to core X_1Y_2 and current must therefore pass through line Y_4 from right to left. From this discussion, we can see that in general current must pass from *left to right* through the *odd*-numbered Y lines, and from *right to left* through *even*-numbered Y lines.

Now, since current must pass from left to right through line Y_1, it is easily seen that current must pass upward through line X_2 in order to set core X_2Y_1. Following an argument similar to that given for the Y lines, we can see that current must pass *downward* through the *odd*-numbered X lines and *upward* through the *even*-numbered X lines.

The matrix shown in Fig. 12-14 has one extra winding which we have not yet discussed. This is the *inhibit wire*. In order to understand its operation and function, let us examine the methods for writing information into the matrix and reading information from the matrix.

In order to write a 1 in any core (that is, set the core to the 1 state), it is only necessary to apply $\frac{1}{2}I_m$ to the X and Y lines selecting that core address. If we desired to write a 0 in any core (that is, set the core to the

0 state), we could simply apply a current of $-\frac{1}{2}I_m$ to the X and Y lines selecting that core address. We can also write a 0 in any core by making use of the *inhibit* wire shown in Fig. 12-14. We assume that all cores are initially in the 0 state. Notice that the application of $\frac{1}{2}I_m$ to this wire in the direction shown on the figure will result in a complete cancellation of the Y line select current (it also tends to cancel an X line current). Thus to write a 0 in any core, it is only necessary to select the core in the same manner as if writing a 1, and at the same time apply an *inhibit* current to the *inhibit* wire. The major reason for writing a 0 in this fashion will become clear when we use these matrix planes to form a complete memory.

To summarize, we write a 1 in any core $X_a X_b$ by applying $\frac{1}{2}I_m$ to the select lines X_a and Y_b. A 0 can be written in the same fashion by simply applying $\frac{1}{2}I_m$ to the *inhibit* line at the same time (if all cores are initially reset).

To read the information stored in any core, we simply apply $-\frac{1}{2}I_m$ to the proper X and Y lines and detect the output on the sense wire. The select currents of $-\frac{1}{2}I_m$ will reset the core and if the core previously held a 1, an output pulse will occur. If the core previously held a 0, it will not switch and no output pulse will appear.

This then is the complete coincident-current selection system for one plane. Notice that reading the information out of the memory results in a complete loss of information from the memory since all cores are reset during the *read* operation. This is referred to as a *destructive readout* or DRO system. This matrix plane is used to store 1 bit in a word, and it is necessary to use n of these planes to store an n-bit word.

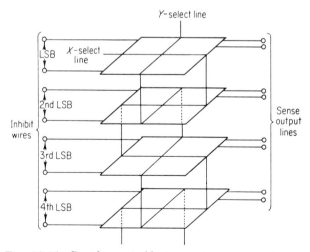

Fig. 12-15 Complete coincident-current memory system.

A complete parallel coincident-current memory system can be constructed by stacking the basic memory planes in the manner shown in Fig. 12-15. All the X drive lines are connected in series from plane to plane as are all the Y drive lines. Thus the application of $\frac{1}{2}I_m$ to lines X_a and Y_b will result in a selection of core $X_a Y_b$ in every plane. In this fashion we can simultaneously switch n cores, where n is the number of planes. These n cores represent one word of n bits. For example, the top plane might be the LSB, the next to the top plane would then be the second LSB, and so on; the bottom plane would then hold the MSB.

To read information from the memory, we simply apply $-\frac{1}{2}I_m$ to the proper address and sense the outputs on the n sense lines. Remember that readout results in resetting all cores to the 0 state, and thus that word position in the memory is cleared to all 0s.

To write information into the memory, we simply apply $\frac{1}{2}I_m$ to the proper X and Y selection lines. This will, however, write a 1 in every core. So for the cores in which we desire a 0, we will simultaneously apply $\frac{1}{2}I_m$ to the *inhibit* line. For example, to write 1001 in the upper four planes in Fig. 12-15, we apply $\frac{1}{2}I_m$ to the proper X and Y lines and at the same time apply $\frac{1}{2}I_m$ to the *inhibit* lines of the second and third planes.

This method of writing assumes that all cores were previously in the 0 state. For this reason it is common to define a *memory cycle*. One *memory cycle* is defined as a *read* operation followed by a *write* operation. This serves two purposes: First it ensures that all the cores will be in the 0 state during the *write* operation; second, it provides the basis for designing a *nondestructive readout* (NDRO) system.

It is quite inconvenient to lose the data stored in the memory every time they are read out. For this reason, the NDRO has been developed. One method for accomplishing this function is to read the information out of the memory into a temporary storage register (flip-flops perhaps). The outputs of the flip-flops are then used to drive the *inhibit* lines during the *write* operation which follows (inhibit to write a 0 and do not inhibit to write a 1). Thus the basic memory cycle allows us to form an NDRO memory from a DRO memory.

Example 12-5

Describe how a coincident-current memory might be constructed if it must be capable of storing 1024 twenty-bit words.

Solution

Since there are 20 bits in each word, there must be 20 planes in the memory (there is one plane for each bit). In order to store 1024 words, we could make the planes square. In this case, each plane would contain 1024 cores, and it would therefore be constructed with 32 rows and 32

Magnetic Devices and Systems   335

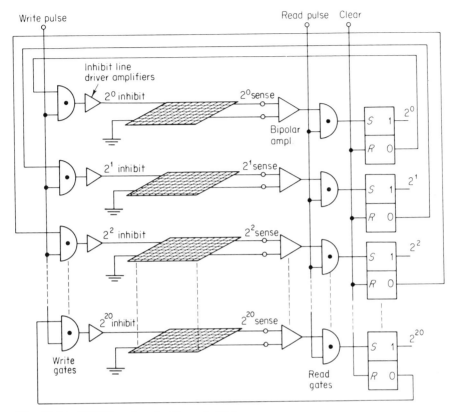

Fig. 12-16 NDRO system for Example 12-6.

columns since $(1024)^{1/2} = (2^{10})^{1/2} = 2^5 = 32$. This memory is then capable of storing $1024 \times 20 = 20{,}480$ bits of information. Typically, a memory of this size might be constructed in a 3-in. cube. Notice that in this memory we have the ability to switch any one of 20,480 cores by controlling the current levels on only 84 wires (32 X lines, 32 Y lines, and 20 *inhibit* lines). This is indeed a modest number of control lines.

EXAMPLE 12-6

Devise a means for making the memory system in the previous example a NDRO system.

SOLUTION

One method for accomplishing this is shown in Fig. 12-16. The basic core array consists of twenty 32-by-32 core planes. For convenience, only the three LSB planes and the MSB core plane are shown in the diagram.

The wiring and operation for the other planes will be the same. For clarity, the X and Y *select* lines have also been omitted. The output sense line of each plane is fed into a bipolar amplifier which will rectify and amplify the output so that a positive pulse will appear any time a set core is reset to the 0 state. A complete memory cycle consists of a *clear* pulse followed by a *read* pulse followed by a *write* pulse. The proper waveforms are shown in Fig. 12-17. The *clear* pulse first sets all flip-flops to the 0 state (this *clear* pulse can be generated from the trailing edge of the *write* pulse). When the *read* line goes high, all the AND gates driven by the bipolar amplifiers are enabled. Shortly after the rise of the *read* pulse, $-\frac{1}{2}I_m$ is applied to the X and Y lines designating the address of the word to be read out. This resets all cores in the selected word to the 0 state, and any core which contained a 1 will switch. Any core which switches will generate a pulse on the *sense* line which will be amplified and will appear as a positive pulse at the output of one of the bipolar amplifiers. Since the *read* AND gates are enabled, a positive pulse at the output of any amplifier will pass through the AND gate and set the flip-flop. Shortly thereafter the half-select currents disappear, the *read* line goes low, and the flip-flops now contain the data which were previously in the selected cores. Shortly after the *read* line goes low, the *write* line will go high, and this will enable the *write* AND gates (connected to the *inhibit* line drivers). The 0 side of any flip-flop which has a 0 stored in it is high, and this will enable the *write* AND gate to which it is connected. In this manner, an *inhibit* current is applied to any core which previously held a 0. Shortly after the rise of the *write* pulse, positive half-select currents will be applied to the same X and Y lines. These select currents will set a 1 in any core which does not have an *inhibit* current. Thus the information stored in the flip-flops is written directly back into the cores from which it came. The half-select currents are then reduced to zero and the *write* line goes low. The fall of the *write* line is used to reset the flip-flops, and the system is now ready for another *read/write* cycle.

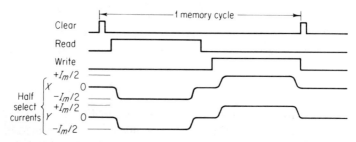

Fig. 12-17 NDRO waveforms for Fig. 12-16 (read from memory).

Magnetic Devices and Systems

The NDRO memory system discussed in the preceding example provides the means for reading information from the system without losing the individual bits stored in the cores. To have a complete memory system, we must have the capability to write information into the cores from some external source (e.g., input data). The *write* operation can be realized by making use of the exact same NDRO waveforms shown in Fig. 12-17. We must, however, add some additional gates to the system such that during the *read* pulse the data set into the flip-flops will be the external data we wish stored in the cores. This could easily be accomplished by adding a second set of AND gates which can be used to set the flip-flops. The logic diagram for the complete memory system is shown in Fig. 12-18. For simplicity, only the LSB is shown since the logic for every bit is identical.

For the complete memory system we recognize that there are two distinct operations. They are *write into memory* (i.e., store external data in the cores) and *read from memory* (i.e., extract data from the cores to be used elsewhere). For these two operations we must necessarily generate two distinct sets of control waveforms. The waveforms for *read from memory* are exactly those shown in Fig. 12-17, and the events are summarized as follows:

1. The *clear* pulse resets all flip-flops.
2. During the *read* pulse, all cores at the selected address are reset to 0 and the data stored in them are transferred to the flip-flops by means of the read AND gates.
3. During the *write* pulse, the data held in the flip-flops are stored back in the cores by applying positive half-select currents (the *inhibit* currents

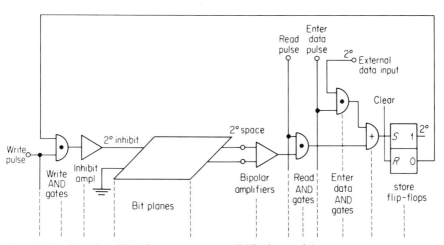

Fig. 12-18 Complete NDRO memory system (LSB plane only).

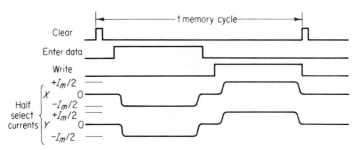

Fig. 12-19 NDRO waveforms for Fig. 12-18 (write into memory).

are controlled by the 0 sides of the flip-flops and provide the means of storing 0s in the cores).

The *write into memory* waveforms are exactly the same as shown in Fig. 12-17 with one exception: that is, the *read* pulse is replaced with the *enter data* pulse. The events for *write into memory* are shown in Fig. 12-19, and are summarized as follows:

1. The *clear* pulse resets all flip-flops.
2. During the *enter data* pulse, the negative half-select currents reset all cores at the selected address. The core outputs are not used, however, since the *read* AND gates are not enabled. Instead, external data are set into the flip-flops through the enter AND gates.
3. During the *write* pulse, data held in the flip-flops are stored in the cores exactly as before.

In conclusion, we see that the *write into memory* and *read from memory* are exactly the same operations with the exception of the data stored in the flip-flops. The waveforms are exactly the same when the *read* and *enter data* pulses are used appropriately, and the same total cycle time is required for either operation.

It should be pointed out that a number of difficulties are encountered with this type of system. First of all, since the *sense* wire in each plane threads every core in that plane a number of undesired signals will be on the *sense* wire. These undesired signals are a result of the fact that many of the cores in the plane receive a half-select current and thus exhibit a slight flux change.

The geometrical pattern of core arrangement and wiring shown in Fig. 12-13 represents an attempt to minimize the *sense*-line noise by cancellation. For example, the signals induced in the *sense* line by the X and Y drive currents would hopefully be canceled out since the *sense* line crosses these lines in the opposite direction the same number of times. Furthermore, the

Magnetic Devices and Systems

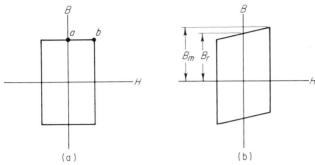

Fig. 12-20 Hysteresis curves. (a) Ideal. (b) Practical (realizable).

sense line is always at a 45° angle to the X and Y select lines. Similarly, the noise signals induced in the *sense* line by the partial switching of cores receiving half-select currents should cancel one another. This, however, assumes that all cores are identical, which is hardly ever true.

Another method for eliminating noise due to cores receiving half-select currents would be to have a core which exhibits an absolutely rectangular **BH** curve as shown in Fig. 12-20a. In this case, a half-select current would move the operating point of the core perhaps from point a to point b on the curve. However, since the top of the curve is horizontal, no flux change would occur and therefore no undesired signal could be induced in the sense wire. This is an ideal curve, however, and cannot be realized in actual practice. A measure of core quality is given by the *squareness ratio*, which is defined as

$$\text{Squareness ratio} = \frac{\mathbf{B}_r}{\mathbf{B}_m}$$

This is the ratio of the flux density at the remanent point $\mathbf{B}_r$ to the flux density at the switching point $\mathbf{B}_m$ and is shown graphically in Fig. 12-20b. The ideal value is, of course, 1.0, but values between 0.9 and 1.0 are the best obtainable.

12-5 Memory Addressing

In this section we will investigate the means for activating the X and Y selection lines which supply the half-select currents for switching the cores in the memory. First of all, since it typically requires 100 to 500 ma in each select line (that is, I_m is typically between 100 and 500 ma), each select line must be driven by a current amplifier. A special class of transistors has been developed for this purpose; they are referred to as core drivers in data sheets. What is then needed is the means for activating the proper core-driver amplifier.

Up to this point, we have designated the X lines as $X_1, X_2, X_3, \ldots, X_n$ and the Y lines as $Y_1, Y_2, Y_3, \ldots, Y_n$. For a square matrix, n is the number of cores in each row or column, and there are then n^2 cores in a plane. When the planes are arranged in a stack of M planes, where M is the number of bits in a word, we have a memory capable of storing n^2, M-bit words. Any two select lines can then be used to read or write a word in memory, and the address of that word is $X_a Y_b$, where a and b can be any number from 0 to n. For example, $X_2 Y_3$ represents the column of cores at the intersection of the X_2 and Y_3 select lines, and we can then say that the address of this word is 23. Notice that in this case the first digit in the address is the X line and the second digit is the Y line. This is arbitrary and could be reversed.

This method of address designation has but one problem: in a digital system we can use only the numbers 1 and 0. The problem is easily resolved, however, since the address 23, for example, can be represented by 010 011 in binary form. If we use 3 bits for the X line position and 3 bits for the Y line position, we can then designate the address of any word in a memory having a capacity of 64 words or less. This is easy to see, since with 3 bits we can represent eight decimal numbers, which means we can define an $8 \times 8 = 64$ word memory. If we chose an 8-bit address, 4 bits for the X line and 4 bits for the Y line, we could define a memory having $2^4 \times 2^4 = 16 \times 16 = 256$ words. In general, an address of B bits can be used to define a square memory of 2^B words where there are $B/2$ bits for the X lines and $B/2$ bits for the Y lines. From this discussion it is easy to see why large-scale coincident-current memory systems usually have a capacity which is an even power of 2.

EXAMPLE 12-7

What would be the structure of the binary address for a memory system having a capacity of 1024 words?

SOLUTION

Since $2^{10} = 1024$, there would have to be 10 bits in the address word. The first 5 bits could be used to designate one of the required 32 X lines and the second 5 bits could be used to designate one of the 32 Y lines.

EXAMPLE 12-8

For the memory system described in the previous example, what is the decimal address for the following binary addresses?
(a) 10110 00101
(b) 11001 01010
(c) 11110 00001

Magnetic Devices and Systems 341

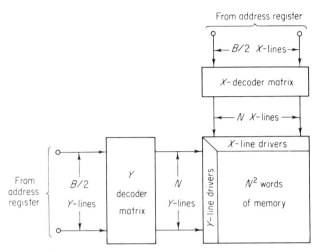

Fig. 12-21 Coincident-current memory addressing.

SOLUTION

(a) The first 5 bits are the X line and correspond to a decimal number 22. The second 5 bits represent the Y line and correspond to the decimal number 5. Thus the address is $X_{22}Y_5$.

(b) $11001_2 = 25_{10}$ and $01010_2 = 10_{10}$. Therefore, the address is $X_{25}Y_{10}$

(c) The address is $X_{30}Y_1$.

The B bits of the address in a typical digital system will be stored in a series of flip-flops called the *address register*. The address in binary form must then be decoded into decimal form in order to drive one of the X line drivers and one of the Y line driver amplifiers as shown in Fig. 12-21. The X and Y decoding matrices shown in the figure can be identical and are essentially binary-to-decimal decoders. You will recall that binary-to-decimal decoding and appropriate matrices were discussed in Chap. 10.

12-6 Apertured Plate

Some of the difficulties inherent in the fabrication of a magnetic-core plane can be overcome by the use of an *apertured plate*. This is a rectangular plate made by molding a ferrite material and drilling a regular array of holes in it. One such plate which has been constructed is shown with its dimensions in Fig. 12-22. The holes are drilled on 50-mil centers and are 25 mils in diameter. Since there are 16 holes on a side, the plate contains a total of 256 holes.

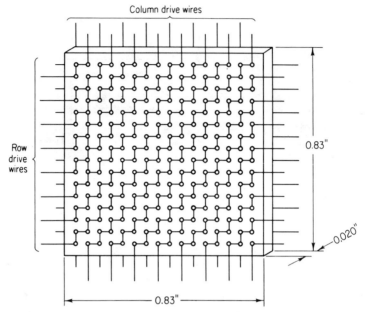

Fig. 12-22 Apertured plate.

Each of the holes in the plate behaves in a fashion similar to an individual magnetic core. Thus if wires are threaded through the holes in the plate it can be used in a fashion similar to a 16-by-16 array of cores. Since the ferromagnetic material from which the plate is constructed is a good insulator, the wiring can be accomplished by printing metallic film strips directly on the plate. The strips are arranged such that they pass through every hole in the plate. There are 16 separate vertical strips each of which passes through 16 holes, and 16 horizontal strips each of which passes through 16 holes. These strips can then be used to select any hole for switching by applying half-select currents just as the X and Y select lines are used in a core array.

There is a critical value of coincident current which will switch the flux direction around a hole in the plate. This value of current cannot be increased without limit since the plate is a singular piece of material and the flux from one hole will spread farther out into the material as the current is increased. Thus an upper limit on the switching current must be observed in order to prevent the switching of flux around one hole from affecting adjacent holes. One interesting difference between a core and the apertured plate is the fact that in the latter an increase in coincident current will increase the flux switched but not the switching time. The switching time

Magnetic Devices and Systems   343

for the plate is relatively constant above a certain value. Typically a switching current of around 300 ma will provide a 30-mv output on the *sense* line with a switching time of 1.5 μsec.

To complete the basic plate, one wire is threaded through all 256 holes on the plate, and this wire is used for both the sensing and the inhibit functions.

One of the problems which arises is the noise currents on the *sense* line present during readout. Because of the pattern of holes it is not possible to thread the *sense* line such that the half-selected hole fluxes will cancel one another. For this reason the noise problem is considerably worse than with cores.

One remedy for the noise due to half-select currents is to use two holes per bit. The method is known in Fig. 12-23 and makes use of two plates placed side by side. The select wires for each plate are connected in series just as the X and Y lines for the core planes are connected. One bit of information is stored by magnetizing a hole in plate A and magnetizing the corresponding hole in plate B in the opposite direction. A stored 1 and a stored 0 are shown in Fig. 12-23b and c, respectively. It is now possible to connect the *sense* wires such that the noise induced because of half-selected holes will nearly cancel out. This, of course, assumes that the plates are identical and uniform to the extent that all holes have exactly the same

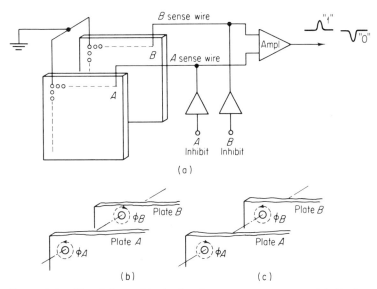

Fig. 12-23 Two-hole-per-bit ferrite-plate memory system. (a) Logic. (b) A stored 1. (c) A stored 0.

switching characteristics. This condition is naturally never reached in actual practice, but nevertheless this method greatly reduces the noise problem.

The method for storing information in the plates is as follows:

1. Assume that a half-select current applied to a vertical line and a horizontal line during the *read from memory* will magnetize both the hole in plate A and the hole in plate B in a counterclockwise direction.

2. Therefore, half-select currents applied during the *write into memory* will magnetize both holes in a clockwise direction.

3. To write a 1 in the memory, we will apply an *inhibit* current to plate A, and the holes will be magnetized as shown in Fig. 12-23.

4. To write a 0 in the memory, we will apply an *inhibit* current to plate B, and the holes will be magnetized as shown in the same figure.

Now, during the *read from memory* operation, there are two possibilities:

1. A 1 was stored in memory. In this case A will not switch, but B will. This will generate a positive pulse at the output of the sense amplifier.

2. A 0 was stored in memory. In this case B will not switch, but A will. This will generate a negative pulse at the output of the sense amplifier. The switching operations of the holes are shown diagrammatically in Fig. 12-24.

It is interesting to note that this is one of four possible combinations. That is, since there are two holes involved and since the flux can be either clockwise or counterclockwise, there are four possible starting points. These are the one discussed, counterclockwise-counterclockwise (CCL-CCL) and three others: CCL-CL, CL-CCL, and CL-CL (see Prob. 12-15).

These basic pairs of plates can then be stacked to form a three-dimensional memory much as the core planes were stacked. There would, of course, have to be n pairs of plates for a memory capable of storing words having n bits.

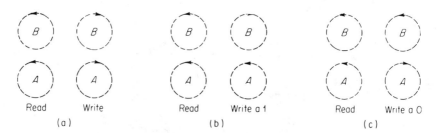

Fig. 12-24 Ferrite-plate recording showing hole flux directions. (*a*) No inhibit. (*b*) A inhibit. (*c*) B inhibit.

Magnetic Devices and Systems

In conclusion, the use of apertured plates eliminates some of the construction problems with threading large arrays of cores, but new problems are at the same time introduced. The first and probably the most serious problem is that of manufacturing plates at a high yield rate. In manufacturing cores, one bad core is simply one bad core; but with apertured plates, one bad hole means the loss of 255 other holes — good or bad. Furthermore, it is considerably more difficult to manufacture plates which have uniform characteristics over the surface of the plate as well as from plate to plate than it is to manufacture individual cores which are all uniform. Finally the magnitudes of the select currents used with apertured plates must be much more carefully controlled than with cores, and there is also the problem of bipolar sensing of the output signals from the sense amplifiers.

12-7 Multiaperture Devices

One of the more common multiaperture devices is a two-holed magnetic core often referred to as a *transfluxor*. The name transfluxor is derived from the fact that the flux in the core is transferred from one leg to the other. A typical transfluxor along with its dimensions is shown in Fig. 12-25. The most important feature of the device is that the flux configuration around one hole can be controlled by the flux configuration around the other hole. Because of this, these devices can be used to form three-dimensional memory systems which have nondestructive readout capabilities. The device is also very useful as a logical element and as a modulator in pulse-modulation systems.

The *block* (reset) and *unblock* (set) windings on the transfluxor can be used to reset and set the device as shown in Fig. 12-26. The two holes in the device form three legs which are labeled 1, 2, and 3 in the figure. The holes are designed so that the cross-sectional areas of legs 2 and 3 are equal and

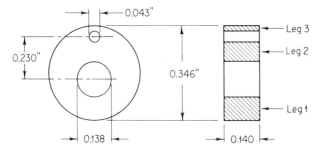

Fig. 12-25 A typical transfluxor element.

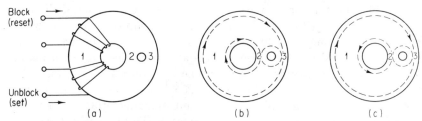

Fig. 12-26 Transfluxor switching. (a) *Reset* and *set* windings. (b) *Blocked* or *reset* state. (c) *Unblocked* or *set* state.

the cross-sectional area of leg 1 is approximately twice that of leg 2 or 3. To reset the device, a current is applied to the block winding such that the flux is entirely clockwise in the device. A 1 is set when a current is applied to the *unblock* or *set* winding. The flux around the larger hole begins to switch to the opposite direction, and when all the flux in leg 2 has been reversed, the *set* operation is complete. The magnetizing force for the *set* operation is limited either by applying a smaller amount of current to the *set* winding or by constructing this winding with fewer turns.

The flux patterns in the transfluxor for a 1 (unblocked) and a 0 (blocked) are summarized in Fig. 12-26c and b, respectively. Notice that in the blocked state the flux in both legs 2 and 3 are in the same direction. In the unblocked state, the flux through these two legs is in opposition. It is the difference between these two states which makes the transfluxor a valuable logical element.

One method of using the transfluxor is to place two more windings on the device. An *input* winding is wound around leg 3 and an *output* winding is wound around leg 2 as shown in Fig. 12-27. The two windings are used as a transformer with the material around the smaller hole acting as the coupling medium between them. An a-c signal is then applied to the input winding. If the core is in a blocked state, no signal will appear at the output winding. But if the core is in the unblocked state, transformer action occurs and an a-c signal appears at the output winding.

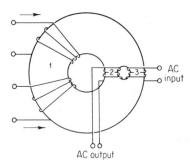

Fig. 12-27 Transfluxor with a-c input-output windings.

Magnetic Devices and Systems

Consider the case when the transfluxor is in the blocked state. When the a-c input signal tends to magnetize the material around the smaller hole in a clockwise direction, it is opposed by the flux through leg 2. When it tries to magnetize around the hole in the opposite direction, it is opposed by the flux in leg 3. Thus no flux reversal takes place because of the input a-c signal and no transformer action occurs. Therefore, no output appears in the output winding.

Now consider the case when the transfluxor is in the unblocked state. When the a-c input signal attempts to magnetize around the hole in a clockwise direction, the fluxes in both legs 2 and 3 are in agreement. Similarly, when trying to magnetize around the hole in the opposite direction, the fluxes in both legs 2 and 3 will switch in the same direction. Therefore, in this state the a-c input signal is capable of switching the flux direction around the smaller hole, resulting in transformer action between the input and output windings. Thus an output signal will appear when the transfluxor is in the set or unblocked state. From the above discussion it should be clear how the transfluxor might be used to provide a modulated pulse train; a pulse represents a 1 and the absence of a pulse represents a 0.

A second method for using the transfluxor as a nondestructive-readout memory device is shown in Fig. 12-28. In this case, two windings are wound through the smaller hole. These are the *prime/drive* winding and the *output* winding. Storing a 1 or a 0 in the transfluxor proceeds in the same manner as before by applying currents to the *block* and *unblock* windings. Reading the stored information out of the device is accomplished by applying first a positive current pulse (prime) and then a negative current pulse (drive) to the *prime/drive* winding. The contents of the transfluxor is then determined by sensing the signal on the output winding.

The *read* operation is summarized in diagram form along with the proper waveforms in Fig. 12-29. The readout operation from the *set* transfluxor is as follows:

1. The prime current sets the flux around the small hole in a counterclockwise direction.

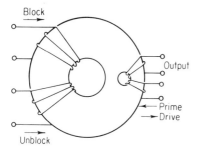

Fig. 12-28 Transfluxor memory element.

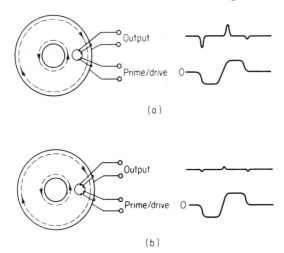

Fig. 12-29 Transfluxor readout. (*a*) Unblocked or set (1). (*b*) Blocked or reset (0).

2. The following drive current switches it back to a clockwise direction. This switching action is possible since the fluxes in legs 2 and 3 in the *set* transfluxor are in opposite directions. Since a switching of flux occurs around the hole, a signal is induced in the output winding and a 1 is detected. It is most important to note that the flux direction around the small hole is returned to its original position after the *read* operation is complete. This means that the transfluxor can be used directly as a nondestructive-readout device without the necessity for all the external logic in the NDRO core memory.

The readout from a *blocked* transfluxor is as follows:

1. The prime current is not capable of reversing the flux around the small hole since the fluxes in legs 2 and 3 are in the same directions.
2. Similarly, the drive current is not capable of switching the flux around the hole. Therefore, there is no net switching of flux and no signal is induced in the output winding.

The transfluxor can be used in a manner similar to cores to construct a three-dimensional NDRO memory system as shown in Fig. 12-30. The X and Y select lines operate in a coincident-current manner and are threaded through the larger holes. The *prime/drive* wires also operate in a coincident-current manner but are threaded through the smaller holes. One *sense* wire is threaded through the smaller hole in every transfluxor in the plane. The *sense* wire is, of course, threaded to minimize pickup from other wires and half-selected devices. Memories constructed using transfluxors generally

Magnetic Devices and Systems

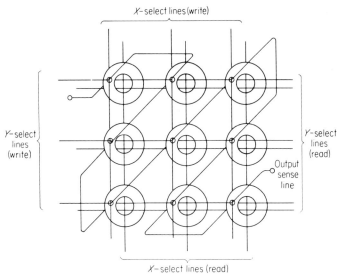

Fig. 12-30 Transfluxor NDRO coincident-current memory.

require larger select currents than those using magnetic cores, but the *sense* output voltages from transfluxors are much greater. You will recall that the typical output from a core is around 100 mv for a 1 with around 30 mv for the 0. A transfluxor might typically have a 1 output of 4 volts and a 0 output of 200 mv. Thus the *sense* amplifier requirements for the transfluxor are somewhat more lenient. Coincident-current memories using cores or transfluxors can be made to operate with about the same overall cycle times.

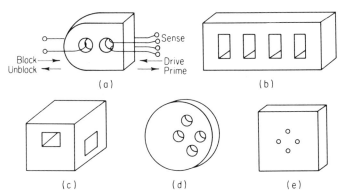

Fig. 12-31 Ohter magnetic logic elements. (*a*) MARS. (*b*) LADDIC. (*c*) BIAX. (*d*) Four-aperture transfluxor. (*e*) Micro-aperture device.

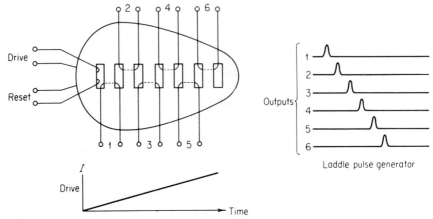

Fig. 12-32 LADDIC pulse generator.

There are a number of other multiapertured devices some of which are shown in Fig. 12-31. The MARS (MultiAperture Reluctance Switch) operates in much the same manner as the transfluxor. The *set* and *block* currents are applied to the same winding but have opposite polarities. The *drive/prime* and *sense* windings operate just as with the transfluxor. Since this device is symmetrical, it is possible to use *block, unblock, prime,* and *drive* currents of the same amplitude. Moreover, these devices have been constructed significantly smaller than the transfluxors and therefore require somewhat smaller *drive* currents.

Laddic elements (from LADDer logIC) of numerous configurations have been constructed and are essentially an extension of the basic ideas involved in the transfluxor. One interesting application of a ladder is the formation of a pulse distributor. This is shown in Fig. 12-32. As the input drive current increases, the area of material switched increases with time. Thus the flux through the legs switches sequentially with time, beginning with leg 1 and then leg 2, etc.

There are numerous other magnetic devices and systems. The reader is referred to articles in technical journals and to texts devoted specifically to magnetics for further information.

12-8 Magnetic-drum Storage

Magnetic cores and multiapertured devices arranged in a three-dimensional form offer great advantages as memory systems. By far the most important advantage is the speed with which data can be written into or read from the memory system. This is called the *access time*, and for core memory systems it is simply the time of one *read/write* cycle. Thus the access time is directly related to the clock, and typical values are from less than 1 to a few microseconds. These types of memory systems are said to be *random-access* since any word in the memory can be selected at random. The primary disadvantage of this type of memory system is the cost of construction for the amount of storage available. As an example of what is meant, you will recall that a magnetic tape is capable of storing large quantities of data at a relatively low cost per bit of storage. A typical tape might be capable of storing up to 20 million characters, which corresponds to 120 million bits (Chap. 10). To construct such a memory with magnetic cores would require about 3 million cores per plane, assuming we use a stack of 36 planes corresponding to a 36-bit word. It is quite easy to understand the impracticality of constructing such a system. What is needed, then, is a system which is capable of storing information with less cost per bit but having a greater capacity.

Such a system is the *magnetic-drum* storage system. The basis of a magnetic drum is a cylindrical-shaped drum, the surface of which has been coated with a magnetic material. The drum is rotated on its axis as shown in Fig. 12-33, and the *read/write* heads are used to record information on the drum or read information from the drum. Since the surface of the drum is magnetic it exhibits a rectangular hysteresis loop property and can thus be magnetized. The process of recording on the drum is much the same as used for recording on magnetic tape as discussed in Chap. 10, and the same methods for recording are commonly used (i.e., RZ, NRZ, and NRZI). The data are recorded in tracks around the circumference of the drum, and there is one *read/write* head for each track. There are three major methods for storing information on the drum surface; they are *bit-serial*, *bit-parallel*, and *bit-serial-parallel*.

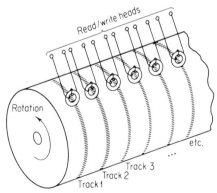

Fig. 12-33 Magnetic-drum storage.

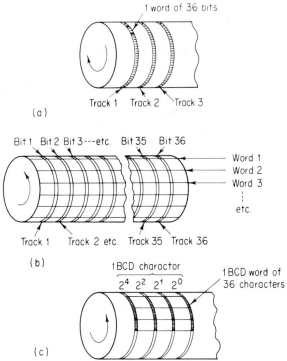

Fig. 12-34 Magnetic-drum organization. (a) Bit-serial storage. (b) Bit-parallel storage. (c) Bit-serial-parallel storage.

In bit-serial recording, all the bits in one word are stored sequentially, side by side, in one track of the drum. Bit-serial storage is shown in Fig. 12-34a. Storage densities of 200 to 1000 bits per inch are typical for magnetic drums. A typical drum might be 8 in. in diameter and thus have the capacity to store $\pi \times 8$ in. $\times 200$ bits per in. $= 5024$ bits in each track. Drums have been constructed having anywhere from 15 to 400 tracks, and a spacing of 20 tracks to the inch is typical. If we assume this particular drum is 8 in. wide and has a total of 100 tracks, we see immediately that it has a storage capacity of 5024 bits per track $\times$ 100 tracks $= 502{,}400$ bits of information. Compare this capacity with that of a coincident core memory, which is 64 cores on a side (quite a large core system), and having 64 core planes. This core memory has a capacity of $2^6 \times 2^6 \times 2^6 = 262{,}144$ bits. The drum described above is actually considered small, and much larger drums have been constructed and are now in use.

Magnetic Devices and Systems

EXAMPLE 12-9

A certain magnetic drum is 12 in. in diameter and 12 in. long. What is the storage capacity of the drum if there are 200 tracks and data are recorded at a density of 500 bits per inch?

SOLUTION

Each track has a capacity of $\pi \times 12$ in. $\times 500$ bits per in. $\cong 18,840$ bits. Since there are 200 tracks, the drum has a total capacity of $18,840 \times 200 = 3,768,000$ bits.

In the preceding example, each track has the ability to store about 18,840 bits. If we use a 36-bit word, it is easy to see that we can store about 523 words in each track. Since the words are stored sequentially around the drum, and since there is only one *read/write* head for the track, it is easy to see that we may have to wait to read any one word. That is, the drum is rotating, and the word we want to read may not be under the *read* head at the time we choose to read it. It may in fact have just passed under the head, and we will have to wait until the drum completes nearly a full revolution before it will be under the head again. This points out one of the major disadvantages of the drum compared with core storage. That is the problem of access time. On the average, we can assume that we will have to wait the time required for the drum to complete one-half a revolution. A drum is then said to have a *restricted access*.

EXAMPLE 12-10

If the drum in Example 12-8 rotates at a speed of 3000 rpm what is the average access time for the drum?

SOLUTION

3000 rpm $=$ 50 rps. Thus the time for one revolution is $1/(50 \text{ rps}) = 20$ msec. Thus the average access time is one-half the time of one revolution, which is 10 msec. Contrast this with a coincident-current core memory which has a direct access time of around a few microseconds.

Notice in the previous example that it will require a short period of time to read the 36 bits of the word since they appear under the *read* head 1 bit at a time in a serial fashion. The actual time required is small compared with the access time and is found to be $(20 \text{ msec/r})/(523 \text{ words per track}) \cong 40$ μsec. This read time can be reduced by storing the data on the drum in a parallel manner as shown in Fig. 12-34b.

The average access time for bit-parallel storage is the same as for bit-serial storage, but it is possible to read and record information at a much faster rate with the bit-parallel system. Let us use the drum in Example 12-9 once more. Since there are 523 words around each track, and since the drum rotates at 50 rps, we can read (or write) 523 words per revolution × 50 rps = 26,150 words per second. If the data were stored in parallel fashion, we could read (or write) at 36 times this rate, or at a rate of 18,840 words per revolution × 50 rps = 942,000 words per second. We would of course arrange to have the number of tracks on the drum an even multiple of the number of bits in a word. For example, with a 36-bit word we might use a drum having 36 or 72 or 108 tracks.

A third method for recording data on a drum is used, and it is called bit-serial-parallel. The method is shown in Fig. 12-34c and is commonly used for storing BCD information. The access and read (or write) times are a combination of the serial and parallel times. One BCD character occupies 1 bit in each of four adjacent tracks. Thus every four tracks might be called a band and each BCD character will occupy one space in the band. If there are 36 BCD characters in a word, we can then store 523 words on the drum of Example 12-9.

Quite often the access time is speeded up by the addition of extra *read/write* heads around the drum. For example, we might use two sets of heads placed on opposite sides of the drum. This would obviously cut the access time in half. Alternatively we might use three sets of heads arranged around the drum at 120° angles. This would reduce the access time by one-third.

Since writing on and reading from the drum must be very carefully timed, one track in the drum is usually reserved as a timing track. On this track, a series of timing pulses is permanently recorded and is used to synchronize the *write* and *read* operations. For the drum discussed in Example 12-9, there are 523 words in each track around the circumference of the drum. We might then record a series of 523 equally spaced timing marks around the circumference of the timing track. Each pulse would then designate the *read* or write *position* for a word on the drum.

SUMMARY

There is a wide variety of magnetic devices which can be used as binary devices in digital systems. By far the most widely used is the magnetic core. Cores can be used to implement various logic functions such as AND, OR, and NOT, and more complicated functions can be formed from combinations of these basic circuits. Magnetic-core shift registers and ring counters can be constructed by using the single-diode transfer loop between cores. Magnetic-core logic is particularly useful in applications experiencing environmental extremes.

Direct-access memories having very fast access times can be conveniently constructed using either magnetic cores or transfluxors. The most popular method for constructing these memories is the coincident-current technique. Memories constructed using cores are inherently DRO-type memories but can be transformed into NDRO memories by the addition of external logic. Memories constructed using transfluxors are inherently NDRO-type memories. There are a number of devices which can be used for both logic and memory functions, among which are the MARS, BIAX, LADDIC, and multiaperture transfluxors. The apertured plate is designed specifically for use in constructing coincident-current memories.

Magnetic drums and disks provide a means for larger storage capacities at a lower cost per bit than core-type memories. They do, however, offer the disadvantage of increased access time.

GLOSSARY

access time For a coincident-current memory, it is the time required for one *read/write* cycle. In general, it is the time required to write one word into memory or to read one word from memory.

address A series of binary digits used to specify the location of a word stored in a memory.

coincident-current selection The technique of applying $\frac{1}{2}I_m$ on each of two lines passing through a magnetic device in such a way that the net current of I_m will switch the device.

DRO Destructive readout.

hysteresis Derived from the Greek word *hysterin*, which means to lag behind.

hysteresis curve Generally a plot of magnetic flux density **B** versus magnetic force **H**. Can also refer to the plot of magnetic flux ϕ versus magnetizing current I.

memory cycle In a coincident-current memory system, a *read* operation followed by a *write* operation.

NDRO Nondestructive readout.

select current I_m The minimum current required to switch a magnetic device.

single-diode transfer loop A method of coupling the output of one magnetic core to the input of the next magnetic core.

squareness ratio A measure of core quality. From the hysteresis curve, it is the ratio of $\mathbf{B}_r/\mathbf{B}_m$.

transfluxor Generally a magnetic core having two holes of unequal diameters. Multiapertured cores are sometimes referred to as multiapertured transfluxors.

REVIEW QUESTIONS

1. Name one advantage of a ferrite core over a metal-ribbon core.
2. Name one advantage of a metal-ribbon core over a ferrite core.
3. Describe the method for detecting a stored 1 in a core.
4. Why is a strobing technique often used to detect the output of a switched core?
5. How is core switching time t_s affected by the switching current?
6. Explain why more complicated logic functions using cores can lead to excessive operating times.
7. What is the purpose of the diode in the single-diode transfer loop?
8. Why is a delay in signal transfer between cores desired?
9. Explain how the R and C in Fig. 12-11 introduce a delay in signal transfer between cores.
10. Explain the operation of the *sense* wire in a magnetic-core matrix plane. Why is it possible to thread every core in the plane with the same wire?
11. Explain how it is possible to store a 0 in a coincident-current memory core using the *inhibit* line.
12. Why is a basic coincident-current core memory inherently a DRO-type system?
13. In the basic memory cycle for a coincident-current core memory system, why must the *read* operation come before the *write* operation?
14. What is the difference between the *write into memory* and the *read from memory* cycles for a coincident-current core memory system?
15. What are some of the advantages and disadvantages of an apertured plate?
16. Explain how a transfluxor can be used to form a NDRO memory.
17. What are some advantages of transfluxors over cores when used in memory systems?
18. Describe the difference between random-access and restricted-access memories.
19. Describe the advantages and disadvantages of using a magnetic-drum storage system.

PROBLEMS

12-1 Draw a typical hysteresis curve for a core and show the two remanent points.

12-2 Show graphically on a ϕI curve the path of the operating point as the core is switched from a 1 to a 0. Repeat for switching from a 0 to a 1.

12-3 Draw the symbol for a magnetic-core logic element and explain the function of each winding.

Magnetic Devices and Systems

12-4 Draw a set of waveforms showing how the exclusive-OR circuit of Fig. 12-8d must operate (notice it requires only two clocks which are spaced 180° out of phase).

12-5 Draw a single-diode transfer loop between two cores and explain its operation (use waveforms if desired).

12-6 Draw a schematic and the waveforms for a core ring counter which will provide seven output pulses.

12-7 Draw a sketch and explain how a core can be switched by the coincident-current method.

12-8 Make a sketch similar to Fig. 12-15 showing a three-dimensional core memory capable of storing one hundred 10-bit words. Show all input and output lines clearly.

12-9 Describe the geometry of a coincident-current core memory capable of storing 4096 thirty-six-bit words (i.e., how many planes, and how many cores per plane, etc.).

12-10 How many bits can be stored in the memory in Prob. 12-9?

12-11 How many control lines are required for the memory in Prob. 12-9?

12-12 Show graphically the meaning of squareness ratio for a magnetic core and explain its importance for magnetic-core memories.

12-13 Describe a structure for the address which could be used for the memory of Prob. 12-9.

12-14 If a certain core memory is composed of square matrices, what is the word capacity if the address is 12 binary digits?

12-15 Show diagrammatically (as in Fig. 12-24) the other three switching possibilities for a two-hole-per-bit apertured plate. Give the polarities of the output signals for each case.

12-16 Describe the method for storing a 1 and a 0 in a transfluxor (use diagrams if desired).

12-17 What is the bit storage capacity of a magnetic drum 10 in. in diameter if data are stored with a density of 200 bits per inch in 20 tracks?

12-18 What would be the diameter of a magnetic drum capable of storing 3140 thirty-six-bit words if there are 10 tracks and data are stored bit-serial at 300 bits per inch?

12-19 What is the average access time for the drum in Prob. 12-18 if it rotates at 36,000 rpm? What could be done to reduce this access time by a factor of 2?

12-20 For the drum in Prob. 12-18, at what bit rate must data be moved (i.e., read or write) if the drum rotates at 36,000 rpm?

CHAPTER 13

DIGITAL ARITHMETIC

In a previous chapter we learned how to perform various arithmetic operations using binary numbers. We learned that binary addition is carried out according to the rules of modulo-2 arithmetic and that binary subtraction can be performed by using 1's and 2's complements. In later chapters various logic circuits were developed which were shown to be capable of carrying out binary arithmetic operations. These logic circuits, namely, half- and full-adders, half- and full-subtractors, and shift registers, form the basis of the arithmetic unit. An arithmetic unit is usually included in a digital system whenever the capability to add, subtract, multiply, divide, etc., is desired. In a small special-purpose digital system the arithmetic requirements may be limited, and the arithmetic unit will then be quite small. On the other end of the scale, a large general-purpose digital computer will have the capability of performing any number of arithmetic operations, and the arithmetic unit will then be quite large and necessarily involved. In this chapter we are interested in investigating some of the methods and techniques for carrying out digital arithmetic using the basic arithmetic logic circuits previously discussed. It should be pointed out that there are almost as many ways of implementing arithmetic

Digital Arithmetic

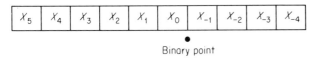

Binary point

Fig. 13-1

functions as there are designers, and this chapter is by no means a comprehensive coverage of the subject. The intent here is to give a solid foundation in the basic principles. The reader interested in advanced techniques is urged to study the references.

13-1 Number Representations

In a previous chapter it was shown that we could store a binary number in a register in two ways, that is, with the MSB on the right or with the MSB on the left. For example, the number 0111_2 when stored in a register with the MSB on the left is equal to 7_{10}. In finding the decimal equivalent of this binary number we simply assumed that the binary point was at the far right. Since no previous mention was made of the binary point, we could just as easily have assumed that it was on the far left. In this case the decimal equivalent would be 0.4375_{10}. For that matter we could have assumed the binary point to be at any number of other places. In any case, it must now be clear that the position of the binary point must be defined.

In trying to decide where to locate the binary point, one suddenly realizes that if we are intending to discuss logic capable of doing arithmetic in general, we must have the capability of handling fractions as well as integers. The most straightforward way of handling this is simply to define the position of the binary point in the word. For example, in Fig. 13-1, the binary-point location is defined. In this particular case, the digits to the left of the binary point represent the integers of the binary number and the digits to the right represent the fractional part of the number. We can then represent numbers between 111111.1111 and 000000.0000, with the smallest fractional part being $0.0001 = 1/16_{10} = 0.0625_{10}$. We can then add or subtract binary numbers in this form and have no difficulties with the location of the binary point. For example, we might add 001110.0011 and 000100.1000 as follows:

$$\begin{array}{rr} 001110.0011 & 14.1875 \\ (+)\ \underline{000100.1000} & (+)\ \underline{4.5000} \\ 010010.1011 & 18.6875 \end{array}$$

Or we might add the following two numbers:

$$\begin{array}{r} 100100.0010 \\ (+)\ 011100.1000 \\ \hline 1000000.1010 \end{array} \qquad \begin{array}{r} 36.125 \\ (+)\ 28.500 \\ \hline 64.625 \end{array}$$

You will notice that in this case we added two numbers the sum of which exceeded the capacity of our register. That is to say, this register is capable of handling numbers only up to $111111.1111_2 = 63.9375_{10}$. The fact that the sum exceeded the capacity of the register can be easily detected by sensing the 1 in the leftmost position of the sum. This 1 is the result of a carry generated when the two MSBs were added, and we will use this fact to detect this condition. Thus when the contents of the register is exceeded, an *overflow* condition exists, and we will sense this by detecting a 1 in the carry from the two MSBs. An overflow must be detected since the contents of the register after this last addition will be $000000.1010 = 0.625_{10}$. This obviously incorrect answer is a result of the fact that the register is capable of storing only 10 bits, and the eleventh bit (i.e., the carry from the two MSB's) is lost during the addition. Thus the overflow is an indication of incorrect operation.

Up to this point, we have shown that the selection of the binary point as shown in Fig. 13-1 is perfectly all right for addition so long as we detect any overflows. Let us see what happens when we multiply two numbers with this system. We might, for example, multiply two numbers as follows:

$$\begin{array}{r} 100110.1001 \\ \times\ 000001.0001 \\ \hline 1001101001 \\ 1001101001 \\ \hline 10100011111001 \end{array} = \text{product}$$

In the product we have no difficulty in finding the proper place for the binary point; we simply count eight places over from the right, and the answer is obviously 101000.11111001. This does not seem to offer any difficulties since there are still only 6 bits to the left of the binary point and we could simply omit the four extra digits to the right of the binary point. However, let us try one more example:

$$\begin{array}{r} 100110.1001 \\ \times\ 100001.0001 \\ \hline 1001101001 \\ 1001101001 \\ 1001101001 \\ \hline 1001111101011111001 \end{array}$$

In this case, we still have no trouble locating the binary point; again we count eight places from the right, giving 10011111010.11111001. We have now obviously greatly exceeded the capacity of the register on both ends.

Digital Arithmetic

Furthermore, the register has no means for counting the number of places from the right to locate the binary point. In this case most machines would simply use the 10 MSBs to give the answer 100111.1101, which is obviously incorrect.

To overcome these difficulties, it is quite common practice to place the binary point to the left of the MSB. We would then represent all numbers as $.X_1X_2 \ldots X_n$, where n is the number of bits. For example, the 10-bit number previously discussed would be $.X_1X_2X_3 \ldots X_9X_{10}$. Let us see how this will solve the problems previously described. In the first place, it makes no difference where the binary point is located for addition (and also therefore for subtraction), and this method is therefore acceptable. We may, however, still have an overflow problem and must still account for this possibility. In the second place, all numbers are represented as fractions since the binary point is to the far left. This means that the product of any two numbers will always be a number less than 1. Thus we can easily locate the binary point by simply using the 10 most significant digits. To clarify this point let us multiply two 10-bit numbers using this system:

$$
\begin{array}{r}
.1101100011 \\
\times\ .1000000010 \\
\hline
0000000000 \\
1101100011 \\
1101100011 \\
\hline
.01101100110011000110
\end{array}
$$

The arithmetic unit will be designed such that it will always generate 20 digits. The 10 most significant will be stored in the register and the other 10 will simply be truncated (omitted). In this way the positioning of the binary point is solved. The fact that all numbers are represented as a fraction is no problem since any number can be represented as a fraction multiplied by a power of 2. For example, $21.75_{10} = 10101.11_2 = .1010111 \times 2^5$. That is, the binary point is simply shifted to the left five places.

Any arithmetic unit which uses the system $.X_1X_2 \ldots X_n$ for representing numbers is called a *binary fractional machine* since all the numbers are fractions. Furthermore this is called a *fixed-point machine* since the location of the binary point is fixed. Using this system we have the capability of storing any number from $.000 \ldots 0$ to $(1 - 1/2^n)$.

EXAMPLE 13-1

What range of numbers can be represented in a binary fractional machine having 4 bits?

SOLUTION

The 4 bits are represented as $.X_1X_2X_3X_4$. The minimum number is of course .0000. The maximum number is $.1111_2 = 0.9375_{10}$. Using the above relationship, the maximum number is $(1 - 1/2^n) = (1 - 1/2^4) = (1 - 1/16) = (1 - 0.0625) = 0.9375$.

With the binary fractional system we have the capability to represent only positive numbers. In order to extend our capability to include negative numbers, it is common practice to include one more bit to the *left* of the binary point. This bit is used exclusively to designate the sign of the number (either positive or negative) and as such is called the *sign bit*. By convention, and also for reasons which will become clear later, a positive number is represented by a 0 in the sign bit and a negative number is represented by a 1 in the sign bit. Thus we can represent a number by the following system:

$$X_0 . X_1 X_2 X_3 \ldots X_n$$

Sign bit   Magnitude

With this system we can then define numbers between $-(1 - 1/2^n)$ and $+(1 - 1/2^n)$.

EXAMPLE 13-2

What is the range of numbers defined by the *sign and magnitude system* if there are a maximum of 5 bits?

SOLUTION

The system is $X_0.X_1X_2X_3X_4$. Notice that for a word of 5 bits, 1 bit is the sign and 4 bits are the magnitude. The most negative number is $1.1111 = -.1111 = -0.9375_{10}$. The most positive number is $0.1111 = +.1111 = +0.9375_{10}$.

EXAMPLE 13-3

Represent the following decimal numbers in sign and magnitude form:
(a) $+0.75$
(b) -0.75
(c) $+0.3125$
(d) -0.4375

SOLUTION

(a) $+0.75 = 0.11$
(b) $-0.75 = 1.11$
(c) $+0.3125 = 0.0101$
(d) $-0.4375 = 1.0111$

The representation of numbers with the system shown in Example 13-2 is more properly called the sign and *true-magnitude* form. You will recall

Digital Arithmetic

from the chapter on binary arithmetic that the addition of a positive binary number and a negative binary number is carried out by adding the former to the complement of the latter (either the 1's or 2's complement). We will find it convenient to represent all negative numbers in complement form. There are then three ways of representing a negative binary number:

1. Sign and true magnitude
2. Sign and 1's complement
3. Sign and 2's complement

EXAMPLE 13-4

Represent the two negative numbers in Example 13-3 in the three different ways described.

SOLUTION

Number	True magnitude	1's complement	2's complement
−.75	1.11	1.00	1.01
−.4375	1.0111	1.1000	1.1001

Notice that in each case the 1 in the sign bit shows that the number is negative. To find the 1's or 2's complement representation only the magnitude portion of the number is complemented.

The next logical question which might arise is how do we represent a number such as 831 in this sign and magnitude system? The answer is quite simple. In a fixed-point system such as this, the operator must ensure that all input information is in the proper form. Thus, in punching cards, paper tape, etc., the operator must change the number 831 to 0.831×10^3. The data which are entered into the machine are simply 0.831 (in binary form) and the operator must remember that the $\times 10^3$ is needed. This process is commonly called "scaling." As an example of this, suppose that one of the parameters to be entered into a system is temperature and the range is from -100 to $+100°C$. The operator would simply enter this parameter as $X_0.X_1X_2X_3$. A value of $-50°C$ would be entered as $-.050$. Output data from the machine representing temperature values would simply have to be multiplied by 1000 to obtain the correct values.

EXAMPLE 13-5

Describe the sign and magnitude system required to represent a parameter whose values vary between ± 200.

SOLUTION

The input data will be scaled by a factor of 10^3. Thus all input data will be between ± 0.200, and all output data will then have to be multiplied by 10^3 for the correct value. We must now make a decision as to the resolution of the data. For example, if we choose to use 6 bits, the LSB is $1/2^6 = 1/64 = 0.0156$. For greater resolution we might choose a system having 8 bits; the LSB is then $1/2^8 = 1/256 = 0.0039$. If we are then satisfied with an 8-bit system, the binary form will be $X_0.X_1X_2X_3X_4X_5X_6X_7X_8$. Notice that it actually requires a 9-bit word, 1 bit for the sign and 8 bits for the magnitude.

EXAMPLE 13-6

In the example above, what is the binary representation and resolution error for a value of $+125$? For a value of -25?

SOLUTION

$+125$ will be represented as $+.125$. The binary word for this value is 0.00100000. There is no resolution error in this case. -25 will be represented as $-.025$. The binary word for this value is 1.00000110 (true magnitude) or 1.11111001 (in 1's complement) or 1.11111010 (in 2's complement). The decimal equivalent of the magnitude is 0.0234375. This represents a resolution error of about $(0.025 - 0.023)/0.025 = 8$ percent. It is quite obvious from this that the resolution error may be very large for small-magnitude numbers.

At this point some of the disadvantages of a fixed-point machine are apparent. To overcome the difficulties of fixed-point arithmetic, most large-scale digital computers use another method called "floating-point arithmetic." In this system numbers are represented with a sign bit, bits for the magnitude and in addition bits for the power of 2. One possible form of the word is shown as

$$X_0.X_1X_2X_3X_4X_5 \ldots X_n$$

| Sign bit | Power of 2 | Magnitude |

The sign and magnitude bits are used just as before. The exponent bits are used as follows: With 3 bits we can represent 8 powers of 2. We would like to have negative powers of 2 as well as positive powers of 2, and we will divide them evenly as shown in Table 13-1. For example, the number $+7.5$ would be represented as

$$7.5 = 111.1_2 = 0.1111_2 \times 2^3$$

The word for $+7.5$ is then 0.111 1111.

Digital Arithmetic

Table 13-1

	$X_1X_2X_3$
Negative powers	$\begin{cases} 000 = 2^{-4} \\ 001 = 2^{-3} \\ 010 = 2^{-2} \\ 011 = 2^{-1} \end{cases}$
Positive powers	$\begin{cases} 100 = 2^0 \\ 101 = 2^1 \\ 110 = 2^2 \\ 111 = 2^3 \end{cases}$

As one more example, the number -6.125 is represented by the word 1.111 110001 (true magnitude).

13-2 Fundamental Processes

In the preceding section we developed two methods for representing numbers in binary form. The first is the binary fractional method which we choose to call sign and magnitude. This method is used in systems employing fixed-point arithmetic. The second method is sign and magnitude and exponent. This method is used in systems employing floating-point arithmetic. In the remainder of this chapter we will discuss arithmetic units employing fixed-point arithmetic using sign and magnitude numbers. Machines using floating-point arithmetic are usually extensions of the basic operations with the added ability to handle exponents.

Using the sign and magnitude system, any number can be represented as $X_0.X_1X_2X_3\ldots X_n$, where n is the number of bits in the magnitude. Since the sign bit X_0 is not actually a part of the magnitude, we need to discover how to handle this bit during simple arithmetic operations. That is, must we treat the sign bit separately or can we simply operate on the entire number $X_0X_1X_2X_3\ldots X_n$ as a whole? As it turns out, we can simply treat the entire number as a whole. This fortunate result is due to our choice of binary-point location and our choice to represent a + with a 0 in the sign bit and a − with a 1 in the sign bit.

Let us first investigate the addition of two numbers using sign and magnitude representation. We will use a 5-bit system for our investigation: $X_0.X_1X_2X_3X_4$. There are four cases to be considered:

1. The addition of two positive numbers. In this case, addition is straightforward; but we must check for overflow. For example:

```
Fractional notation      Computer word
         3/16                 0.0011                9/16       0.1001
   (+)   7/16                 0.0111         (+)   8/16       0.1000
        10/16                 0.1010               17/16      1.0001
```

In adding 3/16 and 7/16, the answer 10/16 is correctly obtained. All three sign bits contain 0, and all is well. However, in adding 9/16 and 8/16, the answer 17/16 exceeds the capacity of our register and we obtain the incorrect answer of $1.0001_2 = -1/16$. This is an overflow error, and it is detected by sensing the 1 appearing in the sign bit.

2. Addition of a positive number and a negative number of smaller magnitude. You will notice first of all that it will be impossible for an overflow to occur. We can then carry out the addition using either the 1's complement or the 2's complement for the negative number. For example,

```
  Fractional
   Notation      True magnitude    1's complement    2's complement
      13/16          0.1101             0.1101            0.1101
(+)   -5/16          1.0101             1.1010            1.1011
      8/16                              ┌0.0111          ┌0.1000
                                        └──→1             ↓
                                         0.1000          omit carry
```

You will notice that we cannot simply add the true-magnitude numbers. In the 1's complement, the end-around carry must be used as previously learned. In the 2's complement, the end-around carry is disregarded as before. Notice in particular that the sign bit was simply used as a part of the number and the proper carry as well as the proper sign in the sum is generated. This will always be the case.

3. Addition of a positive number and a negative number of larger magnitude. As in case 2 above, there cannot be any overflow condition. We therefore proceed with the addition, using complements.

```
  Fractional
   notation     True magnitude    1's complement    2's complement
      7/16         0.0111             0.0111            0.0111
(+)  -12/16        1.1100             1.0011            1.0100
     -5/16                            1.1010            1.1011
```

Notice that there are no end-around carries generated; this will always be the case. The 1's-complement answer is 1.1010, which is −5/16 in 1's-complement form. Similarly, the 2's-complement answer is −5/16 in 2's-complement form.

Notice the result of adding two numbers of the same magnitude but opposite sign.

Digital Arithmetic

```
Fractional
 notation     True magnitude   1's complement   2's complement
    7/16         0.0111            0.0111           0.0111
(+) −7/16        1.0111            1.1000           1.1001
    ─────                          ──────          ┌─0.0000
      0                            1.1111          │
                                                   ↓
                                              omit carry
```

From the 1's complement the sum is 1.1111, which is the 1's complement for −.0. From the 2's complement the sum is 0.0000, which is +.0.

4. The addition of two negative numbers. We note first of all that the sum must necessarily be negative and we must therefore end up with a 1 in the sign bit. Secondly, there is the definite possibility of an overflow. We will be able to detect this overflow by the presence of a 0 in the sign bit of the sum. A few examples will best illustrate the methods.

```
Fractional
 notation     True magnitude   1's complement   2's complement
   −7/16         1.0111            1.1000           1.1001
(+) −8/16        1.1000            1.0111           1.1000
   ──────                         ┌─0.1111         ┌─1.0001
  −15/16                          │      →1        │
                                  ──────           ↓
                                   1.0000      omit carry
```

In the 1's complement, the sum is 1.0000, which is the representation for −15/16. In the 2's complement, the sum is 1.0001, which is the representation for −15/16 in 2's-complement form.

Now let us examine an example where an overflow will occur:

```
Fractional
 notation     True magnitude   1's complement   2's complement
   −9/16         1.1001            1.0110           1.0111
(+) −9/16        1.1001            1.0110           1.0111
   ──────                         ┌─0.1100         ┌─0.1110
  −18/16                          │      →1        │
                                  ──────           ↓
                                   0.1101      omit carry
```

Both sums are indicated to be positive numbers, and thus we know an overflow has occurred giving incorrect results. We detect the overflow by detecting the change of sign in the sign bit.

At this point let us summarize in tabular form the basic addition operations for fixed-point arithmetic. We will use numbers in binary fractional form, and we will use sign and magnitude with either 1's or 2's complements for negative numbers. Table 13-2 summarizes and gives examples of binary-fractional-number representations using 1's and 2's complements.

Table 13-2 Number Representations
$X_0.X_1X_2X_3\ldots X_n$

Fractional notation	True magnitude	1's complement	2's complement
3/16	0.0011		
−3/16	1.0011	1.1100	1.1101
35/64	0.100011		
−35/64	1.100011	1.011100	1.011101

Table 13-3 lists the methods for adding two numbers in the form of Table 13-2.

Table 13-3 Addition $(A > B)$

A	B	
+	+	Add numbers as they are. Overflow indicated by a 1 in the sign bit of the sum
+	−	1. Use 1's complement for B. Add. No overflow possible. Must use carries to correct sum 2. Use 2's complement for B. Add. No overflow possible. Disregard all carries
−	+	1. Use 1's complement for A. Add. No overflow possible 2. Use 2's complement for A. Add. No overflow possible. Disregard all carries
−	−	1. Use 1's complement for A and B. 0 in sign bit of sum indicates overflow. Must use carries to correct sum. 2. Use 2's complement for A and B. 0 in sign bit of sum indicates overflow. Disregard all carries

13-3 Serial Binary Adder

Using the sign and magnitude method developed in the previous sections, we now have the capability to represent any number. The fractional-binary-number representation which we are going to use is characteristic of fixed-point arithmetic units. What we are interested in then is to develop the logic necessary to perform fixed-point binary addition.

In an effort to develop the logic necessary for binary addition, let us consider a binary addition as we would perform it mentally.

Digital Arithmetic

$$\begin{array}{r} 7/16 \\ (+)\ 4/16 \\ \hline 11/16 \end{array} \qquad \begin{array}{r} 0.0111 \\ 0.0100 \\ \hline 0.1011 \end{array}$$

We perform this simple addition in binary form in the following steps:

1. Add the two LSBs.
2. Add the next 2 bits.
3. Add the next 2 bits and place a carry over the MSB column.
4. Add the two MSBs and the carry generated from the previous step.
5. Add the 2 sign bits.

Notice that there are five distinct steps in the complete addition. If we perform the addition according to a time scale, we might do step 1 at time t_1, step 2 at time t_2, ... and step 5 at time t_5.

You will recall that a binary full-adder is capable of adding 2 bits and a carry present at its input terminals. If we could then present the two LSBs of the numbers to be added to the input terminals of a full-adder at time t_1, the LSB of the sum would appear at the output of the adder at time t_1. We would then present the next 2 bits to the adder at time t_2 and the second sum bit would appear at the output of the adder at time t_2. Continuing in this fashion, the third sum bit would appear at time t_3, the MSB at time t_4, and the sign bit at time t_5. A method for causing the bits of a number to appear at a pair of terminals in a time sequence is reminiscent of a shift register. We will then use two shift registers and a full-adder to add two binary numbers in this fashion.

A *serial binary* adder is shown in Fig. 13-2. It will be used to find the sum of two numbers: $A = A_0.A_1A_2A_3A_4$ and $B = B_0.B_1B_2B_3B_4$. It is called a serial adder since the sum is found by adding 2 bits at a time sequentially beginning with the two LSBs. The two numbers are initially stored in the two registers as shown. Each register is shown simply as a series of five boxes, but it is actually five flip-flops forming a shift register. Each register contains one extra flip-flop at the right end. These two flip-flops are used as the inputs to the adder. The carry output of the adder is delayed one bit

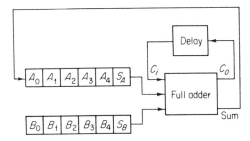

Fig. 13-2 Serial binary adder.

time. Thus, for example, a carry generated at time t_1 will not appear at the input to the adder until time t_2. The operation proceeds as follows:

1. At time t_0: All flip-flops have been previously reset to 0 and the numbers A and B are shifted into the registers from memory. Notice that the numbers can be entered into the registers in either serial or parallel form.

2. At time t_1: The contents of both registers are shifted right one place. Since the output of the adder was initially 0, and 0 is shifted into A_0. The two LSBs are now in the sum flip-flops S_A and S_B (no carry appears) and the LSB of the sum appears at the output of the adder.

3. At time t_2: The contents of both registers are shifted right one place. The LSB of the sum is shifted into A_0. The two second LSBs are now in the sum flip-flops, any carry generated at time t_1 appears at C_i, and the second digit of the sum appears at the output of the adder.

4. This shifting process is continued up to and including time t_6. At the end of time t_6 the sum of the two numbers appears in the upper register. The number A initially stored in the upper register is lost. The number initially stored in the lower register is also lost since it has been shifted out the right end. We could have preserved this number by feeding the output of flip-flop S_B into the first flip-flop in the lower register. In this way, B will remain in the lower register after the addition process is complete.

As an example of the functioning of the adder, the two numbers 0.0110 and 0.0010 are added. The waveforms for the system and the contents of the registers are shown in Fig. 13-3a and b, respectively. The reader should dwell on these two figures until the operation is thoroughly understood. Notice that the 1 in the sum appears at the sum output of the adder at time t_4 (waveforms) and is shifted into bit position A_0 at time t_5 (registers).

We now have the basic requirements to form a serial binary adder. We must, however, consider a few details in order to complete the system. First of all let us consider the problem of overflows. You will recall that an overflow will occur only when adding two numbers of like sign. Furthermore, the overflow is detected when the sign of the sum is different from the signs of the original numbers. One method for detecting overflows is shown in Fig. 13-4. The operation of the circuit is as follows:

1. First of all we will examine for overflow only when the signs of A and B are the same. During time t_0 we will compare the signs of the two numbers. If they are the same, the output of the exclusive-OR gate will be low and the check flip-flop CK will be set. If they are different, the check flip-flop will remain reset.

2. We will circulate the contents of the lower register such that the number B will appear in it at the end of the addition cycle (at time t_6). The sum will be in the upper register at this time.

Digital Arithmetic

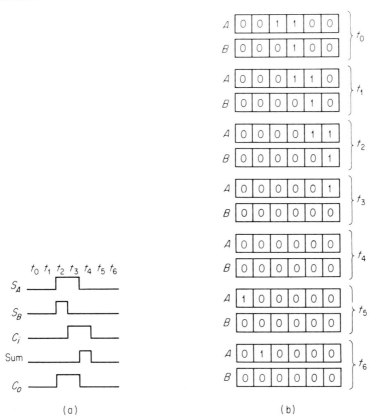

Fig. 13-3 Addition of 0.0110 and 0.0010 in the serial binary adder of Fig. 13-2.

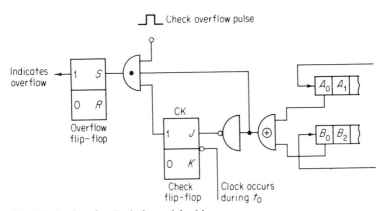

Fig. 13-4 Overflow logic for serial adder.

3. Now during time t_6 a *check overflow pulse* will occur. If flip-flop CK has been set, and if the output of the exclusive-OR is high, the overflow flip-flop will be set. Notice that the output of the exclusive-OR will be high at this time only if the sign bit of the sum A_0 is different from the sign bit of one of the original numbers B_0.

EXAMPLE 13-7

Draw the waveforms for the overflow logic shown in Fig. 13-4 assuming an overflow occurs.

SOLUTION

The waveforms are shown in Fig. 13-5. Notice that at time t_0 the signs of A and B are both the same (indicating positive numbers). The CK flip-flop clock pulse sets this flip-flop during t_0 since the output of the NAND gate is high. At time t_6, the sign of the sum A_0 is now high (representing a negative number and thus an overflow). Thus the output of the exclusive-OR is high, as is CK, and the check overflow pulse will then pass through the AND gate and set the overflow flip-flop. In the usual case, the output of the overflow flip-flop would be used to stop machine operation, since this represents an error condition.

The second detail we must attend to is the handling of signed numbers. If the numbers are stored in memory in sign and true-magnitude form, the 1's or 2's complement of the negative numbers will have to be formed before addition can take place. Forming the 1's complement of a number is quite simple. You will recall from a previous chapter that the 1's complement of the number stored in a register can be formed by simply triggering each flip-flop in the register. Since the trigger pulse will cause each flip-flop

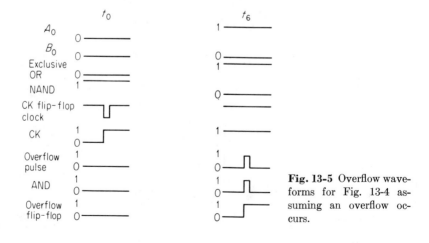

Fig. 13-5 Overflow waveforms for Fig. 13-4 assuming an overflow occurs.

Digital Arithmetic

to change state, the register will then contain the 1's complement of the previously stored number. In the serial adder, we can obtain the 1's complement more easily, however. Since the number appearing at the 0 side of the S_A or S_B flip-flops is the complement of the number appearing at the 1 side, we can simply use the 0 side when we are shifting out a negative number. However, if we choose to use 1's complements for negative numbers we will have to include additional logic to handle the end-around carries. In the serial adder we are discussing, this involves setting a 1 in the lower register (in place of the number B) and going through the add cycle one more time. For this reason, the 1's-complement system is not widely used in serial adders.

We will do better to represent negative numbers by their 2's complements. This will eliminate the problem of end-around carries. On the other hand, it does not seem quite so easy to form the 2's complements. There is, however, a very simple circuit which can be used to form the 2's complement of a number in serial fashion. The circuit is shown in Fig. 13-6. To find the 2's complement of any number it is necessary to complement every bit beginning with the LSB which passes after, but not including, the first 1. The bits are shifted through the 2's complementer serially beginning with LSB. Flip-flop C is initially set and the first bit will pass directly through the upper AND gate. The bits will continue to pass through this AND gate until the first 1 appears. The first 1 will pass through the AND gate unchanged, but it will also reset flip-flop C after a delay of one bit time. Thereafter, the input bits will pass through the lower AND gate after being inverted by the NAND gate and will appear at the output in complement form. The control flip-flop is used to enable or disable the circuit. When it is

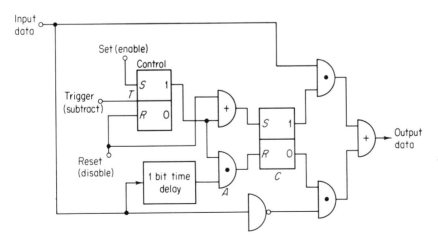

Fig. 13-6 Serial 2's-complement circuit.

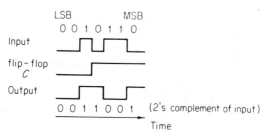

Fig. 13-7 Example 13-8.

set, AND gate A is true and the 2's complement can be formed. When it is reset, AND gate A is disabled and the circuit cannot function. In this case data will simply pass through the upper AND gate and appear at the output unchanged.

EXAMPLE 13-8

Show the waveforms for forming the 2's complement of the number 0110100.

SOLUTION

The waveforms are shown in Fig. 13-7. Notice that the waveforms appear backward from the manner in which we would usually write the number. That is, the LSB is on the left and the MSB is on the right.

We can now use this 2's-complement circuit in our serial adder by placing one in series with each register directly in front of flip-flops S_A and S_B. The complete adder is shown in Fig. 13-8.

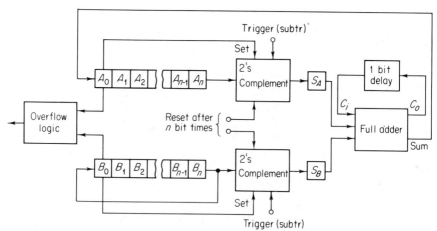

Fig. 13-8 Serial binary adder using numbers stored in sign and true-magnitude form.

Digital Arithmetic

This adder is capable of adding positive or negative numbers which are stored in memory in sign and true-magnitude form. The overflow logic will handle any overflow problems and the 2's-complement circuit will handle any negative numbers. We must take care to ensure that the 2's-complement circuit is allowed to operate for only n bit times (where n is the number of bits in the magnitude). In this fashion we will use the 2's complement of the magnitude of any negative number, but we will not change the sign bit. Thus the control flip-flop in the 2's-complement circuit will be set at time t_0 if a negative sign is detected. It will then be reset at time t_n to ensure that the sign bit passes by unchanged.

A little thought will reveal that the adder shown in Fig. 13-8 can also be used for subtraction. We simply take the four cases of two numbers, $++$, $+-$, $-+$, and $--$, and show the results of subtracting them instead of adding (as was done in a previous section). It can be shown (see Prob. 13-9) that these four cases reduce to the same problems of addition. To use the adder in Fig. 13-8 for subtraction then, it is only necessary to add a trigger input to the control flip-flop in the 2's-complement circuit of Fig. 13-6. Subtraction will then occur if we apply a pulse to this input just prior to time t_1. There are two cases to consider:

1. To subtract a positive number: the 2's-complement circuit will not have been enabled. The subtract control pulse appearing at the T input of the control flip-flop of the 2's-complement circuit will enable it, and the 2's complement will be used, resulting in a subtraction.

2. To subtract a negative number: the 2's-complement circuit will have been enabled and the subtract pulse will disable it. This will then result in a straight addition.

EXAMPLE 13-9

Explain how the serial binary adder in Fig. 13-8 could be used to add or subtract binary numbers stored in the following form: sign and true magnitude for positive numbers, and sign and 2's complement for negative numbers.

SOLUTION

1. Since the negative numbers are stored in 2's-complement form, we will not need to use the 2's-complement circuits during addition. Therefore, the set inputs to the 2's-complement circuits will be removed from the sign bits (A_0 and B_0). The 2's-complement circuits will be held in the reset (disabled) state during addition. The sum will appear in register A.

2. For subtraction we will take the 2's complement of the number to be subtracted and add. Therefore, to subtract B from A we will set the 2's-complement circuit in the B register at time t_0 and add (this 2's-

complement circuit will be reset after n bit times). The 2's-complement circuit in the A register will be held reset (disabled) during this cycle. The difference $(A - B)$ will appear in register A.

3. To subtract A from B, we will activate the 2's-complement circuit in register A for n bit times beginning at t_0 while holding the 2's-complement circuit in register B reset (disabled). The difference $(B - A)$ will appear in register A.

At this point let us examine the total time required to complete an addition or a subtraction using the serial adder of Fig. 13-8. Each word is composed of $(n + 1)$ bits; one sign bit and n bits of magnitude. A little thought will show that it requires $(n + 2)$ bit times to complete an addition (or subtraction) cycle. This time is defined at the *add-cycle time* of the adder. Typically, one bit time is the time of one cycle of the clock since the adder will usually be driven by the clock. Thus, one add-cycle time is equal to $(n + 2)$ clock periods.*

EXAMPLE 13-10

What is the add-cycle time of the serial adder in Fig. 13-8 if each word has 12 bits of magnitude and the clock is 100 kHz?

SOLUTION

One clock period is $1/(100 \times 10^3) = 10$ μsec. The add-cycle time is then $10 \times 10^{-6} (12 + 2) = 140$ μsec.

The length of time required to perform an addition or subtraction using the serial binary adder is one of the major drawbacks to this method. Admittedly a serial adder offers the attraction of a minimum of hardware, but in the interest of accomplishing addition in a shorter period of time we will investigate other methods. The parallel binary adder to be discussed in the next section will offer a substantial reduction in add-cycle time, at the expense, however, of additional hardware.

13-4 Parallel Binary Adder

The primary advantage of a parallel binary adder over a serial adder is the reduction in add-cycle time (the main disadvantage is the increase in hardware required). You will recall that a serial adder requires $(n + 1)$ clock periods to add two binary numbers having one sign bit and n bits of

---

*If the add flip-flops (S_A and S_B) in the registers are omitted, it will require only $(n + 1)$ clock periods for the basic add-cycle time. These two flip-flops are not really necessary and were included in the first place to simplify description of the operation of the adder.

Digital Arithmetic

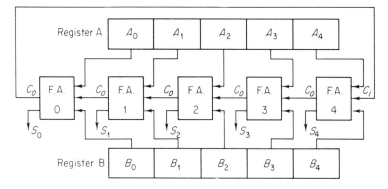

Fig. 13-9 Parallel binary adder.

magnitude. The parallel adder, on the other hand, will provide the sum of two numbers in only one clock period. That is, the sum will appear almost immediately after the two numbers are entered in the registers (allowing a short period of time for settling). Thus the actual addition time for a parallel adder could be considered one clock cycle.

A simple parallel adder is shown in Fig. 13-9 (at this point you might like to review the discussion of the parallel adder given in Chap. 5). This adder has the ability to add two numbers of the form $A = A_0.A_1A_2A_3A_4$ and $B = B_0.B_1B_2B_3B_4$. The number A is stored in register A, the number B is stored in register B, and the sum $S_0.S_1S_2S_3S_4$ appears at the adder outputs. To perform an addition using this adder requires two steps: (1) reset all flip-flops; (2) shift A into register A and B into register B. After a short settling time, the sum will appear at the adder outputs. The add-cycle time can then be considered to be two clock periods. The reason for the short add-cycle time is now clear: The two LSBs are added in full-adder 4 and the carry C_o is connected to the C_i of full-adder 3. Simultaneously, the corresponding pairs of bits from each number are added and each carry is passed on to the next adder. Thus the two numbers are added all at once and the add time is actually the time required for the carries to propagate from one adder to the next down the line. This is the *settling time* required before the addition is complete.

In order to complete the discussion of this parallel adder, we must consider the four cases summarized in Table 13-3. First of all, however, we must decide how the numbers are to be stored in memory. If they are stored in sign and true-magnitude form, we will have to complement the negative numbers in the registers (A or B) before addition can take place. This is easily accomplished for the 1's complement. You will recall that it is only necessary to trigger each flip-flop in the register in order to form the 1's complement. This will, however, add one clock period to our add-cycle time, and we will therefore not choose this representation. The 2's complement is not easily formed, and we will therefore not choose this method. The only method remaining is the most convenient, and we will use it. That is, we will store positive numbers in sign and true magnitude and negative numbers in sign and 1's complement. Recall that the only disadvantage of this method in the serial adder is the end-around carries generated. This is no problem in the parallel adder, however, since the end-around carry (the C_o of adder 0 to the C_i of adder 4) is easily accommodated.

Now assuming we will use sign and true magnitude for positive numbers and sign and 1's complement for negative numbers let us examine the four cases given in Table 13-3:

1. Sum of two positive numbers. The only problem here is the overflow, and we will have to arrange a method for checking this.
2. A positive number added to a negative number of smaller magnitude. No overflow is possible in this case. The end-around carries will be accommodated by connecting the C_o of adder 0 to the C_i of adder 4.
3. A positive number added to a negative number of larger magnitude. No overflow is possible, and there are no end-around carries.
4. The sum of two negative numbers. We must arrange for checking overflows. The end-around carries will be taken care of as in case 2 above.

In summarizing the four cases examined above, it is clear that the end-around carries will be taken care of and it is only necessary to add additional circuitry to check for overflows. A circuit which can be used to detect overflows for this parallel adder is shown in Fig. 13-10. There are two cases where overflows can occur: the addition of two positive numbers and the addition of two negative numbers. In Fig. 13-10, the output of exclusive-OR gate X will be low whenever A_0 and B_0 are the same (i.e., the numbers have

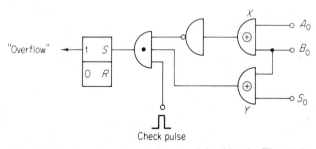

Fig. 13-10 Overflow circuit for parallel adder in Fig. 13-9.

Digital Arithmetic

the same sign). This will cause the top leg of the AND gate to be high. The output of exclusive-OR gate Y will be high whenever B_0 and S_0 are different (indicating an overflow). This will cause the middle leg of the AND gate to be high. Thus when a *check* pulse occurs on the bottom leg of the AND gate, the *overflow* flip-flop will be set only when an overflow condition exists. The check pulse will occur sometime after the addition is complete.

EXAMPLE 13-11

Label the levels present on the gate legs of the circuit shown in Fig. 13-10 for an input of $A_0 = 1, B_0 = 1, S_0 = 0$.

SOLUTION

The levels are shown on the circuit in Fig. 13-11. This case corresponds to the addition of two negative numbers with the sign of the sum indicating a positive sum — an overflow error.

The adder shown in Fig. 13-9 can also be used for subtraction. It can be shown, as in the previous section, that the four cases of subtraction will reduce to the very same four cases of addition listed in Table 13-3. The subtraction process is carried out, however, by complementing the number to be subtracted and then adding. This will simply require one more clock period. There are two cases to consider:

1. $A - B$. During time t_1, the two numbers will be shifted into registers A and B. During time t_2, the contents of register B will be complemented by applying a pulse to the T input of every flip-flop in that register excepting the sign bit. The difference will then appear at the adder outputs (after a small settling time).

2. $B - A$. Similar to the case for $A - B$ except that the contents of register A will be complemented.

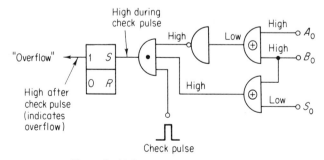

Fig. 13-11 Example 13-9.

It should be noted that the parallel adder discussed could be used to add two numbers having n bits of magnitude by simply constructing the registers with the appropriate number of flip-flops and using $(n + 1)$ full-adders. The total add time is limited by the settling time due to the propagation of the carries from one adder to the next. There are a number of advanced methods for reducing this settling time, and they are discussed in more advanced texts.

13-5 BCD Addition

It is sometimes more desirable to perform digital arithmetic using numbers stored in BCD form. In Chap. 5, we discussed methods for adding numbers which were stored in 8421 BCD code and in excess-3 BCD code. The results of these studies are the 8421 adder and the excess-3 adder developed in that chapter. At this point what we would like to accomplish is the formation of a general BCD adder using these basic adders developed in Chap. 5. We will first of all need to consider the methods of representing numbers in BCD form and the same four cases of addition investigated for adding numbers in straight binary form.

We will soon discover that BCD arithmetic is very similar to arithmetic carried out in straight binary form, and for this reason we will locate the binary point to the left of the MSB as in the previous section. Similarly, we will use a 0 in the sign bit for a positive number and a 1 in the sign bit for a negative number. We will therefore represent numbers in the following form:

$$\underbrace{X_0}_{\text{Sign bit}}.\underbrace{X_1X_2X_3X_4}_{\substack{\text{Tenths}\\\text{digit}}} \underbrace{X_5X_6X_7X_8}_{\substack{\text{Hundredths}\\\text{digit}}}$$

With this system we can represent numbers between $+.99_{10}$ and $-.99_{10}$. The system could be enlarged to include three-digit decimal numbers by adding four more bits to the magnitude, or four-digit decimal numbers by adding eight more bits, etc. Negative numbers can be represented in sign and true magnitude, sign and 9's complement, or sign and 10's complement.

EXAMPLE 13-12

Show the binary representation for the following numbers in 8421 BCD code and in excess-3 BCD code:
(a) 34
(b) −34
(c) −81

Digital Arithmetic

SOLUTION

8421 BCD code

Number	Scaled	Sign and magnitude	True magnitude	9's complement	10's complement
34	+.34	0.34	0.0011 0100		
−34	−.34	1.34	1.0011 0100	1.0110 0101	1.0110 0110
−81	−.81	1.81	1.1000 0001	1.0001 1000	1.0001 1001

Excess-3 BCD code

34	+.34		0.0110 0111		
−34	−.34		1.0110 0111	1.1001 1000	0.1001 1001
−81	−.81		1.1011 0100	1.0100 1011	1.0100 1100

In this example, you will notice that the formation of the 9's and 10's complements of numbers in 8421 BCD code is not easily accomplished electronically. It can be done but it requires additional logic (see Prob. 13-14) and we want to minimize the necessary logic to perform addition. On the other hand, the 9's complements of numbers in excess-3 code is easily formed. It is only necessary to complement the entire magnitude portion of the number. The 10's complement is also easily formed since it involves only the addition of a binary 1 to the rightmost position of the magnitude. With this in mind then, let us consider the four cases of adding two numbers in excess-3 BCD form.

1. Addition of two positive numbers. Overflows can obviously occur, and we will have to check for them. Consider the addition of the following two numbers:

$$
\begin{array}{rccl}
& & True & \\
Numbers & magnitude & Excess\text{-}3 & \\
21 & 0.0010\ 0001 & 0.0101\ 0100 & \\
+\ 13 & 0.0001\ 0011 & 0.0100\ 0110 & \\
\hline
34 & 0.0011\ 0100 & 0.1001\ 1010 \\
& & -11\quad -11 & \} \text{ occurs in BCD adder} \\
& & \overline{0.0110\ 0111} &
\end{array}
$$

The two numbers are first scaled and then changed into sign and true-magnitude form. The sum is then obtained by simply adding these two numbers. Under the excess-3 column, the excess-3 representations of the two numbers are added and the answer (34) appears in excess-3 code. The subtraction of 3 from each digit is accomplished automatically in the excess-3 adder (if you are not sure of the reasoning behind this, you should review the excess-3 BCD adder discussed in Chap. 5).

Let us now consider the addition of two positive numbers when an overflow occurs.

```
                True
  Numbers     magnitude         Excess-3
                                  1↲
     34       0.0011  0100     0.0110 | 0111
  +  86       0.1000  0110     0.1011 | 1001
  ----        ------------     ---------------
    120       0.1011  1010
              +110  +110       1.0001 ↳0000 ⎫
                1↲              +11   +11   ⎬ occurs in BCD adder
              ------------     ---------------⎭
              1.0010 ↳0000     1.0101  0011
```

Under the true magnitude, the answer is −20, which is obviously incorrect. The negative sign (the 1 in the sign bit) is the necessary indication that an overflow has occurred. The answer given under excess-3 is also −20 in excess-3 code. This obviously incorrect answer can also be detected by the 1 in the sign bit.

2. Addition of a positive number and a negative number of smaller magnitude.

```
                                        Excess-3
              True
  Numbers   magnitude   9's complement   10's complement
     21     0.0010 0001  0.0101  0100    0.0101  0100
  - 13      1.0001 0011  1.1011  1001    1.1011  1010
  ----      -----------  ┌0.0000 1101    ┌0.0000 1110⎫ occurs in
      8      Cannot      │ +11    −11    │ +11    −11⎬   adder
               add       │-----------    │-----------⎭
             directly    │0.0011  1010   │0.0011  1011
                         └─────→1         ↓
                          0.0011  1011   Discard carry
```

In this case no overflow can occur. Notice that we use the end-around carry for the 9's complement and discard it for the 10's complement. The correct answer of +8 is given in excess-3 code in both cases.

3. Addition of a positive number and a negative number of greater magnitude.

Digital Arithmetic

		Excess-3	
Number	True magnitude	9's complement	10's complement
−21	1.0010 0001	1.1010 1011	1.1010 1100
13	0.0001 0011	0.0100 0110	0.0100 0110
−8	Cannot add directly	1.1111 0001 −11 +11 1.1100 0100	1.1111 0010 −11 +11 1.1100 0101

Again, no overflow can occur in this case. The answer under 9's complement is −8 in excess-3 code. The magnitude appears in complement form since the answer is negative. Similarly the answer under 10's complement appears as the 10's complement of +8 since it is negative. The 10's-complement answer can be changed to true magnitude by subtracting 1 from the LSB of the magnitude and complementing it.

4. *The addition of two negative numbers.* We must of course examine for any overflow condition. Consider the following two numbers:

		Excess-3	
Numbers	True magnitude	9's complement	10's complement
−21	1.0010 0001	1.1010 1011	1.1010 1100
−13	1.0001 0011	1.1011 1001	1.1011 1010

−34	Cannot add directly	⌐1.0110 0100
		  +11 +11
		  1.1001 0111
		└────→1
		1.1001 1000
			⌐1.0110 0110
			  +11 +11
			↓1.1001 1001
			Omit carry

In this case, no overflow occurs and the correct sum appears in complement form. Now consider a case where an overflow will occur.

		Excess-3	
Numbers	True magnitude	9's complement	10's complement
−51	1.0101 0001	1.0111 1011	1.0111 1100
−51	1.0101 0001	1.0111 1011	1.0111 1100

−102	Cannot add directly	⌐0.1111 0110
		  −11 +11
		  0.1100 1001
		└────→1
		0.1100 1010
			⌐0.1111 1000
			  −11 +11
			↓0.1100 1011
			Omit carry

You will notice in these two cases that the sum appears as +2 in complement form. There is obviously an overflow, and the error can be detected by the 0 in the sign bit of the sum.

We can summarize these four cases of addition by noting the similarity between addition in excess-3 BCD using 9's and 10's complements for negative numbers and addition in straight binary form using 1's and 2's complements for negative numbers. That is to say, we must examine for overflow when adding two numbers of like sign and we must provide for end-around carries when using 9's complements (similar to using the end-around carries when using the 1's complements). With this knowledge we can then proceed with the formulation of a BCD adder using excess-3 code and we can rest assured that the signs will take care of themselves.

13-6 A Parallel BCD Adder

We will now discuss the operation of a parallel BCD adder. This adder will operate in excess-3 BCD code and numbers will be stored in sign and true magnitude for positive numbers and sign and 9's complement for negative numbers.

The basic BCD adder is shown in Fig. 13-12. It is capable of adding two two-digit decimal numbers and has the capacity of handling any numbers between $+99_{10}$ and -99_{10}. The two numbers $A = A_0.A_1A_2A_3A_4A_5A_6A_7A_8$ and $B = B_0.B_1B_2B_3B_4B_5B_6B_7B_8$ are stored in registers A and B, respectively. The sum $S = S_0.S_1S_2S_3S_4S_5S_6S_7S_8$ appears at the outputs of the adders. The add-cycle time is quite similar to that for the parallel binary adder discussed

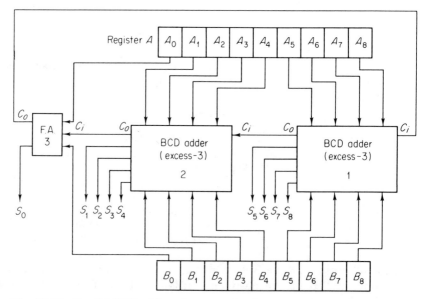

Fig. 13-12 Parallel BCD adder using excess-3 code.

Digital Arithmetic

previously. The numbers A and B are shifted into their respective registers at time t_1 and the sum S appears at the output after a small settling time. Since we are using 9's complements for negative numbers, we must account for the end-around carries, and this is accomplished by connecting the carry out C_o of adder 3 to the carry in C_i of adder 1. Since overflows are detected in the same way as for parallel binary addition, we can use the overflow-detecting circuit of Fig. 13-10. Notice that we must use a BCD excess-3 adder for each 4 bits of the magnitude, but we use a straight binary full-adder for the sign bits. The adder could, of course, be extended to accommodate decimal numbers having n digits by simply using n BCD adders and $4n$ flip-flops for the magnitude portion of the registers.

Since the four cases of subtraction ultimately reduce to the same four cases of addition studied, the BCD adder in Fig. 13-12 can also be used to subtract BCD numbers in excess-3 code. It is only necessary to add one more clock period to the total during which the number to be subtracted is complemented. At this point the great value of using excess-3 code becomes apparent. Since subtraction is performed by finding the complement of a number and then adding, and since excess-3 code is *self-complementing*, we only need to complement the contents of the register containing the number to be subtracted and then execute an addition cycle. The complement operation is, of course, carried out in this case by applying a pulse to the T input of every flip-flop in the register (except the sign bit) containing the number to be complemented. The time required to perform a subtraction using this BCD adder is then two clock periods, and we could define the *subtract-cycle time* as being equal to two clock periods.

Because of the great similarity between this BCD adder and the parallel binary adder previously discussed, we might suppose that there would be a similarity between a serial BCD adder and a serial binary adder. This is indeed the case, and as one might expect it could be most easily accomplished using sign and true magnitude for positive numbers and sign and 10's complements for negative numbers (all in excess-3 code).

EXAMPLE 13-13

Explain how the BCD adder in Fig. 13-12 could be used to add numbers stored in sign and true magnitude (excess-3 code).

SOLUTION

There are three cases to consider:
1. The addition of two positive numbers is carried out by simply shifting the numbers into the registers and reading the output.
2. The addition of a positive number and a negative number is carried out by shifting the two numbers into the registers, complementing the negative number (magnitude portion only), and reading the output.

3. The addition of two negative numbers is carried out by shifting the numbers into the registers, complementing both registers (excepting the signs), and reading the output.

13-7 Binary Multiplication

Binary multiplication and division are both quite complex operations. There are many different methods for performing these operations, and a more comprehensive treatment can be found in texts listed in the bibliography. There is, however, one basic method for multiplication and one basic method for division, which we will discuss at this time. The other methods are variations directed toward shortening the time required to perform a computation. We will discuss multiplication in this section and division in the next.

We can easily discover a method for performing multiplication by carefully examining the multiplication process itself. First consider the multiplication of one digit by another. For example, the multiplication of 8 by 4 is really the same thing as adding the number 8 four times. That is, $8 \times 4 = (8 + 8 + 8 + 8) = 32$ (it is also the same thing as adding 4 eight times). Similarly the product of 5 and 3 could be written as $5 \times 3 = (5 + 5 + 5) = (3 + 3 + 3 + 3 + 3) = 15$. Thus it can be seen that the multiplication process is really the same thing as repeated addition. In the decimal system we may have to add any one number up to nine times. However, in the binary system, multiplication by repeated addition is much simpler since we have only two numbers, 1 and 0. Thus we will have to add any one number only once. For example, consider the binary multiplication of the number 6. There are only two possibilities: (1) $110 \times 0 = 0$, and (2) $110 \times 1 = 110$.

We must now consider the multiplication of some number by a number having more than one digit. The binary multiplication of 6 and 5 would be carried out as follows:

$$\begin{array}{r} 110 \leftarrow \text{multiplicand} \\ \underline{101} \leftarrow \text{multiplier} \\ 110 \\ 000 \\ \underline{110} \\ 11110 \leftarrow \text{product} \end{array}$$

$\leftarrow$ partial products

The first partial product (the top one) is formed by multiplying the multiplicand by the LSB of the multiplier. The second partial product is formed by multiplying the multiplicand by the middle bit of the multiplier, and the third partial product is formed using the MSB of the multiplier.

Digital Arithmetic

7. Shift right

| 0 | 0 | 1 | 1 | 1 | 1 | 0 |

8. Set sign in accumulator

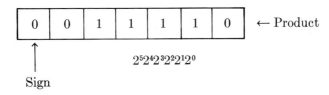

↑
Sign

$2^5 2^4 2^3 2^2 2^1 2^0$

← Product

Only the accumulator and MQ registers were shown since the multiplicand register remains unchanged during the multiply operation. At the end of the eight steps, the product appears in the accumulator and MQ registers.

You will notice that in the preceding example the product of two 3-bit numbers generated a 6-bit product. This, you will recall, is necessary for the proper location of the binary point. We assume, of course, that the numbers are used in binary fractional notation with the binary point between the MSB and the sign bit. In multiplying these two 3-bit numbers a total of eight steps were required after the numbers were shifted into the registers. Two of these steps deal with the sign bit and will always be needed. The remaining steps deal with the magnitudes only, and since there were 3 bits, we had to add and shift right three times for a total of six steps. This technique can, of course, be extended to handle n-bit numbers by simply increasing the size of the registers to $(n + 1)$ bits.

We can now make a general statement regarding the time required to multiply any two numbers using this technique: (1) It will always require two steps to handle the sign bits. (2) It will require $2n$ steps to handle the magnitude, where n is the number of bits in the magnitude. Thus the total time required to perform a multiplication in this way is $(2n + 2) = 2(n + 1)$. Since each step is usually performed in one clock period, we can say that the time required to perform one multiplication is $2(n + 1)$ clock periods.

13-8 Binary Division

Like multiplication, a process for binary division can be best determined by examining the methods used in longhand division. Let us consider the longhand method for carrying out the division $2_{10}/6_{10} = 010/110$. This problem is normally written as

$$\begin{array}{r} 0.0101 \leftarrow \text{quotient} \\ \text{divisor} \rightarrow 110 \overline{\smash{\big)}\ 010.0000} \leftarrow \text{dividend} \\ \underline{1\ 10} \\ 0\ 1000 \leftarrow \text{first remainder} \\ \underline{110} \\ 010 \leftarrow \text{second remainder} \end{array}$$

The steps in this division are:

1. Place the divisor under the first 3 bits of the dividend. We notice that the divisor is too large to be divided into these 3 bits, and we then place a 0 to the left of the binary point. We determine that division is not possible by noting that the magnitude of the first 3 bits is smaller than the divisor.

2. We then move the divisor one place to the right and place it under the second, third, and fourth bits of the dividend. The magnitude of the dividend is still smaller than the divisor; so we place a 0 just after the binary point (in the first binary place).

3. We move the divisor one more place to the right (under the third, fourth, and fifth bits of the dividend). Now the magnitude of the dividend is greater than the divisor, and we can perform a subtraction. We note this by placing a 1 in the second binary place of the quotient. The divisor is then subtracted from the dividend to yield the first remainder.

4. We then bring down a 0 from the dividend, add it to the right side of the first remainder, and place the divisor under the 3 rightmost bits of the first remainder. We note that the magnitude of the first remainder is smaller than the divisor, and we therefore cannot make a subtraction. We note this by placing a 0 in the third binary place of the quotient.

5. We then bring down another 0 from the dividend, add it to the right end of the first remainder, and move the divisor to the right one place. Now the magnitude of the first remainder is larger than the divisor and we can make a subtraction; we note this by placing a 1 in the fourth binary place of the quotient. The subtraction is made, yielding the second remainder.

6. This process is repeated as many times as desired to generate the required number of bits in the quotient.

We notice from the above discussion that division is really a process of repeated subtractions. That is, we repeatedly subtract the divisor from the dividend or the remainders generated. We determine whether or not to subtract by comparing the magnitudes of the divisor and the dividend or remainders. For this reason, this method is sometimes called the "comparison method."

Digital Arithmetic

Division by the comparison method can be accomplished by means of the registers shown in Fig. 13-14. You will notice that this figure is the same as Fig. 13-13 with the exception of the full-subtractors. The divisor is held in the divisor register, the dividend is initially stored in the accumulator, and the quotient appears in the MQ register after the division is complete. The name MQ register is now obvious, since this register holds the *multiplier* at the beginning of a multiplication operation and contains the *quotient* at the end of a division operation. Division is carried out using the registers shown in Fig. 13-14 using the following steps:

1. Shift the divisor into the divisor register; shift the dividend into the accumulator and set the MQ register to 0s.

2. Determine the sign of the quotient by comparing the signs of the divisor and dividend. It is positive if they have like signs and negative if they have unlike signs. Store the sign of the quotient and set the sign bits in the registers to 0s.

3. Compare the magnitudes of the divisor and dividend; if the divisor is greater, proceed; if the dividend is greater, halt. Recall that we are using binary fractional notation and we therefore cannot generate a number larger than $0.111 \ldots 1_n$, where n is the number of bits in the magnitude. In other words, we cannot have a bit to the left of the binary point, other than the sign bit.

4. Shift the contents of the accumulator and MQ registers *left* one place.

5. Compare the magnitudes of the divisor and the accumulator register. If the divisor is smaller, subtract the divisor from the accumulator and place a 1 in the rightmost bit of the MQ register. If the divisor is larger, do not subtract.

6. Shift the accumulator and MQ *left* one place.

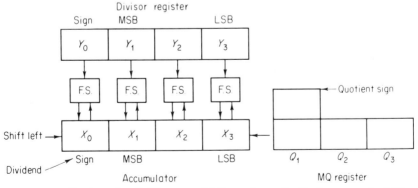

Fig. 13-14 Register for dividing two 3-bit numbers. Quotient appears in MQ register as Q_1, Q_2, and Q_3.

7. Repeat steps 5 and 6 above until n bits have been generated in the MQ register (3 bits in this case).

8. Place the sign in the sign bit of the MQ register. The quotient is now in the MQ register, and the division is complete.

EXAMPLE 13-15

Show the contents of the accumulator and MQ registers while performing the division 010/110.

SOLUTION

The divisor register is shown initially only since its contents remain unchanged during the divide operation.

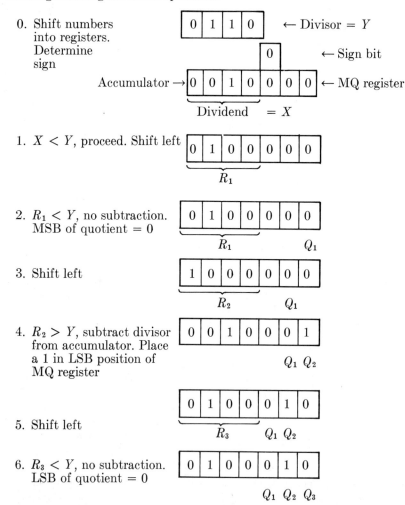

7. Place sign in sign bit of MQ register. Division complete; quotient in MQ register. Quotient = 0.010

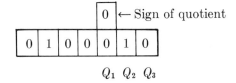

Notice that we cease the division process when we have generated 3 bits of quotient since we have provision for only 3-bit numbers in this system. You will also notice that the answer to this problem is $(1/4)/(3/4) = 1/3 = 0.333 \cdots 3_{10}$. We have ended the division process at three places and thus obtain an approximate answer of $0.010 = 0.25_{10}$. We naturally do not expect great accuracy with a system using only 3 bits. Had we used a 4-bit system we would have obtained an answer of $0.0101 = 0.31_{10}$, which is somewhat closer.

In Example 13-15 it required two clock periods to determine the sign of the quotient. We had to shift left once for each bit in the magnitude; this required three clock periods. We had to make a comparison once for each bit in the magnitude and subtract once; this required four clock periods. Thus the total division time required nine clock periods. It can be easily seen that the division time is variable since it depends on the number of times we have to subtract.

This division process can, of course, be extended to numbers having n bits of magnitude. The total division time is variable, but we can calculate the maximum division time as follows:

1. Two clock periods are required for the sign determination.
2. For a number having n bits of magnitude, we must shift left n times, which requires n clock periods.
3. We must make n comparisons, which may require n clock periods.
4. We will be required to make a maximum of n subtractions, which will require n clock periods.

The total division time is then the sum of these individual times and is $(2 + 3n)$ clock periods.

EXAMPLE 13-16

What is the maximum time required to perform a division using 10-bit numbers if the clock is 1 MHz?

SOLUTION

It will require $(2 + 3n) = (2 + 30) = 32$ clock periods. Each clock period is 1 μsec, and therefore the maximum division time is 32 μsec.

SUMMARY

In this chapter we have discussed the basic methods of performing binary arithmetic. We began by establishing the methods of representing both positive and negative numbers in binary fractional notation. In this system we represent positive numbers in sign and true magnitude and negative numbers in sign and true magnitude, or sign and 1's complement, or sign and 2's complement. These numbers are then used to perform arithmetic in fixed-point machines. We then discussed one method for representing numbers in sign, magnitude, and exponents; this representation is used in machines designed for floating-point arithmetic. The discussion of serial adders and parallel adders clearly shows that parallel addition is considerably faster. By examining the four cases of addition we showed that subtraction reduces to addition and therefore the adders discussed could be easily used for subtraction as well. During the discussion it became clear that the use of 2's complements is more desirable for serial adders while 1's complements are more desirable for parallel addition. The discussion of a parallel BCD adder showed clearly the convenience of using numbers in excess-3 BCD code. The most basic method of multiplication by repeated addition was discussed as well as a simple method of division by repeated subtraction. Using these two methods for multiplication and division explains why these two operations are more complicated than addition or subtraction and therefore require longer execution times.

GLOSSARY

add-cycle time The time required to complete an addition operation.

binary fractional machine A machine designed to perform arithmetic operations using numbers in binary fractional notation.

binary fractional notation The representation of binary numbers by placing the binary point to the left of the MSB.

divide-cycle time The time required to complete a division operation.

fixed-point machine A machine using binary numbers having a fixed position for the binary point and no provision for powers of 2.

floating-point machine A machine which uses binary numbers having a fixed position for the binary point and provision for powers of 2.

multiplication-cycle time The time required to complete a multiplication operation.

overflow The act of exceeding the capacity of a storage register.

parallel binary adder A system which adds all $(n + 1)$ bits of two binary numbers, having n bits of magnitude, simultaneously.

serial binary adder A system which adds two binary numbers by adding 2 bits at a time sequentially beginning with the two LSBs.

Digital Arithmetic 395

settling time The time required for all logic elements in an arithmetic system to reach their final steady-state levels after a change in data.

subtract-cycle time The time required to complete a subtraction operation.

REVIEW QUESTIONS

1. Why is it important to detect overflows? How are they detected?
2. Why is it common practice to locate the binary point to the left of the MSB?
3. Explain the use of the sign bit.
4. What is meant by scaling and where is it used?
5. What are the three ways of representing a negative binary number using sign and magnitude?
6. What is meant by floating-point arithmetic?
7. What are the four cases of addition? When can overflows occur?
8. When adding two numbers, how are overflows detected?
9. Why is the 1's complement for negative numbers not used in a serial adder?
10. What is the major advantage of a parallel adder over a serial adder? What is the major disadvantage?
11. What is the primary cause of settling time in a parallel adder?
12. What are the similarities between binary addition and excess-3 BCD addition (consider the four cases)?
13. Why is excess-3 code more useful in BCD addition than 8421 code?
14. Explain how multiplication is performed by repeated addition.
15. Explain how division is accomplished by repeated subtraction.

PROBLEMS

13-1 What range of numbers can be represented in a binary fractional machine having 8 bits of magnitude?

13-2 Represent the following decimal numbers in sign and true-magnitude form:
  (a) +0.65625.
  (b) −0.65625.
  (c) +0.53125.

13-3 What is the decimal equivalent of the following numbers given in sign and true-magnitude form:
  (a) 1.00101.
  (b) 0.00011.
  (c) 1.01010.
  (d) 0.101010.

13-4 Represent the number -0.625_{10} in sign and true magnitude, sign and 1's complement, and sign and 2's complement.

13-5 What would be the scale factor for the number 4381 if it were used in a machine accepting binary fractional numbers only?

13-6 Make a sketch showing the format of the numbers in a binary fractional machine using 6 bits of magnitude and a sign bit. Using this format, what is the binary number equivalent to 4381? What is the resolution error?

13-7 Make a sketch showing the format of a floating-point number using a sign bit, 4 bits of exponent, and 8 bits of magnitude.

13-8 Using the system of Prob. 13-7, give the binary equivalent of the number 4381_{10}.

13-9 Demonstrate that the four cases of subtraction of two numbers reduce to the same four cases of addition of two numbers.

13-10 What is the add-cycle time of a serial binary adder having 12 bits of magnitude if the system uses a 1-MHz clock?

13-11 Estimate the settling time for a parallel binary adder having 12 bits of magnitude if the delay through each full-adder is 0.1 μsec. Assume the flip-flop delay times are negligible.

13-12 Make a table showing the eight possible input conditions to the overflow detector in Fig. 13-10 and determine which cases give an overflow indication.

13-13 Show the binary representation of the following numbers in 8421 code and excess-3 code:

(a) 67_{10}.

(b) -34_{10}.

(c) -93_{10}.

13-14 Make a truth table and verify the operation of the 9's-complement circuit shown in Fig. 13-15.

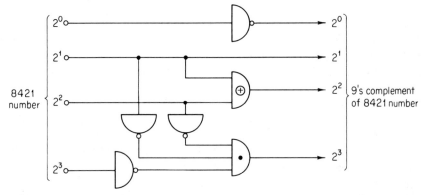

Fig. 13-15 Circuit to form the 9's complement of a 8421 BCD number (Prob. 13-14).

Digital Arithmetic

13-15 Draw a logic diagram for an excess-3 parallel BCD adder capable of handling numbers between ± 0.999.

13-16 What multiply-cycle time is required for a system using numbers having 12 bits of magnitude and having a 2-MHz clock? What is the time required for numbers having 30 bits of magnitude?

13-17 What is the maximum division time required for two numbers having 12 bits of magnitude if the clock is 2 MHz? What is the maximum time for two numbers having 30 bits of magnitude?

13-18 Make a sketch similar to Example 13-14 showing the multiplication of $+.111$ and $-.111$ (these numbers are sign and true magnitude).

13-19 Make a sketch similar to Example 13-15 showing the division of $+.101$ by $-.111$ (these numbers are sign and true magnitude).

13-20 What happens if you attempt to divide $+.111$ by $+.111$?

CHAPTER 14

CLOCK AND CONTROL

The operation or control of a digital system can be classified into two general categories — synchronous and asynchronous. In a *synchronous system* the flip-flops are controlled by the system clock and can therefore change states only when the clock changes state. Therefore, all the flip-flops and logic gates will change levels in time (or in synchronism) with the clock. An example of such a synchronous system is the parallel counter constructed using the *master/slave* clocked flip-flops. In this counter, the flip-flops can change state only when the clock goes low and at no other time (notice that a system could be constructed such that the flip-flops would change state when the clock goes high). On the other hand, in an *asynchronous system* the flip-flops are controlled by events which occur at random times. Thus the flip-flops may change states at random and are not in synchronism with any timing signal such as a clock. An example of such a system might be the operation of a push button by a human operator. Depression of the push button will cause a flip-flop to change state. Since the operator can depress the button at any time he desires, the flip-flop will change states at some random time, and this is therefore an asynchronous operation. Most large-scale digital systems operate in the synchronous mode,

Clock and Control

and if one gives a little thought to the checkout and maintenance of such a system it is easy to see why. Any synchronous system must have a system clock (some systems have more than one clock), and we will therefore devote this chapter to basic clocks and the generation of control signals.

14-1 Basic Clocks

Since all logic operations in a synchronous machine occur in synchronism with a clock, the system clock becomes the basic timing unit. The system clock must then provide a periodic waveform which can be used as a synchronizing signal. The square wave shown in Fig. 14-1a, is a typical clock waveform used in digital system. It should be noted that the clock need not be a perfectly symmetrical square wave as shown. It could simply be a series of positive pulses (or negative pulses) as shown in Fig. 14-1b. This waveform could, of course, be considered as an asymmetrical square wave. The main requirement is simply that the clock be perfectly periodic. Notice that the clock defines a basic timing interval during which logic operations must be performed. This basic timing interval is defined as a *clock cycle time* and is equal to one period of the clock waveform. Thus all logic elements, flip-flops, counters, gates, etc., must complete their transitions in less than one clock cycle time.

EXAMPLE 14-1

What is the clock cycle time for a system which uses a 500-kHz clock? A 2-MHz clock?

SOLUTION

A clock cycle time is equal to one period of the clock. Therefore, the clock cycle time for a 500-kHz clock is $1/(500 \times 10^3) = 2$ μsec. For a 2-MHz clock, the clock cycle time is $1/(2 \times 10^6) = 0.5$ μsec.

EXAMPLE 14-2

The total propagation delay through a *master/slave* clocked flip-flop is given as 100 nsec. What is the maximum clock frequency that can be used with this flip-flop?

Fig. 14-1 Basic system clock.

Solution

An alternative way of expressing the question is, how fast can the flip-flop operate? The flip-flop must complete its transition in less than one clock cycle time. Therefore, the minimum clock cycle time must be 100 nsec. So, the maximum clock frequency must be $1/(100 \times 10^{-9}) = 10$ MHz.

In many digital systems the clock is used as the basic standard for measurement. For example, the accuracy of the A/D converter discussed in Chap. 11 is related directly to the frequency of the clock used to drive the counter. If the clock changes frequency, the accuracy is reduced. For this reason, it is necessary to ensure that the clock maintains a stable and predictable frequency. In many digital systems only short-term stability

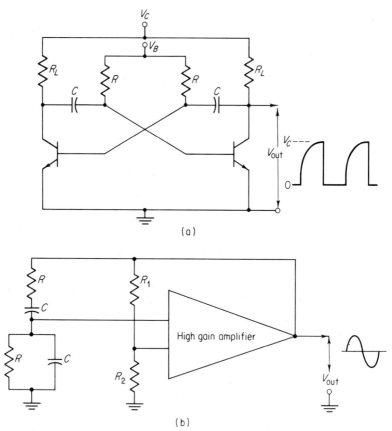

Fig. 14-2 Basic clock circuits. (a) Free-running multivibrator. (b) Wien-bridge oscillator.

Clock and Control

is required of the clock. This would be the case in a system where the clock could be monitored and adjusted periodically. For such a system, the basic clock might be derived from a free-running multivibrator or a simple sine-wave oscillator as shown in Fig. 14-2a and b. For the free-running multivibrator the clock frequency f is given by

$$f \cong \frac{1}{2RC \ln(1 + V_C/V_B)} \qquad (14\text{-}1)$$

From Eq. (14-1) it can be seen that the basic clock frequency is affected by the supply voltages as well as the values of the resistors R and the capacitors C. Even so, it is possible to construct multivibrators such as this which have stabilities better than a few parts in 10^3 per day. The frequency of oscillation f for the Wien-bridge oscillator is given by

$$f \cong \frac{1}{2\pi RC} \qquad (14\text{-}2)$$

Again it is not difficult to construct these oscillators with stabilities better than a few parts in 10^3 per day. If greater clock accuracy is desired, a crystal-controlled oscillator such as that shown in Fig. 14-3 might be used. This type of oscillator is quite often housed in an enclosure containing a heating element which maintains the crystal at a constant temperature. Such oscillators can be obtained having accuracies better than a few parts in 10^9 per day.

EXAMPLE 14-3

The multivibrator shown in Fig. 14-2a is being used as a system clock and operated at a frequency of 100 kHz. If its accuracy is better than

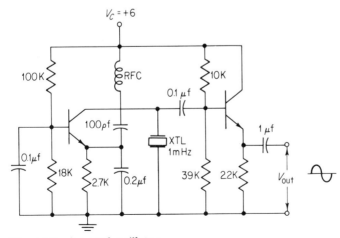

Fig. 14-3 A crystal oscillator.

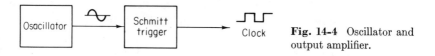

Fig. 14-4 Oscillator and output amplifier.

±2 parts in 10^3 per day, what are the maximum and minimum frequencies of the multivibrator?

SOLUTION

One part in 10^3 can be thought of as one cycle in 1000 cycles. Two parts in 10^3 can then be thought of as two cycles in 1000 cycles. Since the multivibrator runs at 100 kHz, 2 parts in 10^3 is equivalent to 200 cycles. Thus the maximum frequency would be 100 kHz + 200 cycles = 100.2 kHz, and the minimum frequency would be 100 kHz − 200 cycles = 99.8 kHz.

None of the oscillators shown in Figs. 14-2 and 14-3 has a square-wave output waveform, and it is therefore necessary to convert the basic frequency into a square wave before use in the system. The simplest way of accomplishing this is to use a Schmitt trigger on the output of the basic oscillator as shown in Fig. 14-4. This will provide two advantages:

1. It will provide a square wave of the basic clock frequency as desired.
2. It will ensure that the clock output amplifier (the Schmitt trigger in this case) has enough power to drive all the necessary circuits without loading the basic oscillator and thus changing the oscillating frequency.

14-2 Clock Systems

Quite often it is desirable to have clocks of more than one frequency in a system. Alternatively, it might be desirable to have the ability to operate a system at different clock frequencies. We might then begin with a basic clock which is the highest frequency desired and develop other basic clocks by simple frequency division using counters. As an example of this suppose we desire a system which will provide basic clock frequencies of 3, 1.5, and 1 MHz. This could be accomplished by using the clock system shown in Fig. 14-5. We begin with a 3-MHz oscillator followed by a Schmitt trigger to provide the 3-MHz clock. The 3-MHz signal is then fed through one flip-flop which divides the signal by 2 to provide the 1.5-MHz clock. The 3 MHz is also fed through a divide-by-3 counter, which provides the 1-MHz clock. Systems having multiple clock frequencies can then be provided by using this basic method.

Clock and Control

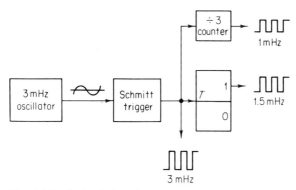

Fig. 14-5 Basic clock system.

Example 14-4

Show a clock system which will provide clock frequencies of 2 MHz, 1 MHz, 500 kHz, and 100 kHz.

Solution

The desired system is shown in Fig. 14-6. Beginning with a 2-MHz oscillator and a Schmitt trigger, the 2-MHz clock appears at the output of the Schmitt trigger. The first flip-flop divides the 2 MHz by 2 to provide the 1 MHz clock. The second flip-flop divides the 1-MHz clock by 2 to provide the 500-kHz clock. Dividing the 500-kHz clock by 5 provides the 100-kHz clock.

It is sometimes desirable to have a two-phase clock in a digital system. A *two-phase clock* simply means we have two clock signals of the same frequency which are 180° out of phase with one another. This can quite easily be accomplished by taking the outputs of a flip-flop. The 1 output will be one phase of the clock and the 0 output will be the other phase. These two signals are clearly 180° out of phase with one another since one

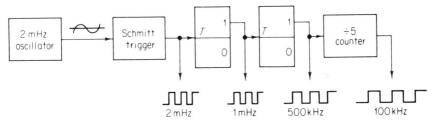

Fig. 14-6 A clock system.

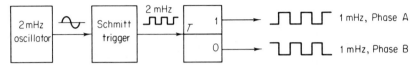

Fig. 14-7 A 1-MHz two-phase clock.

is simply the complement of the other. A system for developing a two-phase clock of 1 MHz is shown in Fig. 14-7. For distinction, the two clocks are sometimes referred to as phase A and phase B. You will recall that one use for a two-phase clock system is to drive the magnetic-core shift register discussed in Chap. 12 (Fig. 12-10). It is interesting to note that the two-phase clock system can be used to overcome the race problem encountered with the basic parallel counter discussed in Chap. 8 (Fig. 8-5). The race problem is solved by driving the *odd* flip-flops (i.e., flip-flops A, C, E, etc.) with phase A of the clock and the *even* flip-flops (i.e., flip-flops B, D, F, etc.) with phase B of the clock (see Prob. 14-12).

The race problem as initially discussed in Chap. 8 can occur any time two or more signals at the inputs of a gate are undergoing changes at the same time. The problem is therefore not unique in counters and can occur anywhere throughout a digital system. For this reason, a *strobe pulse* is quite often developed using the basic clock. This strobe pulse will be used to interrogate the condition of a gate at a time when the input levels to the gate are not changing. If the gate levels render the gate in a true condition, a pulse will appear at the output of the gate when the strobe pulse is applied. If the gate is false, no pulse will appear. In Fig. 14-8, a strobe pulse is used to interrogate the simple three-input AND gate. The waveforms clearly show that outputs appear only when the three input levels to the gate are true. It is also quite clear that no racing can possibly occur since the strobe

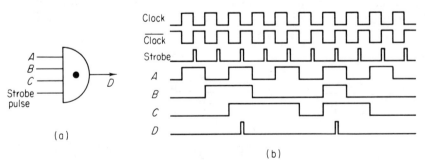

Fig. 14-8 The use of a strobe pulse. (*a*) A three-input AND interrogated by a strobe pulse. (*b*) Waveforms for the AND gate.

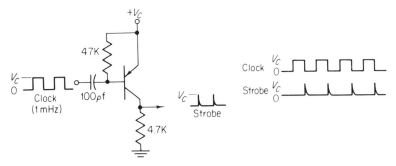

Fig. 14-9 Developing a strobe pulse.

pulses are placed exactly midway between the input-level transitions. The strobe signal can be developed in a number of ways. One way is to differentiate the complement of the clock, $\overline{\text{clock}}$, and use only the positive pulses. A second method would be to differentiate the clock and feed it into an "off" transistor as shown in Fig. 14-9.

14-3 D/A Converter Control

The D/A converter shown in Fig. 11-13 can be redrawn as shown in Fig. 14-10a. In this figure we have expanded the basic 4-bit converter to a 10-bit converter. This simply means that we need 10 flip-flops in the converter and the ladder is expanded to accommodate 10 inputs. The storage register represents the flip-flops holding the data to be converted to analog form. Let us suppose that data to be converted appear in the storage register once every 10 μsec and we wish to shift these data from the storage register into the D/A converter every 10 μsec. Thus we will be performing a D/A conversion every 10 μsec. We need to develop the proper *read in* or strobe pulse for the D/A converter. Let us assume that the storage register is composed of master/slave flip-flops which are driven by a clock of 1 MHz. The desired strobe pulse can be developed using the logic shown in Fig. 14-10b. The basic 1-MHz clock is divided by 10 using a five-flip-flop shift counter. We can then form the control waveform A shown in Fig. 14-10c by decoding one of the 10 states of the shift counter (decoding of the shift counter is discussed in Chap. 9). This might at first glance seem to be all we need, since this waveform A is a pulse which occurs once every 10 μsec as we desire. We must, however, consider the race problem which might occur if we use this waveform. It is clear that the control waveform A changes state during the time the clock goes low; however, the storage register also changes state at this time and we do not therefore want to

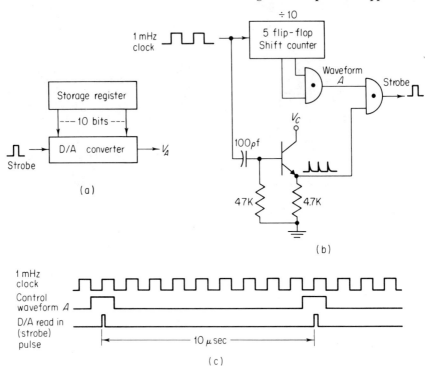

Fig. 14-10 The use of a strobe pulse with a D/A converter. (a) 10-bit D/A converter. (b) Logic to develop a strobe pulse. (c) Control waveforms.

initiate a shift at this time since we would certainly run the risk of racing at the shift-gate inputs. We need then to develop a strobe pulse to initiate the shift of data from the storage register into the converter. A convenient time for this to occur would be exactly in the middle of the positive portion of the control waveform A since the clock is going high at this time and therefore the storage-register flip-flop outputs are static. The desired strobe pulse can be easily formed by ANDing the differentiated clock with the control waveform A as shown in Fig. 14-10b. It is clear that the D/A strobe pulse occurs at 10-μsec intervals and at a time when the storage-register flip-flop outputs are static.

14-4 Multiple D/A Conversion

In Chap. 11 we discussed two methods for obtaining the analog equivalent of a number of digital signals. The two methods involve channel selection using a separate D/A converter for each signal to be decoded. The second

Clock and Control

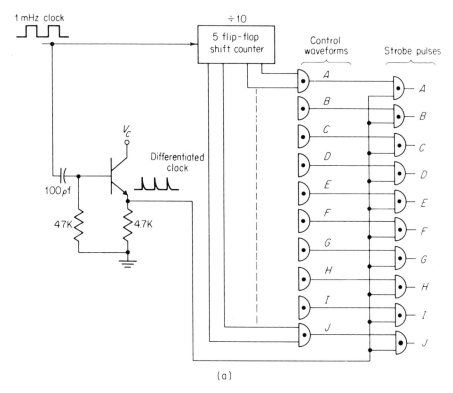

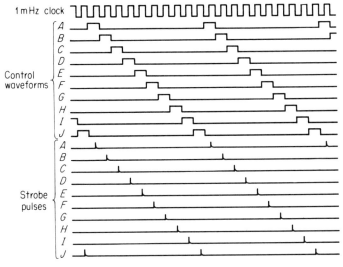

Fig. 14-11 Multiple D/A conversion by channel selection. (*a*) Logic diagram. (*b*) Waveforms.

method involves the use of only one D/A converter and multiplexing to switch the output of the converter to a number of sample-and-hold amplifiers. These two methods are summarized in Fig. 14-11.

In the channel-selection method shown in Fig. 11-14a, all the D/A converter inputs are connected in parallel with the input storage register. Let us assume that we have 10 such converters, each of which is similar to the converter discussed in the previous section and shown in Fig. 14-10. The 10 signals to be converted will appear in the storage register in a time sequence and will remain in the register for 1 μsec. In order to decode these signals we must then develop a series of 10 strobe pulses similar to the one developed in Fig. 14-10. The proper strobe signals can be quite easily developed by decoding all 10 states of the five-flip-flop shift counter in this figure. This will then provide 10 control waveforms (A, B, C, D, E, F, G, H, I, and J) as shown in Fig. 14-11b. The correct strobe pulses can then be formed by differentiating the clock and ANDing the signals as before. The complete logic diagram and the resulting waveforms are shown in Fig. 14-11. It is clear from the strobe waveforms developed that the data will be shifted into the converters at 1-μsec intervals and that each converter will be selected every 10 μsec in a time sequence.

EXAMPLE 14-5

What would be one method for altering the system shown in Fig. 14-11 to accommodate five D/A converters?

SOLUTION

If symmetry is desired, it would be quite easy to form the desired control signals by retaining only the gates required to form the strobe pulses A, C, E, G, and I.

14-5 A/D Converter Control

The operation of the A/D converter discussed in Chap. 11 is summarized by the waveforms for one conversion cycle shown in Fig. 11-23. These waveforms are self-contained in the converter with the exception of the clock and the *start* pulse. The *start* pulse could be formed by a push-button switch and the converter would make one conversion each time the push button is depressed. More often, however, it is desirable to have the converter perform conversions at a predetermined rate. The *start* pulse could then be derived from the output of a free-running multivibrator, and a conversion would begin each time the multivibrator output goes high. It is only necessary to ensure that the period of the multivibrator waveform is greater than the maximum conversion time of the converter. If the fre-

Clock and Control   409

quency of the multivibrator is variable (as in Fig. 14-2a) the number of conversions per second can be varied. This principle is quite often used in commercial counters and digital voltmeters.

EXAMPLE 14-6

If the D/A converter previously discussed operates with a basic system clock of 1 MHz, what is the maximum frequency that the *start* pulse multivibrator can operate if the full-scale count of the converter is 500 counts?

SOLUTION

With a 1-MHz clock, the basic clock cycle time is 1 μsec. Therefore, the full-scale count will require 500 μsec, which corresponds to the maximum conversion time. The *start* pulse multivibrator must then have a period greater than 500 μsec. Therefore, the multivibrator must operate at a frequency less than $1/(500 \times 10^{-6}) = 2$ kHz.

EXAMPLE 14-7

If the output of the D/A converter in the previous example is used to drive the grids of nixie tubes, and if the *start* pulse multivibrator runs at 10 Hz, for what percentage of time will the nixie tubes be illuminated?

SOLUTION

The converter conversion time is 500 μsec, or 0.5 msec. During the conversion time, the converter output is changing and the nixie tubes will not respond properly. However, a conversion is initiated only once every tenth of a second, or once every 100 msec. Thus the output of the converter is static with the desired output signal for $100 - 0.5 = 99.5$ msec out of every 100 msec. Therefore, the nixie tubes are illuminated with the proper output signal for $(99.5/100) \times 100 = 99.5$ percent of the total operating time.

14-6 Ring-counter Control

You will recall the operation of the basic ring counter discussed in Chap. 9 and shown in Fig. 9-3. The proper operation of this counter depends on the fact that one and only one of the flip-flops can be set at any one time and all other flip-flops must be in the *reset* state. In other words, the counter can contain one and only one 1. When power is first applied to the system, the initial states of the flip-flops are unpredictable, and therefore some external circuitry must be used to ensure that the counter operates in the proper mode.

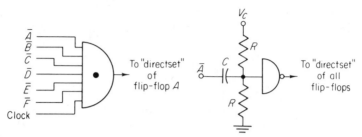

Fig. 14-12 Logic to provide automatic presetting of the ring counter in Fig. 9-3.

One possible state which could occur when power is first applied is for all flip-flops to turn on in the *reset* state. In this case, the counter contains all 0s and will not function at all. One method of accounting for this condition is to use the six-input AND gate shown in Fig. 14-12. If all flip-flops are in the 0 state, it is clear that the output of the AND gate will be true and flip-flop A will be set immediately. The counter will then be ready for operation. We can account for any other condition by applying a positive pulse to the *reset* inputs of every flip-flop except flip-flop A whenever A goes high. This can be accomplished by means of the NAND gate shown in Fig. 14-12. Notice that the output of the NAND gate must be a pulse since most integrated-circuit *master/slave* flip-flops have d-c-coupled *direct set* and *direct reset* inputs.

EXAMPLE 14-8

Define the logic necessary to preset the ring counter in Fig. 9-3 automatically with a 1 in flip-flops A and D and a 0 in all other flip-flops.

SOLUTION

The same logic circuits shown in Fig. 14-12 can be used. However, the output of the AND gate must now be connected to the *direct set* of both flip-flops A and D. The output of the NAND gate must be connected to the *direct reset* of flip-flops B, C, E, and F and the *direct set* of flip-flop D.

14-7 MPG Computer

Up to this point we have covered quite a wide variety of topics generally encountered in the study of digital systems. Some of the topics have been discussed in great detail while others have been treated in a more general way. In any case the reader should now have the necessary background to study any digital system with good comprehension and a minimum of effort. Even so, the reader may still be somewhat unsure about the overall organization of a digital system. In an effort to overcome this feeling and to attempt to tie together many of the topics discussed in the previous chapters, we will at this time consider the implementation of a small special-purpose digital computer.

The special-purpose computer which we will consider will be used to calculate the *miles per gallon* of a motor vehicle, thus the name *MPG computer*. It is termed a *special-purpose* computer since this is the only use for which it is intended. A *general-purpose* computer would be a more complicated machine which might be used for a number of different applications.

The first step in considering the design of the MPG computer must necessarily be the determination of the system performance requirements. The first requirement might be that the system must be capable of operating from a supply voltage of ± 6 or ± 12 volts d-c since the machine will be operated in a motor vehicle. The second requirement might be that the readout of the computer be in decimal form. Nixie tubes might be good for the readout but they will require an additional power supply of around $+100$ volts to operate the tubes. Digital modules are commercially available which provide decimal readout and they operate on $+6$ or $+12$ volts d-c. These modules are somewhat more expensive, but might provide a better choice in this case. The final decision will be one of economics. The third requirement is that the computer calculate the miles per gallon used by the vehicle to an accuracy of ± 1 mile per gallon. The fourth requirement which we will impose is that the computer perform a calculation at least once every 15 sec when the vehicle is traveling at a speed greater than 10 mph. In other words, we would like to sample the mileage performance of the vehicle at least once every 15 sec (faster sampling rates are acceptable). The fifth requirement is that the computer be capable of operating in vehicles using fuel at rates between 10 and 40 miles per gallon. We can now summarize the five basic requirements of the MPG computer as follows:

1. Power-supply voltage is either ± 6 or ± 12 volts d-c.
2. The computer must provide a decimal readout in miles per gallon.
3. The computer must provide the readout to an accuracy of ± 1 mile per gallon.
4. The computer must provide a readout of miles per gallon at least once every 15 sec when the vehicle is traveling at a speed greater than 10 mph.
5. The computer must be capable of calculating miles per gallon between the limits of 10 and 40 miles per gallon.

It should be noted that the system requirements for the computer under study here are quite simple and somewhat less stringent than in the usual case. The requirements here are intentionally made simple in order to simplify the discussion. Nevertheless the principles applied are the same regardless of the severity of the system specifications, and the study is therefore instructive.

We will assume that we have available two transducers which are to be used as an integral part of the MPG computer. The first transducer will be used to measure the volume of fuel flowing into the engine. This flow transducer will provide an electrical pulse each time $\frac{1}{1000}$ of a gallon of fuel passes through it. The second transducer will be used to measure the distance traveled and will be driven by the speedometer cable. This distance transducer will provide an electrical pulse each time the vehicle has traveled a distance of $\frac{1}{1000}$ of a mile.

Now in order to implement the necessary logic for the computer, let us examine the outputs of the flow and distance transducers. Let us begin by assuming that we have a flow transducer which gives an output pulse each time 1 gallon is used, and we have a distance transducer which gives an output pulse each time the vehicle has traveled 1 mile. If our vehicle is obtaining a mileage slightly better than 10 miles per gallon, the transducer waveforms will appear as shown in Fig. 14-13. Notice that the number of distance pulses appearing between two flow pulses is exactly equal to the miles per gallon we desire. Thus we can *calculate* the miles per gallon by simply counting the number of distance pulses occurring between two flow pulses. We can check this by noting that, if the vehicle were operating at 20 miles per gallon, there would be 20 distance pulses between two flow pulses. Notice that if the flow transducer supplied 10 pulses per gallon, and at the same time the distance transducer provided 10 pulses per mile, the basic waveform in Fig. 14-13 would remain unchanged. That is, the number of distance pulses appearing between two flow pulses would still be equal to the number of miles per gallon. From this it should be clear that we can choose any number of pulses per gallon from the flow transducer so long as we choose the same number of pulses per mile from the distance transducer. The transducers we are going to use in the MPG computer provide 1000 pulses per gallon of flow and 1000 pulses per mile of distance. Therefore, the number of miles per gallon can be obtained by simply counting the number of distance pulses between consecutive flow pulses.

The reason for using the transducers given can be determined by examining the time between flow pulses. Let us first consider the flow transducer having one pulse per gallon and the distance transducer having one pulse per mile. If the vehicle were obtaining a rate of 10 miles per gallon, one flow pulse

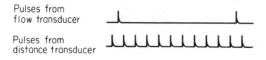

Fig. 14-13 Transducer pulses for the MPG computer when the rate is 10 miles per gallon.

would occur every 10 miles. If the vehicle were traveling at a speed of 10 mph, the flow pulses would occur at a rate of one per hour. This is clearly not a fast enough sampling rate. On the other hand, with the specified transducers, the flow pulses occur at a rate of 1000 pulses per gallon or at the rate of 1000 pulses per hour under the same conditions. Thus the flow pulses will occur every 1 hr/1000 = 3.6 sec. This sampling time is clearly within the specified rate. The worst case occurs when the vehicle obtains the maximum miles per gallon. At 40 miles per gallon and 10 mph the flow pulses will occur every $3.6 \times 4 = 14.4$ sec. We have therefore met the minimum-sampling-time requirements.

The logic diagram for the MPG computer can now be drawn and is shown in Fig. 14-14 along with the complete waveforms. The flow pulses are fed

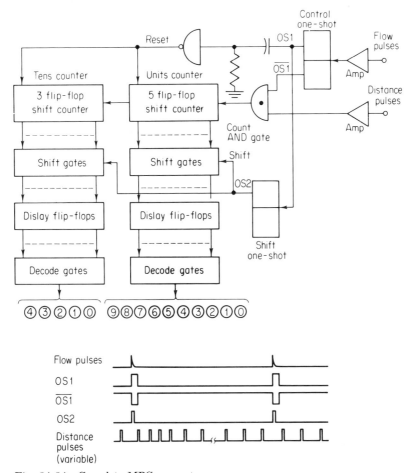

Fig. 14-14 Complete MPG computer.

into a conditioning amplifier and then into a one-shot to develop the waveform $OS1$ and $\overline{OS1}$. The distance pulses are also fed into a conditioning amplifier. Since we desire to count the number of distance pulses occurring between two flow pulses, we will use the distance pulses as one input to the *count* AND gate. If $\overline{OS1}$ is used as the other input to this AND gate, it will be enabled between flow pulses and the distance pulses will appear at its output. We will use the pulses appearing at the output of the *count* AND gate to drive a counter. Since we desire to display the miles per gallon between the limits of 10 and 40 we will use a five-flip-flop shift counter for the *units* digits and a three-flip-flop shift counter for the *tens* digits of miles per gallon.

One conversion time is the time between two flow pulses, and we want to shift the accumulated count into the display flip-flops at the end of each conversion cycle. Notice first of all that, when $\overline{OS1}$ is low, the *count* AND gate is disabled and therefore the *units* and *tens* counters cannot change states. It is during this time that we must shift the contents of these counters into the display flip-flops. We will use the leading edge of $OS1$ to trigger the *shift* one-shot and develop the shift waveform $OS2$. The falling edge of $OS2$ will be applied to the shift gates, and at this time the count stored in the *units* and *tens* counters will be shifted into the display flip-flops. The falling edge of $OS1$ will then be used to reset all flip-flops in the *units* and *tens* counters. The contents of the display flip-flops are then decoded and used to illuminate the indicator lights. In this system, the distance pulses can be considered to be the basic system clock. The flow pulses form a variable control gate by means of the *control* one-shot which determines the period of time that the *count* AND gate is enabled and therefore the number of distance pulses counted. The output of the *shift* one-shot $OS2$ can be considered as a strobe pulse which shifts data from the counters into the display flip-flops in such a way that racing is avoided. The system will clearly have an accuracy of ± one count, which corresponds to ±1 mile per gallon.

SUMMARY

Any digital system can be considered synchronous or asynchronous in operation (there are systems which are both). Any synchronous system must have a basic system clock which synchronizes the operations and defines a basic cycle time. The system clock may be derived from any oscillator, multivibrators and crystal oscillators being the most popular. The stability of the system clock has a direct bearing on the system accuracy since it is often used as a measurement standard. Crystal oscillators provide a considerably more stable clock frequency than do multivibrators

Clock and Control

or other sinusoidal oscillators. Many systems operate with more than one clock. Secondary clock frequencies can be derived from the basic clock by simple division using counters. In this way, the secondary clocks are always in synchronism with the basic clock. The basic clock is quite often used to develop strobe pulses. These strobe pulses are used to interrogate the condition of a gate or to shift data from one register to another at a time which will avoid racing.

GLOSSARY

asynchronous system A system in which logic operations and level changes occur at random times.

clock cycle time One clock period; the reciprocal of clock frequency.

general-purpose computer A computer designed to accomplish a number of tasks. For example, all the arithmetic operations as well as decision making (i.e., equal to, greater than, less than, go, no go, etc.).

oscillator stability The stability of the frequency of oscillation; usually expressed in parts per thousand or parts per million for a period of time.

secondary clock A clock of frequency lower than the basic system clock which is derived from the basic system clock.

special-purpose computer A computer designed to accomplish only one task, for example, the MPG computer in this chapter.

strobe pulse A pulse developed to interrogate gates or to shift data at a time such that racing is avoided.

synchronous system A system in which logic operations and level changes occur in synchronism with a system clock.

two-phase clock The use of two clock waveforms of the same frequency which are 180° out of phase with one another, for example, the 1 and 0 outputs of a flip-flop.

REVIEW QUESTIONS

1. Explain why a clock must be perfectly periodic.
2. How can the clock cycle time be found from the clock frequency?
3. Why must flip-flops have a delay time less than one clock cycle time?
4. What factors affect the oscillating frequency of the multivibrator in Fig. 14-2?
5. What is the purpose of the Schmitt trigger in Fig. 14-4?
6. Explain one method for obtaining a two-phase clock.

7. What is the main purpose for developing a strobe pulse?
8. Why is it advantageous to develop the strobe pulse in Fig. 14-9 by turning the transistor on rather than off?
9. Why is it necessary to develop the strobe pulses for the D/A converters shown in Fig. 14-10 or 14-11?

PROBLEMS

14-1 Beginning with a symmetrical square wave, show a method for developing a clock consisting of a series of positive pulses. A series of negative pulses.

14-2 What is the clock cycle time for a system using a 1-MHz clock? A 250-kHz clock?

14-3 What is the maximum delay time for a flip-flop if it is to be used in a system having an 8-MHz clock?

14-4 At what frequency will the multivibrator in Fig. 14-2a oscillate if $R = 100$ kilohms, $C = 100$ pf, $V_C = 20$ volts d-c, and $V_B = 10$ volts d-c?

14-5 What will be the frequency of the multivibrator in Prob. 14-4 if V_B is changed to 20 volts d-c?

14-6 What value of C is required for the multivibrator in Fig. 14-2a if $V_C = V_B$, $R = 47$ kilohms and the desired frequency is 100 kHz?

14-7 What is the oscillating frequency of the Wien-bridge oscillator in Fig. 14-2b if $R = 47$ kilohms, and $C = 100$ pf?

14-8 If the crystal oscillator in Fig. 14-3 has a stability of ± 3 parts in 10^7 per day, what are the maximum and minimum frequencies of the oscillator?

14-9 Show the logic necessary to develop clock frequencies of 5 MHz, 2.5 MHz, 1 MHz, and 200 kHz.

14-10 The 5-MHz oscillator in Prob. 14-9 has a stability of ± 1 part in 10^6 per day. What will be the maximum and minimum frequency of the 1-MHz clock?

14-11 What would be the maximum and minimum frequency of the 200-kHz clock in Prob. 14-10?

14-12 Draw the waveforms for a parallel binary counter being driven by a two-phase clock. Show that this will result in a solution to the race problem. Remember that each flip-flop has a finite delay time.

14-13 If the D/A converter in Example 14-6 has a clock of 100 kHz, what is the maximum frequency of the *start* multivibrator?

14-14 If the D/A converter in Prob. 14-13 samples at a rate of ten times per second, for what percentage of the time will the nixie tubes be illuminated?

14-15 How could the MPG computer be modified to give a solution to the nearest 1/10 mile per gallon?

BIBLIOGRAPHY

Bartee, T. C.: "Digital Computer Fundamentals," 2d ed., McGraw-Hill Book Company, New York, 1966.
Blitzer, R.: "Basic Pulse Circuits," McGraw-Hill Book Company, New York, 1963.
Burroughs Corporation: "Digital Computer Principles," McGraw-Hill Book Company, New York, 1962.
Chu, Y.: "Digital Computer Design Fundamentals," McGraw-Hill Book Company, New York, 1962.
Flores, I.: "The Logic of Computer Arithmetic," Prentice-Hall, Inc., Englewood Cliffs, N.J., 1963.
Gray, H. J.: "Digital Computer Engineering," Prentice-Hall, Inc., Englewood Cliffs, N. J., 1963.
Ketchum, D. J., and E. C. Alvarez: "Pulse and Switching Circuits," McGraw-Hill Book Company, New York, 1964.
Nashelsky, L.: "Digital Computer Theory," John Wiley & Sons, Inc., New York, 1966.
Phister, M., Jr.: "Logical Design of Digital Computers," John Wiley & Sons, Inc., New York, 1960.
Quartly, C. J.: "Square-loop Ferrite Circuitry," Prentice-Hall, Inc., Englewood Cliffs, N.J., 1962.
Richards, R. K.: "Arithmetic Operations in Digital Computers," D. Van Nostrand Company, Inc., Princeton, N.J., 1957.
Shannon, C. E.: A Symbolic Analysis of Relay and Switching Circuits, *Transactions of the AIEE*, vol. 57, 1938.

APPENDIX A

STATES AND RESOLUTION FOR BINARY NUMBERS

Word length in bits n	Max number of combinations 2^n	Resolution of a binary ladder ppm
1	2	500 000.
2	4	250 000.
3	8	125 000.
4	16	62 500.
5	32	31 250.
6	64	15 625.
7	128	7 812.5
8	256	3 906.25
9	512	1 953.13
10	1 024	976.56
11	2 048	488.28
12	4 096	244.14
13	8 192	122.07
14	16 384	61.04
15	32 768	30.52
16	65 536	15.26
17	131 072	7.63
18	262 144	3.81
19	524 288	1.91
20	1 048 576	0.95
21	2 097 152	0.48
22	4 194 304	0.24
23	8 388 608	0.12
24	16 777 216	0.06

APPENDIX B

DECIMAL-OCTAL-BINARY NUMBER CONVERSION TABLE

Decimal	Octal	Binary	Decimal	Octal	Binary
0	00	000 000	32	40	100 000
1	01	000 001	33	41	100 001
2	02	000 010	34	42	100 010
3	03	000 011	35	43	100 011
4	04	000 100	36	44	100 100
5	05	000 101	37	45	100 101
6	06	000 110	38	46	100 110
7	07	000 111	39	47	100 111
8	10	001 000	40	50	101 000
9	11	001 001	41	51	101 001
10	12	001 010	42	52	101 010
11	13	001 011	43	53	101 011
12	14	001 100	44	54	101 100
13	15	001 101	45	55	101 101
14	16	001 110	46	56	101 110
15	17	001 111	47	57	101 111
16	20	010 000	48	60	110 000
17	21	010 001	49	61	110 001
18	22	010 010	50	62	110 010
19	23	010 011	51	63	110 011
20	24	010 100	52	64	110 100
21	25	010 101	53	65	110 101
22	26	010 110	54	66	110 110
23	27	010 111	55	67	110 111
24	30	011 000	56	70	111 000
25	31	011 001	57	71	111 001
26	32	011 010	58	72	111 010
27	33	011 011	59	73	111 011
28	34	011 100	60	74	111 100
29	35	011 101	61	75	111 101
30	36	011 110	62	76	111 110
31	37	011 111	63	77	111 111

ANSWERS TO SELECTED ODD-NUMBERED PROBLEMS

CHAPTER 1

1-1 0.5 ma **1-3** On the positive half-cycle, it is clipped at the zero level. On the negative half-cycle it is sinusoidal with a peak of -16 volts **1-5** Negatively clipped at the zero level **1-7** Positive half-cycle: sinusoidal with a peak of 19.3 volts. Negative half-cycle: sinusoidal with a peak of -14 volts **1-9** Waveform is positively clipped at the 0.7-volt level **1-11** $I_{max} = 8$ ma and $I_{min} = 4$ ma **1-13** 5 **1-15** $I_B = 0.03$ ma, $I_C = 3$ ma, and $V_{CE} = 10$ volts for the ideal approach. $I_B = 0.023$ ma, $I_C = 2.3$ ma, and $V_{CE} = 13.5$ volts for the second approximation **1-17** About 4.06 volts **1-19** 12.5 MHz

CHAPTER 2

2-1 1, 10, 11, 100, 101, 110, 111, 1000, 1001, 1010, 1011, 1100, 1101, 1110, 1111, 10000, 10001, 10010, 10011, 10100, 10101, 10110, 10111, 11000, 11001, 11010, 11011 **2-3** (a) 10100 (b) 11000110 (c) 11000111 **2-5** 11000, 1000001, 1101010 **2-7** 10111.01110 plus a remainder **2-9** 25.375 **2-11** (a) -011 (b) 10000 (c) -10111 **2-13** 1000. No **2-15** (a) 0100 (b) 001110 (c) -010111 **2-17** (a) 100011 (b) 10000.1110 (c) -10000010.1 **2-19** 110.11 **2-21** 53 **2-23** (a) 1216 (b) 215 **2-25** (a) 257 (b) 15.334 (c) 123.55

CHAPTER 3

3-1 (a) 0101 1001 (b) 0011 1001 0101 1000 0100 **3-3** (a) 387 (b) 967,873 **3-5** (a) 0111 1001 1010 (b) 1000 1011 0110 1100 **3-7** (a) 846 (b) 75,320 **3-9** 1010 1000 0011 **3-11** (a) 0100 0011 1000 (b) 0010 1011 1000 **3-13** (a) 952 (b) 047 **3-15** (a) 10110010 (b) 1100011000110 (c) 101011100111000 **3-17** 0, 1, 1, 0, 0, 0, 0, 1, 1, 0 **3-19** (a) 11000 01111 (b) 10000 00111 00011 (c) 11111 01111 11000 00011 **3-21** (a) 01 00100 10 00001 (b) 01 10000 10 00001 01 00100 **3-23** (a) 11101 (b) 110100110101 (c) 11111011001010

420

Answers to Selected Odd-Numbered Problems

CHAPTER 4

4-1 The ABx entries are 3-3-3, 3-12-12, 12-3-12, 12-12-12 **4-3** The $ABCx$ entries are 1-1-1-1, 1-1-10-1, 1-10-1-1, 1-10-10-1, 10-1-1-1, 10-1-10-1, 10-10-1-1, 10-10-10-10 **4-5** The $ABCx$ entries are 0000, 0010, 0100, 0111, 1001, 1011, 1101, 1111 **4-7** $x = (\overline{A + B})B$. The ABx entries are 000, 010, 100, 110 **4-9** The ABx entries are 000, 011, 101, 110 **4-11** Construct the truth tables to show equality **4-15** $A \cdot \overline{B}$ **4-23** AB

CHAPTER 5

5-11 1 half-subtractor and 34 full-subtractors

CHAPTER 6

6-1 250 kHz **6-3** 4 kHz **6-5** 0.1 ma, 2 volts **6-9** (a) 11, 01, 00, 10, 11, ... (b) 00, 10, 11, 01, 00, ... **6-11** 0.147 msec **6-13** 70 kHz to 700 kHz

CHAPTER 7

7-1 The truth-table entries are 5-bit numbers following a straight binary count from 00000 to 11111 **7-3** 0010, 0011, 0100, 0101, 0110, 0111, 1100, 1101, 1110, 1111, 0010, ... **7-5** 2,480 **7-9** $7.5(10^6), 2.23(10^6)$

CHAPTER 8

8-1 (a) 3 (b) 4 (c) 4 (d) 5 (e) 5 **8-3** Mod-7
8-5

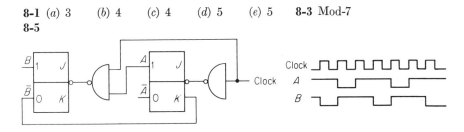

8-7 Yes, cured

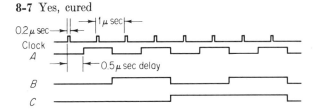

8-9

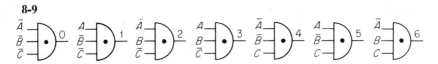

8-11 The difficulty is in decoding

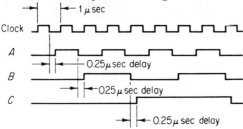

8-13 $2 \times 3 \times 3, 2 \times 9, 3 \times 6$

CHAPTER 9

9-1 The only restriction on RS flip-flops is that both the R and S inputs cannot be high at the same time. Since it cannot possibly occur in Fig. 9-2, the logic diagram and waveforms are exactly as shown in Fig. 9-2 **9-3** Apply a positive pulse to *direct reset* of flip-flops A, B, C, D, and E and *direct set* of F **9-5** Set 1s in any two adjacent flip-flops **9-7** Yes **9-9** Yes, 5 **9-11** Corrected on counts 2 or 10 or 18 or 26. Worst case is 9 clock periods. $\bar{C}$ input to gate is not necessary

9-13

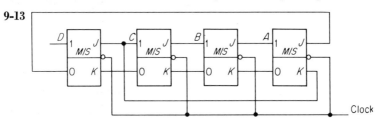

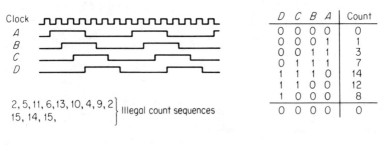

D	C	B	A	Count
0	0	0	0	0
0	0	0	1	1
0	0	1	1	3
0	1	1	1	7
1	1	1	0	14
1	1	0	0	12
1	0	0	0	8
0	0	0	0	0

2, 5, 11, 6, 13, 10, 4, 9, 2, 15, 14, 15, } Illegal count sequences

Cure

Answers to Selected Odd-Numbered Problems

9-15 Skips counts 16, 17, 48, and 49. E is symmetrical. Would not matter if E were not symmetrical

9-17
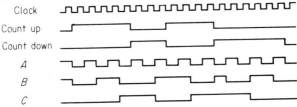

CHAPTER 10

10-7 120 in.   **10-9** 1.37×10^6 **10-11** $F = 5/11$, $L = 275$ in.   **10-13** No savings in diodes   **10-15** Eight three-input AND gates using the three LSBs of the binary number will form the units digit of the octal number. The fourth bit of the binary number will form the other digit of the octal number (1 or 0).

CHAPTER 11

11-1 $1/63$, $2/63$, $4/63$, $8/63$, $16/63$, $32/63$ **11-5** 51.2 ma   **11-7** (a) 0.641 volt (b) 0.923 volt   (c) 0.766 volt   **11-9** 1 part in 4096; 2.44 mv   **11-11** 31 **11-13** (a) 4.096 msec   (b) 2.048 msec   (c) approximately 500 conversions per second   **11-15** 12 μsec, not counting delay and control times   **11-17** Resolution = 0.0244 percent; accuracy of about 0.05 percent

CHAPTER 12

12-1
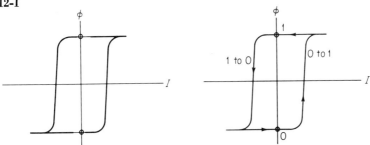

12-3 Symbol shown in Fig. 12-7b. Pulse at 0 input will reset; pulse at 1 will set; pulse at advance will cause a 1 output pulse if core previously held a 1
12-5 Fig. 12-9

12-7

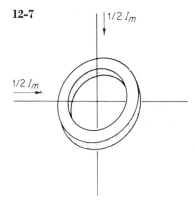

12-9 36 planes; each plane a square array of 4096 cores, 64 cores on each edge
12-11 64 X lines, 64 Y lines, 36 *inhibit* lines   **12-13** 12 bits in the address word; 6 bits for X and 6 bits for Y
12-15

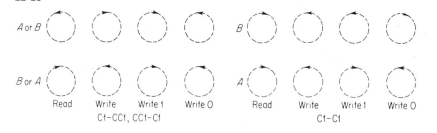

12-17 125,600 bits   **12-19** 0.833 msec; use two sets of *read/write* heads

CHAPTER 13

13-1 0 to 0.996094   **13-3** (a) -0.156 (b) $+0.0938$ (c) -0.312 (d) $+0.656$
13-5 10^4 **13-7** $X_0.X_1X_2X_3X_4 \ \ X_5X_6X_7X_8X_9X_{10}X_{11}X_{12}$
13-9 $(+x) - (+y) = \ \ x - y, (+x) - (-y) = \ \ x + y$
            $(-x) - (+y) = -x - y, (-x) - (-y) = -x + y$
13-11 Approximately 1.2 μsec
13-13

	8421	True	9s	10s
67_{10}	0.0110 0111			
-34_{10}	1.0011 0100	1.0110 0101	1.0110 0110	
-93_{10}	1.1001 0011	1.0000 0110	1.0000 0111	

Answers to Selected Odd-Numbered Problems

	Excess-3 True	9s	10s
67_{10}	0.1001 1010		
-34_{10}	1.0110 0111	1.1001 1000	1.1001 1001
-93_{10}	1.1100 0110	1.0011 1001	1.0011 1010

13-15 Same as Fig. 13-12 with four more bits in each register and one more BCD adder   **13-17** $19\,\mu\text{sec}; 46\,\mu\text{sec}$

CHAPTER 14

14.1

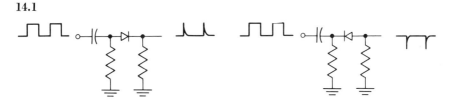

14-3 125 nsec   **14-5** 722 kHz   **14-7** 33.9 kHz

14-9

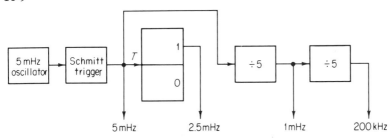

14-11 200 kHz $\pm$ 0.2 Hz   **14-13** 200 Hz   **14-15** Use transducers having 10^4 pulses per mile and 10^4 pulses per gallon

INDEX

Abacus, 73
Access time, 351, 355
Accumulator register, 387
A/D converter, 275–313
  accuracy, 307
  continuous type, 302
  control, 408
  conversion rate, 299
  conversion time, 298, 306
  counter method, 297
  electromechanical, 309
  signal reconstruction, 301
  simultaneous, 294
  successive-approximation, 305
Add-cycle time, 376, 394
Adder, BCD, 380, 384
  8421, 127
  excess-3, 129, 384

Adder (*cont.*)
  full, 123
  half, 121
  parallel, 376, 394
  serial, 368, 394
Address, 330, 355
Address register, 341
Addressing, 339
  capacity, 340
Advance pulse, 322
Alphanumeric, 250, 271
Analog error, 307
AND gate, 85, 112
  symbol, 93
AND multiplication, 92
Apertured plate, 341
Arithmetic units, 359
Asynchronous system, 398, 415

Index

Avalanche diode, 9
Average conversion time, 298

Balanced multiplicative decoder, 268
Base (number), 33, 54
Base (transistor), 11
BCD, 57, 78
BCD adder, 127, 129, 380
Beta, 13
Biasing, 261
BIAX, 349
Binary, 30, 54
  addition, 35
  division, 48, 389
  multiplication, 48, 386
  subtraction, 43
  weight, 38, 54
Binary addition, rules for, 368
Binary codes, biquinary, 73, 78
  8421, 57, 78
  excess-3, 60, 78
  4-bit, 59
  Gray, 74, 78
  possible number of, 59
  unweighted, 62
  weighted, 59, 78
Binary counter, 146
Binary division, shift register, 244
Binary equivalent weight, 277, 312
Binary fractional, capacity, 361
  machine, 361, 394
  notation, 361, 394
Binary fractions, 39
Binary ladder, 281
  output, 284
  termination, 286
Binary multiplication, shift register, 243
Binary multivibrator, 146
Binary point, 39, 359
Binary-to-Gray conversion, 75
Binary-to-Gray converter, 119

Bit, 42, 54, 271
Bit-parallel, 351
Bit-serial, 351
Block, 345
Boole, George, 81
Boolean algebra, 81
  duality, 109
  laws and theorems, 105, 112
Boolean expressions, 97
Breakdown point, 2
Breakdown voltage, 2

Capacitor delay, core, 328
Capacity of a register, 360
Card punch, 252
Card reader, 251
Cascade, 167
Cathode-ray tube, 264
Centering, on LSB, 303
Character, 271
Character-check bit, 258
Charactron, 264
Circulating register, 217
Clock, 151, 159, 163, 167, 181
  multivibrator, 400
  secondary, 402
  stability, 401, 415
  systems, 402
  two-phase, 403, 415
Clock cycle time, 399, 415
CML, 136
Code wheel, 309
  ambiguity, 310
Coincident-current drive, 330, 355
Coincident-current memory system, 329, 334
  noise in, 338
Collector, 11
Comparator, 294
Comparison method, 390
Complement, 44
Complementary circuit, 90

Computer, 411, 415
  MPG, 411
Conventional current, 1, 25
Conversion, binary-to-decimal, 37
  binary-to-Gray, 75
  binary-to-octal, 52
  decimal-to-binary, 40
  Gray-to-binary, 76, 312
  octal-to-binary, 51
  time, 298
Converter, binary-to-gray, 119
Counter, 167, 183
  BCD, 208
  binary, 169, 181
  capacity, 169
  combination, mod-5, 198
    mod-7, 200
  decade, 171, 181, 202, 224
    decoding, 203
  even modulus, 224
  feedback, 171,
    mod-5, 187
    mod-10, 188
  higher modulus, 206
  illegal states, 199
  modified, 186
  odd modulus, 225
  parallel, mod-5, 192
    mod-6, 191
    mod-7, 190
    mod-8, 189
  permuting, 187
  ripple, 183, 211
  settling time, 189
  synchronous, 189
Crystal oscillator, 401
Current hogging, 134, 139
Current source, 12, 25
Cutoff, 19

D/A converter, 275, 287, 312
  accuracy, 292

D/A converter (cont.)
  control, 405
  resolution, 292
Data density, 255
DCTL logic, 133
DCU, 175
Decade counter, 171, 181, 202, 224
Decimal counter (see Decade
    counter)
Decimal encoder, 269
Decoding, 173
Decoding gates, 203
Decoding matrix, 265, 271
  minimizing, 268
Delay time, 23, 194
De Morgan's theorems, 98
Density of data, 255
Destructive readout (see DRO)
Differential linearity, 308, 312
Differentiate (square wave),
    145, 163
Digit, 31, 54
Digit punch, 249
Digital clock, 228
Digital error, 307
Digital plotter, 265
Digital recording, 260
Digital voltmeter, 179, 245, 307
  gate time of, 180
Diode, avalanche, 9
  ideal, 3, 25
  second approximation, 5
  semiconductor, 1
  zener, 9
Diode AND gate, 85
Diode OR gate, 82
Direct access, 351
Divide-cycle time, 393, 394
Divide-by-two, 146
Dividend, 390
Divisor, 390
Double complement, 102
Double-dabble method, 40

Index

DRO, 333, 355
DTL, 134

Eccles-Jordan multivibrator, 146
Eight-hole code, 254
8421 adder, 127
8421-code addition, 59
8421 decade counter, 172
Emitter, 11
Emitter follower, 24
Encoding matrix, 265, 271
End-around carry, 45, 54
Equivalent binary weighted voltage, 276
Error detection, 68
Excess-3, addition, 62
  subtraction, 64
Excess-3 adder, 129, 384
Exclusive-OR addition, 75
Exclusive-OR gate, 116, 139
  symbol, 119

Fall time, 23
Fan-in, 133, 139
Fan-out, 133, 139
Ferrite core, 317
Ferrite-plate memory, 343
  capacity, 344
  critical current, 342
  two-hole-per-bit, 343
Ferromagnetic, material 316
5-bit codes, 72
Fixed-point machine, 361
Flip-flop, clocked, 196
  JK, 151, 197
  master/slave, 196, 211
  RS, 148, 197
  RST, 148
  T, 142
Floating-point arithmetic, 364, 394

Flux, 317
Forbidden combinations, 8421, 58
  excess-3, 61
Forward characteristics, 2
4-bit BCD codes, 65
Free-running multivibrator, 159
Frequency counter, 177
Frequency divider, 146
Full-adder, 123, 139
Full-subtractor, 131, 139

Gating, 176, 181
Gating pulse, 176
Gray code, 74, 310
Gray-code wheel, 310
  resolution, 311
Gray-to-binary conversion, 76, 312
Guard time, 194

Half-adder, 121, 139
Half-select current, 330
Half-subtractor, 131, 139
Hollerith code, 249, 272
Hundreds, 176
Hysterein, 317
Hysteresis, 156
Hysteresis curve, 318, 339, 355

i-v characteristic, diode, 5
  transistor, 14
  zener diode, 9
Inhibit wire, 332
Input-output, 248
Integer, 54
Integrated circuits, 138
Interrecord gap, 258, 272
Inversion, 95
Inverter, 90, 113

J input, 151
JK flip-flop, 151
  symbol, 151
  truth table, 152
Johnson counter, 221
"JR. is 11," 250

K input, 151
Key punch, 251
Knee voltage, 2, 25

LADDIC, 349
Least significant digit, 38
  (*see also* LSB)
Level amplifiers, 287
Light pen, 264
Limit points, 305
Logic gate, 82, 112
Longitudinal parity, 258, 272
Look-ahead logic, 210
Lower trip point, LTP, 156
LSB, weight, 280

Magnetic core, 316
  sensing, 320
  switching time, 321
Magnetic-core logic, 321
  basic functions, 323
  symbol, 322
Magnetic-core shift register, 325
Magnetic drum, 350
  access time, 353
  storage density, 352
Magnetic flux density, 317
Magnetic force, 317
Magnetic recording, 260
Magnetic tape, 255
  capacity, 256, 259
  code, 256
  density, 258

Magnetic tape (*cont.*)
  parity, 258
  recoding, reading, 257
Man-machine interface, 248
MARS, 349
Master/slave flip-flop, 196, 211
  symbol, 196
Matrix, 331
Memory cycle, 334, 355
Memory matrix currents, 332
Metal ribbon core, 317
Millman's theorem, 278, 313
Modified count, 184
Modulo-2 addition, 75, 118, 139
Modulus, 184, 211
Monotonicity test, 290
MQ register, 387, 391
Multiaperture devices, 345
Multiplex rate, 290
Multiplexing, 289
Multiplicand, 386
Multiplication-cycle time, 389, 394
Multiplier, 386
Multivibrator, 142, 163
  astable, 159, 163
  bistable, 143, 146, 163
  frequency, 159
  monostable, 161, 163

NAND gate, 100, 113
  symbol, 100
NAND/NOR gate, ORing, 239
Natural count, 184, 211
NDRO, 334, 355
Negation, 95
Negative code weights, 66
Negative logic, 89, 113
9 edge, 249
9's compliment, 47
Nixie tube, 174, 203
Nondestructive readout (*see* NDRO)

Index

NOR gate, 103, 113
NOT circuit, 90, 113
  symbol, 95
NOT operation, 94
NRZ, 260, 262, 272
NRZI, 263, 272
Number representations, 361–368
Number systems,
  binary, 30, 31
  decimal, 30
  octal, 30, 49

Octal, 30, 49, 54
One-core-per-bit, 328
One-shot, 162
1's complement, 44, 243
Optical encoder, 309
OR addition, 91
OR gate, 82, 112
  symbol, 92
Oscillator, 401
Overflow, 243, 360, 394
  detection, 366, 370, 378

Paper tape, 253
  eight-hole code, 253
  end-of-line, 254
  guide holes, 254
  parity, 254
Parallel binary adder, 125, 376, 394
Parallel binary subtraction, 126
Parallel counter, 189
Parity, double, 70
  even, 69
  odd, 69
Parity bit, 69
Parity checker, 119
Partial products, 48, 386
Period, 178
Peripheral equipment, 263

Permuting counter, 187
Positive clipper, 5, 25
Positive logic, 89, 113
prime/drive winding, 347
Printer, 264
Punched cards, 249
  capacity, 251

Quantization error, 307, 313
Quantum, 307
Quinary, 30, 54
Quotient, 390

Race problem, 192, 211
Radix, 33, 54
Random-access, 351
RCTL, 134
Read from memory, 336
Read operation, 334
Read/write cycle, 336
Read/write head, 257
Record, 258
Recording density, 255, 272
Rectilinear information, 309
Reflected binary system, 74
Regenerate, 145
Remainders, 390
Remanence, 318
Remanent flux, 318
Remanent point, 318
Reset, 148
Resistive divider, 278
  drawbacks, 281
Restricted access, 353
Reverse resistance, 7, 25
Ring counter, 216, 245
  control of, 409
  magnetic core, 329
  presetting, 218
Ring-counter code, 73
Ripple counter, 169, 181, 211

Rise time, 23
Rotational information, 309
RS flip-flop, 148
   symbol, 150
RST flip-flop, 148
   symbol, 150
RTL, 134
RZ, 260

Sample-and-hold amplifier, 290
Saturation, 19, 20
Saturation current, base, 19
   collector, 20
Scaler, 147, 163
Scaling, 363
Schmitt trigger, 154
Section counter, 306
Select current, 325, 355
Self-complementing code, 62, 385
Sense wire, 331
Serial binary adder, 368, 394
Set, 148
Settling time, 290, 377, 395
Shannon, Claude, 81
Shift counter, 219, 245
   illegal states, 3-stage, 221
      4-stage, 223
      5-stage, 224
Shift-counter code (Johnson code), 72
Shift-counter decoding, 226
Shift left, 243
Shift register, 213, 245
   capacity, 216, 245
   cross-coupled, 245
   feedback, 216
   inverse feedback, 220
   parallel, 241
   serial, 213
   shift time, 216, 241
   6-bit, 215

Shift right, 243
Sign and magnitude, 362
Sign bit, 362
Single-diode transfer loop, 325, 355
Speed-up capacitor, 144
Squareness ratio, 339, 355
Squaring circuit, 158
Staircase waveform, 290
Steady-state accuracy, 290
Steering diodes, 144
Storage time, 23
Strobe, 261, 321
Strobe pulse, 404, 415
Subtract-cycle time, 385, 395
Subtractor, 131
Switching time, 21
Synchronous counter, 189
Synchronous system, 398, 415

T flip-flop, 142
   symbol, 146
Table of combinations, 83
Tape utilization factor, 259, 272
Tens, 175
10's complement, 47
Thévenin's theorem, 283
Toggle, 146, 163
Trailing-edge logic, 194
Transfluxor, 345, 355
   ac, 347
   NDRO, 347
Transistor, characteristics, 11
   ideal, 14
   model, 15
   n-p-n, 11
   p-n-p, 11
   second approximation, 17
   as a switch, 18
   switching time, 21
   symbols, 11
Transistor AND gate, 88

Index

Transistor OR gate, 85
Tree decoding, 266
Trigger, 142, 163
Trip point, 159
True magnitude, 362
Truth table, 83, 113
   combinations, 84
TTL, 136
Turn-off time, 23, 25
Turn-on time, 23, 25
12 edge, 249
2-out-of-5 code, 72
two-phase clock, 327, 403, 415
2's complement, 44
2's complement circuit, 373

Unblock, 345
Unit record, 251, 272
Units, 175
Universal building block, 103

Up-down counter, 245
   parallel, 239
   serial, 234
Upper trip point (UTP), 156

Variable reference voltage, 297
Variable resistor network, 276
Vigesimal, 30, 54
Voltage comparator, 156, 164
Voltage regulation, 10

Weight, 54
Wein-bridge oscillator, 401
Word, 68
Write into memory, 337
Write operation, 334

Zener diode, 9
Zener voltage, 9
Zone punch, 249